INSECTICIDE AND FUNGICIDE HANDBOOK

# INSECTICIDE AND FUNGICIDE HANDBOOK

## FOR CROP PROTECTION

issued by the

### BRITISH CROP PROTECTION COUNCIL

edited by

## HUBERT MARTIN

D.Sc. A.R.C.S. F.R.I.C.

## THIRD EDITION

BLACKWELL SCIENTIFIC PUBLICATIONS
OXFORD AND EDINBURGH

*Printed in Great Britain by*
ADLARD AND SON LTD, DORKING
*and bound by*
WEBB SON & CO LTD., LONDON & FERNDALE

# Members of the Recommendations Committee of the British Insecticide and Fungicide Council

| | |
|---|---|
| Dr H. Martin | *Chairman* |
| Mr D.J. Higgons | *Secretary* |
| Mr A.L. Abel | *Association of British Manufacturers of Agricultural Chemicals* |
| Mr J.L. Hunt | |
| Dr W. Linke | |
| Mr G. Culpan | |
| Mr H.J. Terry | |
| Mr T. Manns | |
| Mr R.D. Wiseman | |
| Mr W.H. Read | |
| Mr G. Stell | *Ministry of Agriculture, Fisheries and Food, Plant Pathology Laboratory* |
| Dr H.H. Glasscock | *National Agricultural Advisory Service* |
| Mr M.S. Bradford | *National Association of Corn and Agricultural Merchants* |
| Dr W.G. Keyworth | *National Vegetable Research Station* |

## Co-opted members

| | |
|---|---|
| Dr R.A. Dunning | *Broom's Barn Experimental Station* |
| Dr R. Hull | |
| Mr J.A.R. Bates | *Ministry of Agriculture, Fisheries and Food, Plant Pathology Laboratory* |
| Mr K.S. George | |
| Dr Joan Moore | |
| Mr G.H. Brenchley | *National Agricultural Advisory Service* |
| Mr R. Gair | |
| Mr B.C. Knight | |
| Mr B.D. Moreton | |
| Mr E.T. Roberts | |
| Mr J.B. Byass | *National Institute of Agricultural Engineering* |
| Dr J.A. Dunn | *National Vegetable Research Station* |
| Miss A.V. Brooks | *The Royal Horticultural Society* |
| Mr K.M. Harris | |
| Dr D. Bevan | *Forest Research Station* |
| Miss P.M. Smith | *Glasshouse Crops Research Institute* |

# CONTENTS


1  THE BIOLOGICAL BACKGROUND
*Pests*
  1.1  Insect life histories                                                                   1
  1.2  Destructive insects                                                                     3
  1.3  Useful insects                                                                          4
  1.4  Control measures                                                                        6
  1.5  Economics of treatment                                                                 9
  1.6  Development of resistance to pesticides                                                11
  1.7  Increasing importance of previously non-damaging organisms  12
  1.8  Effects on soil                                                                         12
  1.9  Phytotoxicity                                                                           13
  1.10 Conclusions                                                                             13

*Plant diseases*
  1.11  Factors influencing disease incidence                                                16
  1.12  Control measures                                                                       18

2  THE CHEMICAL BACKGROUND
  2.1  Insecticides                                                                            23
  2.2  Fungicides                                                                              24
  2.3  Formulation                                                                             25
  2.4  The active components                                                                  28

3  THE APPLICATION OF PESTICIDES
  3.1  Spraying
         3.1A  Methods and terminology                                                       62
         3.1B  Equipment                                                                      63
  3.2  Specification of main components                                                       65
  3.3  Operation                                                                              69
  3.4  Nozzle arrangements and procedures for particular crops                               75
  3.5  Dusts and dry applications                                                             78

4  THE SAFE AND EFFICIENT USE OF PESTICIDES
  4.1A  The Pesticides Safety Precautions Scheme                                              83
  4.1B  The Agriculture (Poisonous Substances) Act, 1952                                      86

| | | |
|---|---|---:|
| 4.2 | The Agricultural Chemicals Approval Scheme | 93 |
| 4.3 | Pesticides—Code of Conduct | 95 |

**5 PEST AND DISEASE CONTROL IN CEREALS**

| | | |
|---|---|---:|
| 5.1 | Wheat | 98 |
| 5.2 | Barley | 104 |
| 5.3 | Oats | 107 |
| 5.4 | Rye | 109 |
| 5.5 | Stored cereal pests | 110 |

**6 PESTS AND DISEASES OF POTATOES**

| | | |
|---|---|---:|
| 6.1 | Pests | 111 |
| 6.2 | Diseases | 117 |

**7 PESTS AND DISEASES OF SUGAR BEET, FODDER BEET AND MANGOLDS**

| | | |
|---|---|---:|
| 7.1 | Pests | 125 |
| 7.2 | Diseases | 132 |

**8 PEST AND DISEASE CONTROL IN GRASS AND FODDER CROPS**

*Pests*

| | | |
|---|---|---:|
| 8.1 | Cereals | 138 |
| 8.2 | Grasses | 139 |
| 8.3 | Legumes (excluding Field Beans) | 142 |
| 8.4 | Field Beans | 144 |
| 8.5 | Cruciferous crops | 145 |

*Diseases*

| | | |
|---|---|---:|
| 8.6 | Cereals | 149 |
| 8.7 | Grasses | 149 |
| 8.8 | Leguminous crops | 153 |
| 8.9 | Field Beans | 156 |
| 8.10 | Brassica crops | 156 |

**9 PESTS AND DISEASES OF FRUIT AND HOPS**

| | | |
|---|---|---:|
| 9.1 | Apple | 160 |
| 9.2 | Pear | 186 |
| 9.3 | Quince | 192 |
| 9.4 | Apricot, Nectarine and Peach | 192 |
| 9.5 | Cherry | 194 |

9.6   Plum and Damson                                      197
9.7   Hazelnut, Cobnut and Filbert                         201
9.8   Currant and Gooseberry                               201
9.9   Blackberry, Loganberry and Raspberry                 210
9.10  Strawberry                                           214
9.11  Grape vine                                           221
9.12  Hop                                                  221


## 10   PEST AND DISEASE CONTROL IN GLASSHOUSE CROPS


10.1  General                                              228
   10.1A   Treatment of Border soils                    229
   10.1B   Treatment of soil for propagating and potting  230
   10.1C   Susceptibility of plants to chemicals         231
   10.1D   Atomized fluids (aerosols) and smokes         231
   10.1E   Glasshouse hygiene                            231
10.2  Specific crops
   Arum (Cally) lily                                     232
   Asparagus fern                                        232
   Azalea                                                233
   Beans (French and climbing French)                    234
   Begonia                                               234
   Calceolaria                                           235
   Carnation                                             236
   Chrysanthemum                                         239
   Cineraria                                             245
   Cucumber and melon                                    245
   Cyclamen                                              250
   Freesia                                               251
   Fuchsia                                               252
   Geranium and Pelargonium                              252
   Grape (vines)                                         253
   Hippeastrum (Amaryllis)                               254
   Hydrangea                                             255
   Lettuce                                               256
   Mushroom                                              257
   Orchid                                                261
   Poinsettia                                            263
   Primula                                               263
   Rose                                                  264
   Saintpaulia (African violet)                          266
   Sweet Pea                                             266
   Tomato                                                267
10.3  Miscellaneous commercial glasshouse plants           273

11  PESTS AND DISEASES OF VEGETABLE CROPS
    11.1  General pests and diseases — 275
    11.2  Specific crops
        Broad Bean — 277
        French and runner beans — 278
        Beetroot (Red beet) — 279
        Brassica crops — 280
        Carrot — 286
        Celery — 288
        Lettuce — 289
        Mint — 291
        Onion — 292
        Pea — 293
        Spinach — 295
        Tomato (outdoor) — 295
        Watercress — 296

12  PEST AND DISEASE CONTROL IN OUTDOOR ORNAMENALS AND IN TURF
    Acer — 298
    Antirrhinum — 299
    Azalea — 299
    Beech — 300
    Begonia (outdoor bedding types) — 300
    Camellia — 300
    Carnation — 301
    China aster — 301
    Chrysanthemum — 302
    Clarkia — 303
    Clematis — 303
    Cornflower — 304
    Cotoneaster — 304
    Crocus — 304
    Dahlia — 305
    Daphne — 306
    Delphinium — 306
    Euonymus — 306
    Forsythia — 307
    Fuchsia — 307
    Gladiolus — 307
    Godetia — 308
    Hawthorn — 309
    Hellebore — 309

Holly 309
Hollyhock 310
Hyacinth 310
Hydrangea 311
Iris 311
Juniper 312
Larch 312
Lilac 312
Lily 313
Lobelia 313
Mahonia 313
Malus 313
Meconopsis 314
Michaelmas daisy 314
Narcissus 314
Oak 317
Paeony 317
Pansy 318
Pelargonium 318
Petunia 319
Phlox 319
Plane 319
Polyanthus 320
Poplar 320
Privet 321
Prunus (Ornamental) 321
Pyracantha 322
Pyrethrum 322
Rhododendron 322
Rose 324
Snowdrop 326
Spruce 327
Stock 328
Sweet pea 328
Sweet William 328
Sycamore 328
Syringa 329
Thuja 329
Tulip 329
Turf 330
Violet 332
Wallflower 333
Willow 333

Yew 334
Yucca 334
Zinnia 334

13 PESTS AND DISEASES OF FOREST TREES
13.1 Forest nurseries 337
13.2 Forest plantations 341

APPENDIX Properties of Insecticidal and Fungicidal Compounds 348
Index of Scientific Names 363
General Index 371

# NOTES FOR READERS

In addition to commoner abbreviations, the following have been used:

| | |
|---|---|
| a.i. | active ingredient |
| A.L. | Advisory Leaflet (M.A.F.F.) |
| BSI | British Standards Institution |
| CT | Concentration × time product |
| e.c. | emulsifiable concentrate |
| F.C.L. | Forestry Commission Leaflet |
| g.p.h. | gallons per hour |
| H.V. | High Volume (the use of over 100 gal/acre of spray on bushes and trees; of over 60 gal/acre on ground crops) |
| L.V. | Low Volume (the use of less than 50 gal/acre of spray on bushes and trees; of less than 20 gal/acre on ground crops) |
| M.A.F.F. | Ministry of Agriculture, Fisheries and Food |
| m.p.h. | miles per hour |
| N.A.A.S. | National Agricultural Advisory Service |
| N.I.A.B. | National Institute of Agricultural Botany |
| ppm | parts per million |
| p.s.i. | pounds per square inch |
| PVC | polyvinyl chloride |
| r.p.m. | revolutions per minute |
| s.c. | soluble concentrate |
| s.e. | stock emulsion |
| s.p. | soluble powder |
| STL | Short-term Leaflet (M.A.F.F.) |
| V/V | volume per volume |
| w.g. | water gauge |
| W.H.O. | World Health Organization |
| w.p. | wettable powder |
| W/W | weight per weight |

In the description of the diseases and pests of crop plants, the significance of each is indicated by asterisks thus:

*** of major importance

** of local importance

* of minor importance

# CONTRIBUTORS

1 Mr K.S. George and Dr Joan Moore

2 Dr H. Martin

3 Mr J.B. Byass

4 Mr J.A.R. Bates and Mr G. Stell

5 Dr D.W. Empson (the late) and Mr B.C. Knight

6 Mr G.H. Brenchley and Mr R. Gair

7 Dr R.A. Dunning and Dr R. Hull

8 Mr R. Gair and Mr E.T. Roberts

9 Dr H.H. Glasscock and Mr B.D. Moreton

10 Mr W.H. Read and Miss P.M. Smith

11 Dr W.G. Keyworth and Mr J.A. Dunn

12 Miss A.V. Brooks and Mr K.M. Harris

13 Dr D. Bevan and Mr D.A. Burdekin

# FOREWORD TO FIRST EDITION

It is now generally accepted that, in the present state of knowledge, pesticides are indispensable for food production. By definition, their use is detrimental to certain forms of life—the pest and pathogen—and the danger is that this toxicity may be exerted in unwanted directions, to the detriment of the user, the consumer and to the biological environment in which the materials are used. These hazards are scrutinized by appropriate authorities who report to the Ministers concerned. If necessary, safeguards are put into effect, either by Statutory Regulations or by advisory means. The adequacy of these safeguards is obviously dependent on a strict adherence, by the user, to the rules and recommendations proposed. The main purpose of this Handbook is to place before the grower and adviser the information necessary for the correct use of pesticides in crop protection.

The chemist has been successful, in recent years, in finding new compounds of more selective toxicity and, yearly, the list of hazardous pesticides is being reduced by the introduction of less noxious alternatives. A second purpose of this Handbook is to record this progress and to advise the grower on better methods of crop protection. In some instances, the new methods have been described in advance of their inclusion in official recommendations; if so, they must be regarded as tentative and for the information of the keen grower.

The material of the Handbook has been assembled by the Recommendations Committee of the British Insecticide and Fungicide Council, with the unstinted help of colleagues within the National Agricultural Advisory Service or at the state-aided Research Stations. In this first edition it has been found necessary, for reasons of space, to limit the text to the subject of crop protection, but the inclusion, in future editions, of farm storage may become possible. Suggestions for improvement, and correction of any errors or omissions, would be most welcome and should be sent to the Secretary of the Recommendations Committee, Lenton Experimental Station, Lenton House, Nottingham.

Every effort has been made to ensure that the recommendations are correct, but the British Insecticide and Fungicide Council cannot accept responsibility for any loss, damage or accident arising from carrying out the methods advocated in this Handbook.

H.G. SANDERS M.A. PH.D

*Ministry of Agriculture, Fisheries and Food*
*July* 1963

# CHAPTER I

# THE BIOLOGICAL BACKGROUND

Agricultural and horticultural practices aim to establish reasonably uniform crops over relatively restricted areas. These conditions provide a potentially favourable environment for the development of pest and disease epidemics. But many other factors play their part both in determining the likelihood of an attack and the rate at which the infestation or infection may increase and become of economic significance. The influence of these factors depends to a large extent on the biology of the insects and disease organisms concerned, and varies greatly from one to another. An appreciation of the interreactions of these factors is essential if control measures are to be effective. The various recommendations given in later chapters of this book have been determined with the biological background of the pest or disease concerned constantly in mind.

The approach to pest and disease problems is essentially different, and hence the two parts of this chapter have been treated separately. When insects become pests on crop plants, they can be readily seen and destroyed: if an attack cannot be entirely prevented it can often be cured. On the other hand a disease only becomes apparent when the symptoms on the plant indicate that the pathogen is already established within it: prophylactic methods of control are therefore necessary and prevention rather than cure is a more practical possibility.

# PESTS

## 1.1 INSECT LIFE HISTORIES

Most insect species lay eggs although, in certain groups, living young may be produced at some stage in the life cycle. Young insects hatch from the egg in several ways; for example, the young of moths and butterflies eat their way out while many of the grasshoppers, crickets and related forms apply pressure to the egg from within by an everted, blood-filled sac behind the head; many other insects break through the egg with a spine which is an outgrowth from the cuticle of the head. The newly-hatched insect is outwardly different from the adult and becomes successively closer to the adult form by a series of moults during each of which the old, tight skin is cast off leaving the insect with a new skin previously secreted below the old one.

There are two major divisions in insect classification depending on the degree of transition which occurs between the immature and mature forms. In the first, the newly hatched insect is similar to the adult but has undeveloped reproductive organs and wings. Immature stages are known as nymphs and increase in size by a number of moults until they become adults. The new skin is soft but soon hardens. Growth in size therefore only occurs on moulting. During this period of growth the genitalia and wings develop, becoming functional when the insect reaches maturity. The well-known aphids typify this form of development and adults and nymphs at all stages can often be seen on an infested plant. The other type of transition occurs in insects such as moths and butterflies, beetles and two-winged flies where the juvenile form is a larva which differs fundamentally from the adult, usually having different mouthparts and a totally different feeding habit. When the larva is fully grown it changes to the adult form by passing through a physically quiescent but physiologically active stage, the pupa or chrysalis.

In the first group, the insects which feed on plants do so both in the adult and nymphal stages. For example, aphids at all stages of growth may be found feeding on plum and apple, soft fruit, beans, brassicae and lettuce. Thrips feeding on peas and capsids on fruit trees are also included in this category.

In the second group most plant destruction and damage occurs as a result of feeding by the larval stages. Caterpillars of moths and butterflies are voracious feeders, larvae of codling moth feed inside apples, cabbage caterpillars on brassicae and cutworms on the roots and shoot bases of various plants. Among the beetles, Colorado beetle larvae attack potato foliage, raspberry beetle larvae feed inside the developing fruits of raspberries and loganberries, wireworms or larval click beetles feed on the underground parts of such plants as cereals and potatoes and grain weevil larvae destroy individual cereal grains by feeding within them. Among the two-winged flies, cabbage root fly larvae feed on brassica plants especially plants in the seedling stage; larvae of carrot fly, celery fly and onion fly feed on appropriate vegetables while wheat bulb fly larvae and frit fly larvae are common pests of grain crops. A general feeder on roots is the larval crane fly or leatherjacket. The larvae of narcissus flies feed in daffodil and narcissus bulbs, and secondary rots frequently destroy the bulbs. Among adult insects in this group, beetles and weevils frequently feed on plant tissue. Weevils feed on pea, bean and clover plants, apple blossom and grain, Colorado beetles destroy potato foliage and flea beetles eat holes in seedling brassica and other leaves. Other insects worthy of mention are wasps which damage ripe fruit and leaf-cutter bees which do not feed on plant tissue but make their nests of segments of leaves.

Of the other creatures which are allied to insects, those which damage plant growth include millepedes, symphylids and spider mites; slug and nematode damage also falls within the province of the entomologist. Millepedes at all stages in the life cycle feed on vegetable matter such as roots, bulbs and tubers and they are often abundant at the sites of primary damage caused by other pests such as slugs. Symphylids most commonly attack glasshouse crops such as tomatoes, cucumbers and lettuce, feeding on the root systems until the plants wilt and die. Red spider mites are most destructive, feeding on glasshouse crops and also attacking outdoor crops especially when the weather is hot and dry. The mites suck sap from leaves and sometimes spin webs over the foliage causing it to wither and die.

## 1.2 DESTRUCTIVE INSECTS

The foregoing paragraphs show that insect damage to crops is mainly associated with feeding, and reference must now be made to various types of feeding. Some insects feed by chewing plant tissue and others by sucking sap from the plant. Secondary damage occurs when waste products such as excreta and honeydew spoil the produce.

(*a*) **Chewing insects** damage crops in many ways according to the form of the mouth-parts and the part of the plant attacked. *Leaf eaters* include various caterpillars such as those of cabbage white butterflies, cabbage moths and diamond back moth which cut away and eat sections of leaf tissue. Adult weevils attack peas, beans and clover by eating U-shaped notches in the leaf margins, often considerably reducing the leaf area. *Stem borers* such as the larvae of frit fly and wheat bulb fly have the mouth-parts reduced to a pair of hook-like organs which rasp away the internal conducting tissue of cereal plants. Shoots are frequently killed in this way and the crop loss is often extensive. *Miners* tunnel and feed between the upper and lower epidermis of the leaf. Damage is caused mainly when the plant is attacked in the seedling stage as in tomato. On a crop of chrysanthemums, leaves may be disfigured to such an extent that they must be removed before the flowers are sold. *Root-eaters* diminish the vigour of the plants by eating and severing the roots. Wireworms, cutworms and leather-jackets occur generally on farm and market garden crops. Clover weevils attack roots just below the soil surface often severing the main root, while pea and bean weevil larvae destroy root nodules. Some insects such as swede midges lay their eggs near the growing point of their host plants, and subsequent feeding by the larvae may destroy the shoot, causing such damage as blindness in cauliflowers. Many insects destroy fruits and seeds; blossom weevils lay their eggs in the flower buds of strawberries and then puncture the stalks so that the buds wither. Raspberry beetles, codling moth and wasps have already been noted as other pests of fruit. Although insect damage to a plant may be relatively insignificant in itself, the damaged part of the plant often allows entry of disease organisms which cause the plant to decay or rot.

(*b*) **Sucking insects** have mouth-parts adapted for piercing plant tissue and sucking out sap. Within the needle-like proboscis, or beak, are two fine tubes through one of which saliva from the insect may pass into the plant and through the other sap from the plant into the insect.

In temperate areas of the world, great damage is done by aphids. Their capacity for rapid build-up on crops is brought about by the fact that they bear living young which results in a telescoping of generations and enables them to produce more generations in a year than exclusively egg-laying insects. Apart from the debilitating effect on the plant of large numbers of aphids feeding on the sap, otherwise profitable parts of the plant are often rendered unsaleable because of physical deformation or fouling by cast skins and honeydew. The latter is made up of the unused part of the sap which has passed through the aphid body and been ejected. Moulds grow freely on these sugary excreta which become even more objectionable.

Although the entry of fungi and bacteria into the punctures made by these insects can cause secondary damage, the greatest potential danger lies in the transmission of viruses by the aphids to the plants. Viruses such as raspberry mosaic, sugar beet yellows, barley yellow dwarf and those which attack brassicas and potatoes are spread within and between crops by winged aphids. Virologists recognize two main types of virus, the persistent and the non-persistent. To acquire and transmit the former, an aphid must feed on an infected plant. The sap containing the virus passes from the plant into the insect gut. From here, the virus migrates to the salivary glands from where it can be injected with the saliva into a healthy plant when the insect is feeding. The migration of the virus from the gut to the salivary glands of the insect may take hours or even days. On the other hand, non-persistent viruses are easily acquired by the aphid merely probing within the epidermal cells of the plant, a common procedure before feeding commences. The virus is carried on the mouth-parts and is thus readily transmitted from one site to another in a few minutes.

The general effect of virus infection is to reduce the growth and yield of the plant and to reduce the vigour of the vegetative offspring.


## 1.3 USEFUL INSECTS

(*a*) **Pollinators.** Flower-visiting insects, including hive bees, are of great importance in the pollination of seed-producing crops and fruit. The use of insecticides when bees are working the crop can seriously reduce their numbers and possibly the yield of the crop. The residual effects of insecticides vary but it is usually safe for bees to work in an area 48 hours after an insecticide has been applied. Different formulations of the same insecticide, particularly chlorinated hydrocarbons, vary in their toxicity to bees. Some honeybee colonies are known to have developed a partial resistance to some insecticides, for example, chlordane, DDT and methoxychlor, but there is no evidence of resistance to others such as gamma-BHC, dieldrin, aldrin, heptachlor, malathion, parathion, and carbaryl.

(b) **Parasites and predators.** Insects become pests usually because they grow, feed and breed quickly on their host plants. Numbers usually fluctuate, a decline being generally associated with unfavourable weather factors such as cold or heavy rain over a period of time, a dwindling supply of food as the crop is progressively destroyed or becomes too small to support the rapidly increasing numbers of insects, or with the activities of insect enemies. The latter may be predators of the pest, using it as a source of food, or parasites which use the pest insect as food and shelter for their developing young. The interrelationships between pests and their enemies are complex and vary from example to example. But in general, as the pest insects increase in density, so the available enemies are able to find them more quickly, with the result that the populations of parasites and predators increase. Eventually there are so many individuals searching for prey that the pest numbers decline and shortly thereafter the parasites and predators are themselves restricted for want of hosts. Thus, many pest fluctuations may be due to the effect of their insect enemies, although it must be noted that often a pest insect can reach epidemic proportions before its enemies are sufficiently numerous to effect any worthwhile degree of control.

A good example of the value of predatory and parasitic insects can be seen on almost any large colony of aphids. Observation will show that ladybirds of various species eat these pests and also lay their eggs on the infested plants. When the larvae hatch they thrive on the ample food supply with which they are surrounded. Hover fly adults feed on nectar but lay their eggs on infested plants; the larvae move over the plant searching for aphids and may eat 400–500 before they pupate. Other larvae commonly found feeding on aphid colonies are those of certain midges and lacewings. In addition, minute wasps oviposit into aphids and the larvae feed on, and then pupate within the host body. Parasitized aphids are easily recognized for they are swollen, shiny brown and attached to the plant by silk produced by the developing parasite. This is not the whole story for there are 'lesser fleas', hyperparasites which lay their eggs on the developing parasite larvae within dead aphids. These are obviously not useful insects inasmuch as they destroy the potentially valuable parasites.

## 1.4 CONTROL MEASURES

Any insect which destroys part of a pest population is exerting some measure of biological control; known pests can be influenced by the cultural operations imposed on the crop by the grower; an insect which is a pest in one country may be prevented from establishing itself in another by legislative action; finally, if an insect does increase in numbers to the stage when it is doing an economically significant amount of damage, it can usually be controlled by chemical means.

(*a*) **Biological control.** There have been cases where serious pest problems have been solved by the introduction of parasites and predators but these are relatively rare (about 20 cases in the last 50 years). In Britain, two pest insects may be controlled by parasites. Woolly aphid on apple trees is attacked by the parasite *Aphelinus mali*; parasitized aphids may be overwintered in cold stores and put out in the following early summer when the pest is beginning to build up. Those which overwinter in the open are unaffected by normal tar oil and DNOC, and early sprays of DDT, BHC, nicotine and lime sulphur have little effect on subsequent emergence of the parsites from their dead hosts. Once the parasites are active, however, insecticides have a lethal effect.

In glasshouses, the whitefly found on plants such as cucumbers and tomatoes may be controlled by the introduction of the parasite *Encarsia formosa* The parasite attacks the immature stages of the whitefly and completes its life-cycle in approximately 5 weeks. As each parasite may lay up to 50 eggs, a severe whitefly infestation can be controlled in a few months. To ensure that the parasite does not die out, a small reservoir of whiteflies is kept by growing one or two plants which are attacked by whitefly but disliked by the parasite, e.g. Nicotiana.

More details of these and other examples of biological control can be found in M.A.F.F. Bull. 20, 'Beneficial Insects'.

(*b*) **Legislative control.** The Destructive Insects and Pests Acts, 1877–1927, form the basis of plant health legislation in Britain, the object being to prevent the introduction or spread of the more important foreign pests and diseases, and to minimize the planting of diseased material. Two examples will suffice to illustrate this aspect of control.

Colorado beetle, which spread from the U.S.A. into Canada and thence to Europe in 1922 is not present in Britain. Despite stringent precautions at all ports and airfields, there is a potential danger of the importation of plants carrying larvae, pupae or adult beetles. Anyone finding this pest is required to inform the Ministry of Agriculture; if the presence of the insect is confirmed, control measures are undertaken at public expense, the Ministry's policy being completely to eradicate the infestation. During the years 1946–1952, a number of breeding colonies of beetles were found and subsequently exterminated. Since then, there have been no cases of infestations developing in the field although each year live beetles and larvae are intercepted on imported plants of foodstuffs.

Beet eelworm attacks sugar beet and can cause severe crop losses. To prevent this happening on a large scale, the growing of susceptible crops on land known to be infested is prohibited under the Beet Eelworm Orders and the sale of plants grown on infested land is forbidden. A proper crop rotation does much to prevent a rapid build-up of this pest, and is the best possible insurance against trouble on farms where sugar beet is grown extensively.

(c) **Cultural control.** Crop rotations, normal farm management and hygiene, and the use of healthy planting material do much to prevent unnecessary losses due to insect pests. Other practices such as manuring and irrigation, which assist plants to grow strongly and healthily also help to prevent undue damage as a result of pest attack.

Cultivation of the soil at certain times of the year exposes many soil-living insects to predation by birds. Thus, populations of leatherjackets and wireworms can be reduced by birds following the plough in autumn and spring. Using the roller on crops attacked by wheat bulb fly restricts the ease with which the larvae move through the soil in search of host plants, may crush larvae within the stems and encourages tillering.

Changing an accepted sequence of cropping is often limited by practical considerations, but where heavy infestations of wheat bulb fly occur regularly, wheat should not be grown after a crop which provides the uncovered earth in July–September which is necessary for egg-laying by the adult flies.

The rotation of crops is a normal practice and is essential to prevent the build-up of eelworms which attack crops such as potatoes, cereals and bulbs.

Badly drained fields provide favourable habitats for slugs especially when the humus content of the soil is high. Damage consists of hollowing of cereal grains a short time after sowing, holing in plant stems, shredding of leaves, and irregular holes in tubers and roots. As these pests are difficult to control chemically, breeding sites and ideal conditions for the slugs should be minimized, for example by drainage.

Attention to the sowing date of a crop will often ensure that plants are not in a susceptible stage when a particular pest is ready to lay eggs. In areas where carrot fly is common, risk of infestation can be lessened by delaying sowing until the end of May. The crop will largely escape damage by the first generation, and the second generation of flies will be correspondingly smaller.

At the other end of the season, crops should not be left in the ground when they have completed their growth if they are susceptible to pest attack. For example, wireworms attack mature potatoes causing a serious reduction in quality. As this damage increases the longer the crop is left in the soil, lifting should begin as soon as possible after maturity is reached.

(*d*) **Chemical control.** There can be little doubt that the use of chemicals for the control of insect pests is of primary importance in agriculture, and later chapters in this handbook gave details of the wide variety of crop pests which can be controlled with these materials.

It is inherent in a competent grower's technique that his normal management minimizes the risk of insect attack as far as this is possible. His resort to chemical control means that additional money must be spent on the crop, but this may mean the difference between an economic return and a loss.

Earliest records of insect control refer to the use of materials which were so noxious to humans that their effects on insects were unquestioned. But by the second half of the nineteenth century, substances were being used which are still in use today; arsenical compounds were proving reliable against Colorado beetle and Gypsy moth in the United States of America and nicotine, which had been prepared for years as a simple infusion of tobacco leaves, effectively controlled such insects as aphids, thrips and mites. In the 1920's pyrethrum, produced from the powdered flowers of the pyrethrum plant, entered the insecticide market for the first time. It was the principal insecticide of medical importance during World War II until 1942, when the use of DDT became widespread. But pyrethrum, because of its rapid paralysing effect on insects and its relative harmlessness to mammals, is still extensively used, especially in fly sprays. The other principal insecticide available prior to 1939 was derris, the ground root of species of the genera *Derris* and *Lonchocarpus*.

Since 1945, the number of synthetic organic insecticides available for agricultural and horticultural use has increased enormously, and the two main groups which we shall consider are the chlorinated hydrocarbons such as DDT, and the organophosphorus insecticides such as parathion. DDT, probably the best-known of the chlorinated hydrocarbons, was developed during the early 1940's and used mainly against disease-carrying insects: it was immediately successful because of its effectiveness at very low concentrations and its usefulness against a wide range of insects. Its introduction was rapidly followed by the discovery of the high insecticidal activity of such compounds as aldrin, dieldrin, gamma-BHC and others of the chlorinated hydrocarbon group.

The general action of the chlorinated hydrocarbons depends on the insects eating treated vegetation or walking over a deposit of the insecticide applied to the plant or other surface. To be effective, careful application is required to ensure that the feeding site, or the area over which the insect will walk is covered with insecticide.

Phosphorus compounds are used extensively as insecticides. They are inhibitors of enzymes which occur in insects, particularly in the tissues of the nervous system. The actual chain of events by which an insect is killed is not completely clear but the nervous system is certainly damaged beyond recovery. In general, organophosphorus insecticides such as parathion act very rapidly but they are not so persistent as chlorinated hydrocarbon insecticides. In this respect, and to a certain extent, they are useful in that they overcome the problem of residues which remain active for a long time, either in the soil or the plant.

Several of these insecticides are systemic, that is, they are absorbed into the sap stream of the plant. The insecticide may be applied as spray or granules to the soil from which it is taken up by the plant roots, or by spraying the plant, when it is absorbed through the leaves. Sap-feeding insects, for example aphids, are more readily killed by systemic insecticides than by those which have a contact action. Parasites and predators are not affected unless they come into contact with the insecticide, that is, if the plants are sprayed.

Some insecticides of this type are potentially dangerous to man, having a damaging effect on the nervous system, but others are now available such as malathion, menazon and trichlorphon, which have a reduced toxicity to man while still being effective against insects.


## 1.5 ECONOMICS OF TREATMENT

From the foregoing sections it is apparent that insects attack a large number of crop plants and that vast sums of money may be spent annually in attempts to control them. Quite casual observation may show that damage is occurring but intensive work is necessary to show the extent of the damage, either in terms of loss of crop or lowering of profit obtained from selling the produce. The majority of the estimates of crop losses quoted early in this century were made as a result of intelligent observation by field entomologists. More recent work has aimed at determining these losses by critical field experimentation followed by systematic survey work. Crop losses may be expressed as the average loss of crop in a particular year, or the area of land equivalent to that from which the entire crop was lost, or in terms of financial returns. The last is the most difficult to evaluate because prices fluctuate with the supply/demand ratio, and when a commodity is scarce it is more expensive than when it is freely available. Thus, a 15% loss of crop might not lead to a proportionate drop in income from that crop.

A further difficulty which applies to all loss assessment is that although the individual effects of various pests and diseases of a crop may be determined, they may not be additive. Thus, an estimated loss of crop due to attacks by several pests may be much greater than the actual loss. In addition, the compensating factor which results in healthy plants adjacent to dead or dying ones giving more than the average yield can upset predictions of loss of yield. If a crop of uniform size is required for some particular market the last factor can be of importance.

When the economic aspects of crop loss are being considered the over-riding question is whether the value of the extra crop to be expected as a result of pest control is greater than the cost of chemicals and applications. On quality crops such as fruit, protection is nearly always worthwhile for the slightest blemish reduces the quality rating on the open market and the value is correspondingly decreased. On the other hand, there is little point in protecting fodder crops from insect attack unless this results in a substantial increase in yield. Between these extremes comes a host of problems associated with the limits to which pest control should be taken. One insecticidal application may give only 70% control of a pest; is it worthwhile making a second application to control a further 20%, and another to obtain almost complete control? Definite answers to such questions relating to specific pests and crops are difficult to give, mainly because there are few results available of actual damage assessment. Obviously too, conditions from field to field, and season to season vary enormously. Usually, the grower relies on his experience, or that of his advisers, in deciding whether and when to spray. But often by the time an insect has reached pest status, the damage to the crop has been done. For example, by the time colonies of wingless aphids are noticed feeding on crops, their winged parents have possibly spread virus to several localities in the area. Extensive work on the relationship between cabbage aphid attack on Brussels sprouts and yield has indicated that early infestations of a few aphids per plant are responsible for as much damage as that caused by the large numbers of aphids seen on plants during August and September. Obviously, early spraying, when the grower may not even be aware of the presence of the pest, is worthwhile in many cases. It has been estimated too, that frit fly damage to oats causes an annual loss in the region of £10 million, but about half the loss is caused by low pest densities which it is not economic to control.

Alternatively, it has been shown experimentally that an insect attack which looks dangerous may be causing little damage; extensive defoliation and loss of stand in the seedling stages of some crops has very little subsequent effect on yield. In experiments, loss of 50% of the leaf area of sugar beet seedlings at the 4–8 leaf stage, simulating the type of damage caused by such pests as mangold fly, reduced the yield of roots by only 5% while destruction of up to one half of the initial plant population resulted in a 10% loss of yield. In other experiments on peas, loss of 12·5% of the leaf area when the plants were at the 4 expanded leaflet stage led to a loss of 8% of the yield of shelled green peas. It was suggested that damage of this level by the weevil *Sitona lineatus* was the exception rather than the rule and that, if done at all, spraying should be restricted to the headlands of fields.

Finally, spraying a crop against a particular pest may result in a profit margin that is so small that it is essential that the material be applied at the critical time. Work in East Anglia on dry harvesting peas has shown how vital is the proper timing of spray applications.

This brief account, although restricted to a few examples, is sufficient to show that the economics of chemical pest control is a very complicated subject and the grower needs to be up-to-date in his approach to the problem, seeking unbiassed advice whenever possible before committing himself to heavy expenses which may not be recovered.

## 1.6 DEVELOPMENT OF RESISTANCE TO PESTICIDES

A population of insects on a crop is largely made up of individuals which are essentially the same genetically, a few differing slightly from the rest. When repeated insecticidal application results in the appearance of resistance among the population, the process has been one of preferential selection. Those insects with a genetic make-up favouring survival under the conditions which destroy the main part of the original population become dominant and subsequent populations are made up of this strain. Such resistance may be due to the insect not being affected by the insecticide itself or to its behavioural pattern effectively isolating it from sites which have received the insecticide. Such resistance can be seen when applications of insecticides which once controlled a pest no longer do so, and most frequently on crops on which routine control measures are the rule rather than the exception. Thus, 'insurance' spraying of orchards and the use of persistent insecticides as soil treatments may eventually produce resistant populations of the insect species they initially controlled.

There are many examples of resistance among insects in the United States of America and in the tropical regions of the world, where insecticides are used more freely than they are in Britain and where insect life cycles are shorter due to the more favourable climatic conditions. Codling moth in America, and mosquitoes and flies in the tropics are but three examples of insects in which resistance is found. Examples in Great Britain are the red spider mite on fruit trees and in glasshouses, *Myzus persicae* on chrysanthemums in glasshouses, carrot fly on carrots, cabbage root fly on brassicas and bean seed fly. Resistant races can be controlled by using insecticides of a different biochemical type: substituting the organophosphorus insecticides for DDT will destroy insects resistant to this material.

## 1.7 INCREASING IMPORTANCE OF PREVIOUSLY NON-DAMAGING ORGANISMS— FRUIT TREE RED SPIDER MITE

One of the drawbacks of the insecticide approach to pest control is the general effect on the whole fauna of the plant and soil. Although pest insects are destroyed, so too are many parasites and predators living within the field of application of the insecticide. This effect was known when tar oil washes for fruit trees were first introduced, but another unforeseen result soon produced a problem as great as the one which had been solved. Various insects, and the red spider mite began to appear in large numbers on the trees. With the advent of DDT, the mite began to occur in numbers never seen before. The killing of predatory and parasitic insects which previously kept the mite at unimportant levels was a disaster which serves to underline possible side-effects of spraying with non-selective insecticides. From its role as a relatively insignificant member of the fruit tree fauna, red spider mite is now one of the most troublesome pests with which the grower has to contend.

## 1.8 EFFECTS ON SOIL

Some insecticides are applied directly to the soil to control insects attacking plants at, or below, the soil surface. The insecticides used are aldrin and dieldrin and, where taint is not a problem, gamma-BHC. Before 1964, aldrin and dieldrin were widely used both as seed dressings and as directly applied soil insecticides. In 1964 however, the Advisory Committee on Poisonous Substances used in Agriculture and Food Storage recommended, and the Government accepted, restrictions in the use of aldrin and dieldrin. The crops which may be treated with these insecticides are listed in the current copies of the Agricultural Chemicals Approval Scheme booklet. Other chemicals, mainly in the organophosphorus group, are available, for example, disulfoton and phorate for the control of carrot fly, and diazinon for the control of carrot fly and cabbage root fly.

Recent surveys have shown that, in England and Wales, about one quarter of the potato crops for human consumption have been grown on land treated with aldrin against wireworm attack. Other insecticides, notably DDT, drop to the soil as run-off from treated vegetation, substantial amounts being found beneath treated potatoes, peas, soft and orchard fruits. These insecticides remain active in the soil for periods running into years and there are many problems associated with this persistence. It has been observed that the incidence of cabbage root fly on brassica crops grown on previously treated land has risen due to the adverse effects of aldrin on the beetle predators of this pest. In addition, it is apparent that there will be long-term effects on the soil fauna through these have not yet been critically evaluated. However, preliminary results suggest that beetle and fly larvae, and some mites are reduced in numbers on insecticide treated soil; and that aldrin reduces the springtail populations whereas DDT has the opposite effect. The mechanics of these population fluctuations, and the inter-actions between predators, prey and insecticides are currently under investigation.

Another aspect of the problem is the extent to which plant tissues become contaminated by insecticide passing from the soil into the crop. Very little is known about the theory of plant absorption and retention of insecticides and the measurement of residues is a time-consuming task; there are, however, recommendations concerning minimum periods which must elapse between treatment and harvesting of various crops. Obviously, all the problems associated with insecticide residues both in the soil and in crops will attract much attention and research in the next few years.

## 1.9 PHYTOTOXICITY

Chemicals applied to plants to control pests and diseases sometimes have a harmful phytotoxic effect. This may amount to no more than a temporary check in growth, as may occur where seed dressings retard germination but, in severe cases, extensive necrosis of leaf tissues and possible death of plants may result.

Knowledge of possible phytotoxic hazards is obviously greatest for those materials which have been widely used over a long period of time and it is well known for example, that some varieties of black and red currant are sulphur shy; that DDT is toxic to cucumbers and related plants, and has a harmful effect on certain barley varieties; that dimethoate and formothion should not be used on chrysanthemums. Where appropriate, such information is given in the text of this book, but it should be noted that new cases of phytotoxicity are likely to occur, particularly where newer materials and formulations are used.

## 1.10 CONCLUSIONS

At present, control of many pests can only be established by chemicals although the conditions which lead to their application need much more critical appraisal. Faced with secondary effects such as the development of resistance, residues in soil and plants and the effects on insect parasites and predators, there needs to be more discrimination as to when insecticides should be applied. More detailed ecological investigations need to be made to determine the value of beneficial insects; and possibilities should be explored of changing accepted cultural techniques to produce habitats less acceptable to pests. At the same time, more selective insecticides are required so that pests may be destroyed without altogether upsetting the attendant complex of non-damaging insects. The development and production of more specific insecticides implies higher costs; these might well be absorbed by the grower and cost/benefit ratios remain acceptable, if less spraying were done on an insurance basis or in an attempt to keep a crop completely clear of pests.

The application of pesticides should never be a hit or miss affair; this aspect of pest control should be critically timed and conscientiously carried out. In most cases it is necessary to obtain skilled entomological advice on the time at which the pest should be sprayed, but this is of little use if only a small proportion of the insects is touched by the pesticide. In the case of control of virus vectors, the treatment must be as near completely effective as possible.

# PLANT DISEASES

Most plants are liable to be attacked by several and sometimes many different pathogens: usually only a few are of major significance and the others occasionally of importance under certain conditions. Over the years, changes, often associated with changes in agricultural or horticultural practices, may alter considerably the pattern of predominant troubles. Most pathogens are restricted to relatively few hosts and these are often fairly closely related botanically. There are however several pathogens with a wide host range: the ubiquitous grey mould can be found on many plants, the Verticillium wilt fungus and cucumber mosaic virus have a very wide host range and Armillaria root rot can attack most woody plants. Some pathogens can be recognized by the characteristic symptoms that they produce and the common name for the diseases is often descriptive of this symptom: red core of strawberries and eyespot of wheat are two good examples. But in general plants often react similarly to invasion by different pathogens. Attack on the roots results in the yellowing of leaves and gradual death of the plant and only laboratory examination will reveal the culprit. Sometimes the causal fungus can be seen on the roots: Armillaria root rot and take-all of cereals both produce characteristic mycelium on the surface of affected roots. The fungi attacking through the roots and causing wilts are often identifiable only after they have been isolated in the laboratory from the affected stems. For all the many fungi causing leaf spots, stem cankers and fruit rots, the presence of characteristic spores or other structures enables the causal pathogen to be identified. For both bacterial and virus diseases much reliance has to be placed on the symptoms produced: the identification of the pathogens often involves the inoculation of plants or parts of plants.

Most plant pathogens belong to one of the three groups: fungi, bacteria and viruses. A brief outline of the main characteristics of these groups and their sub-groups may help those who are unfamiliar with them to appreciate the salient characters.

The fungi causing plant diseases can be grouped into four sub-groups. These sub-groups are distinguished from one another by the microscopic characters of the mycelium, of the spores and the organs on or in which the spores are developed.

The first sub-group are known as the Phycomycetes: they typically produce both motile spores, which depend on water for their spread, and long-lived resistant resting spores. In this group are the soil-borne organisms causing club root of brassicae, wart disease of potatoes and crook root of watercress; also included are the downy mildews with their best-known representative potato blight and the damping-off fungi belonging to *Phytophthora* and *Pythium*.

The second group consists of the Ascomycetes, again typically producing two different types of spore, one produced in vast numbers on the affected parts and the other developing either in a thick-walled perithecium, usually formed in or on dead host tissue, or a cup-shaped apothecium, normally arising from a resistant structure, the sclerotium. Included in this group are the perithecia-forming powdery mildews and the fungi causing apple scab, blackcurrant leaf spot and rose black spot: Sclerotinia disease and the closely allied clover rot belong to those producing apothecia.

The third group are the Basidiomycetes. They also often produce two or more kinds of spores but one is always formed on a characteristic structure known as a basidium. Into this group fall the smuts represented by loose smut of wheat and barley, and the rusts, species of which can be found on most plants. The smuts and rusts share with the powdery mildews, the characteristic that they are all obligate parasites: they survive only on living plants and have not yet been grown on artificial media in the laboratory. The so called 'higher' fungi also come into this group and are represented by silver leaf of fruit trees producing small purple brackets on dead branches and Armillaria root rot with its clumps of tawny toad-stools.

The fourth group is known as the Fungi Imperfecti; a miscellaneous group for most of which only one kind of spore is known and some have apparently none. Included here are leaf mould of tomato, leaf spot of celery and white rot of onion.

There are relatively few bacteria causing plant diseases in this country and they can conveniently be divided into five main groups corresponding to the genera of bacteria. These genera are differentiated largely on biochemical, physiological and to some extent also on pathological grounds. In the genus *Agrobacterium* are included crown gall and leafy gall: in *Corynebacterium*, silvering of beet: in *Pseudomonas*, bacterial canker of cherry and plum, halo blight of beans, and halo blight of oats: in *Erwinia*, fireblight and some of the important soft rotting organisms: and in *Xanthomonas*, hyacinth yellows and begonia bacterial blight.

The viruses, perhaps most aptly described as possibly the simplest type of organism, have so far defeated attempts to classify them satisfactorily. The common names given to them are usually descriptive of the symptoms on one of the host plants, often the one on which the virus was first found. Viruses are normally systemic throughout infected plants and consequently are readily transmitted in cuttings, layers, tubers, etc., of all vegetatively propagated plants. Unlike fungi and bacteria, viruses can only be spread from plant to plant by some external agency; insects, such as aphids, thrips and leaf hoppers, are the most common vectors, though some viruses are spread by nematodes.

## 1.11 FACTORS INFLUENCING DISEASE INCIDENCE

There are three primary requirements for the development of a disease: firstly the pathogen must be present or be introduced; secondly the host plant, or certain parts of it, must be susceptible to the particular pathogen; thirdly the environment including soil and weather conditions must be favourable for infection to take place and for spread to occur.

Knowledge of the likely source of each pathogen is of fundamental importance for the successful application of any control measure. Most pathogens need to survive for a considerable period of the year or sometimes longer in the absence of either their host or the tissues which they attack. They may be soil-borne, they may survive in debris from the previous crop, they may be seed-borne or perennate in buds or rootstocks or other parts of perennial plants. They may find volunteer seedlings or other hosts and then spread into the new crop from nearby, or occasionally far distant, sources as airborne spores, or for the viruses with their insect vector. Each of these sources will be considered in greater detail in the section below on control measures.

As has already been mentioned, most pathogens are restricted in their ability to attack plants and even the rusts, powdery mildews and downy mildews, which as groups cause rather similar symptoms on very many plants, are often individually specific and varieties of the same plant may or may not be susceptible. The factors responsible for these differences in resistance to attack are varied and often obscure but such resistance can be controlled by the genetic make-up of the plant and plant breeders have been able to incorporate resistance into new varieties of crop plants. In this country most success has been obtained with potatoes resistant to wart disease, with wheat resistant to yellow rust, with strawberries to red core and hops to Verticillium wilt. The appearance of races of fungi able to attack such resistant varieties is a constant source of concern to plant breeders.

Most fungal and bacterial pathogens attack only certain parts of plants: root pathogens rarely extend to above-ground parts and pathogens attacking the leaves, stems, flowers and fruits are often most damaging to one of these parts and only occasionally occur on the others. Many of these pathogens infect the tissues when they are at a certain stage of growth or maturity. For instance, the apple scab fungus infects the very young leaves: the typical greenish brown patches may take up to a month to become visible and the disease then appears on old leaves. It is thus understandable that the critical time for the establishment of scab on apple trees is in the spring when for some weeks, particularly the period from bud burst to green cluster, much of the leaf area on the tree consists of young leaves. For some storage diseases too, the age of the tissues they infect is of significance. The fungi causing dry rot and gangrene of potato can infect tubers at lifting time but they normally do not invade the tubers until later in the storage season when the tissues become more susceptible. A similar effect occurs with *Gloeosporium* rots of apple which become increasingly important causes of loss during storage: the commercial prolongation of the storage season can result in greater losses due to such storage diseases.

The nutrition of the host can influence the susceptibility of plants to disease. It is sometimes difficult to dissociate the effects of nutrition from the inevitable modifications in the microclimate and other environmental factors but there is no doubt that nutrients, particularly nitrogen, phosphorus and potassium can affect the incidence of some diseases. In general excesses or deficiencies often lead to greater susceptibility. Nutritional effects can alter symptom expression of virus diseases.

The environment plays an all-important part in determining the incidence of soil- and air-borne pathogens. Many soil-borne pathogens are sensitive to soil reaction. For example, alkaline conditions are unfavourable for the club root fungus. It can survive and attack under these conditions but a narrower range of other factors needs to be satisfied if the disease is to occur; the temperature becomes more critical and the spore load in the soil needs to be higher than in a more acid soil. On the other hand an alkaline soil reaction favours attacks of potato common scab and cereal take-all, which tend to be most troublesome in chalkland areas. The temperature of the soil can also affect the incidence of diseases. Under cold, wet conditions damping-off fungi can cause serious losses of germinating seedlings. Early sown peas are often severely attacked by soil-borne fungi unless protected by seed treatment with a fungicide. Potatoes planted into cold, dry soils can be attacked by the black scurf fungus often present on the surface of the tuber or in the soil: under good growing conditions the sprouts normally grow away unaffected.

For all fungi and bacteria the weather, both directly and indirectly in its influence on the microclimate around the plants, inevitably affects profoundly the occurrence and development of all the diseases that these organisms cause. Weather conditions affect their take-off and spread from the source, their transport to the host plant, whether they infect or not and, by affecting the subsequent local spread, determine the development of the epidemic. For most fungi and bacteria, temperature and humidity or wetness are the two most important factors. Temperature is often the factor limiting the geographical range of fungi and bacteria, whereas humidity determines their incidence in a particular crop in any one year. This close association of diseases with weather conditions is widely appreciated but the actual conditions that individual pathogens respond to vary greatly and are often critical. For most diseases the symptoms appear after the pathogen has become well established on its host and the conditions leading to the initial development took place weeks or months previously and subsequent favourable weather conditions will promote its rapid spread. Knowledge of the specific conditions enabling early infections to take place and subsequent spread to occur provide the basis for forecasting schemes such as those developed in this country to help growers combat apple scab and potato blight. Much more needs to be discovered about the life histories and weather relationships of our other important diseases.

Cultural operations can also influence the occurrence of diseases. The date of sowing can be a significant factor in the development of barley powdery mildew: late sown barley crops are often growing rapidly in early summer when spores are available on overwintered plants and the weather often suitable for disease development and spread. On some other crops it is the early sown ones that are likely to be the most affected. Wounding by pruning operations or during harvesting and subsequent handling can also predispose plants to infection. The silver leaf fungus invades pruning wounds on fruit trees and many storage diseases and particularly soft-rotting bacteria follow damage to the tissues.

## 1.12 CONTROL MEASURES

With very few exceptions the pathogens causing plant diseases penetrate into the host tissue and once they have done so are very difficult to control by the application of chemicals: measures can be taken to prevent spread but even these are unlikely to be effective unless they are applied at an early stage of the development of the disease. As fungi are themselves plants, the control measure that will kill the pathogen and leave the host unaffected is not always easy to find and phytotoxicity is a constant problem. For a few diseases heat treatment will eradicate the pathogen: hot water treatment of celery seed for the control of celery leaf spot, and of brassica seed for the control of canker affords effective disinfection but the temperature of the bath needs to be regulated carefully to prevent serious loss in germination. Heat therapy of chrysanthemum plants for the elimination of a virus disease is now being used commercially.

The powdery mildews are exceptional in growing superficially on leaf surfaces with only small 'pegs' penetrating into the cells below. It is possible, though not easy, to eradicate such infections with appropriate spray materials. It is also possible to eradicate early infections on other plants but, with apple scab for instance, sprays must be applied within a few days of infection: only occasionally can the fungus in established lesions be killed.

For the vast majority of diseases prevention rather than cure is essential if losses are to be avoided. Whenever research is focused on to a disease detailed studies of the life history of the pathogen become necessary and they almost invariably reveal that the adoption of careful cultivation practices must form the basis of any general protective programme for its control: the judicious use of chemical treatments can then be expected to play a decisive part. Assessment of the likelihood of a disease appearing and developing severely enough to justify a control programme is one of the many difficult problems that a grower has to contend with. Frequently it is not until a severe attack has been experienced that some action becomes necessary and then drastic measures may need to be taken and followed up the next season. The most important protective measure is the prevention of the introduction of the pathogen into the crop: the most dangerous source of infection in any crop is a diseased plant within it. It is perhaps unnecessary to stress the importance of planting only healthy planting stock in land in which soil pathogens have not been allowed to increase by repeated successive cropping. The following very brief summary of the control measures available will show that variations on this general theme, depending upon the source of the particular pathogen, form the basis for the control of all diseases.

Contaminated soil harbouring soil-borne pathogens causing root rots and wilts poses many problems. Under field conditions the careful selection of cropping sequences with appropriate intervals between susceptible crops will keep, or reduce, the pathogen to a low level insufficient to cause serious losses to a subsequent host crop. The length of the interval varies from pathogen to pathogen and depends on the form in which it survives.

It may persist on roots and stubble and die out as these disintegrate: hence for some cereal foot rots up to three years is needed. Other pathogens survive as resting spores or resistant sclerotia and for club root and clover rot an interval of up to eight years may be necessary. No methods of eliminating such pathogens from large areas of soil are available but for some soil-borne diseases there are chemical control measures which effectively protect plants planted out in the soil: young brassica plants can be grown in club root contaminated soil after they have been dipped in a fungicide and onion seedlings can be protected from attack by white rot and smut by seed treatment. Under glasshouse conditions the build-up of soil-borne pathogens can be dangerously rapid and the economics of the industry are such that rotation of crops is impracticable. Routine soil disinfection has become an essential part of the intensive cropping systems associated with the production of crops of tomatoes, lettuces, carnations and other plants in glasshouses. But treated soil may be quickly invaded by reintroduced pathogens and even greater care is needed to ensure that planting stock is healthy.

Infected debris is another common source of infection and much can be done by general hygienic measures to eliminate this source. The effective disposal of tomato haulms, which can become a prolific source of Didymella stem rot infection, and of hop vines from gardens affected by Verticillium wilt should be routine procedures. Blighted potato tubers discarded at riddling near clamp sites can become the starting point for early local spread of the disease. For some diseases for which infected debris constitutes a major source of infection, no commercial methods of destruction have as yet been devised. Overwintering dead leaves containing the perithecia of apple scab and black currant leaf spot, on the orchard and plantation floor, are not easy to destroy and spraying of the young susceptible growth is at the moment the only practicable control measure.

Infection within the tissues of planting material is much more common than is generally appreciated: seeds, runners, tubers or cuttings all frequently carry unseen pathogens. The importance of maintaining stocks of healthy material for planting cannot be over-emphasized. For virus diseases of those plants that are vegetatively propagated, this is of the utmost importance and the Ministry of Agriculture, Fisheries and Food operates certification schemes for potatoes, hops, strawberries, raspberries and fruit tree rootstocks to ensure that a supply of healthy planting material is available for growers. Methods for testing cuttings of carnations for freedom from wilt pathogens have been devised and are being used by some commercial raisers of cuttings. A number of pathogens are commonly seed-borne, either occurring as contaminants on the surface of the seed or penetrating into the internal tissues of the seed. Seed of cereals and sugar beet can be effectively disinfected by seed treatment with mercury compounds and this is normally recommended as a routine treatment: hot water treatment for the control of the internally-borne loose smut of cereals, leaf spot of celery and canker of broccoli has been referred to earlier; and a thiram soak treatment is very effective for the control of some internal fungal pathogens. Seed potato tubers may carry blight and the fungus growing up into shoots from these tubers can be responsible for early local outbreaks from which spread will occur. Tulip fire is carried on the bulbs and infection of the young shoots as they emerge from the soil provides a source of spores which in their turn infect the leaves and flowers of nearby plants. Many other bulb diseases are also carried over from season to season in this way and very careful sorting of the bulbs before they are planted to eliminate those that show signs of a rot or indeed any signs of infection will help to prevent an attack on the new crop. The hop downy mildew fungus overwinters in the rootstock and appears in the spring on the infected 'spikes'. Apple powdery mildew overwinters in buds which became infected the previous summer: the mildewed trusses which appear in the spring are prolific sources of spores which give rise to infections of the new leaves and buds to complete the cycle. Systematic cutting of the mildewed trusses in the spring together with a routine spray programme with a suitable fungicide will reduce incidence of this disease. The fungus causing peach leaf curl also overwinters in the buds or on the stem. Some fungi and bacteria establish themselves as cankers on fruit trees and sometimes cutting-out can be a very useful aid to their control.

All these sources of initial infection are often few and far between and difficult to detect but, when conditions are favourable, the speed with which they can build-up, spread and cause widespread damage is remarkable. The application of an appropriate protective fungicide before infection is noticed and has made an appreciable headway can often prevent the development. But to ensure that this protective treatment is effective many other factors need to be considered. The timing of the application of the fungicide, both for the first application and for subsequent treatments to cover new growth and to renew deposits that have been washed off by weathering, can play a very significant part. Experiments and experience have established useful routine spray programmes for several diseases and new information is continually being obtained which enables some of the more empirical programmes to be replaced by more concise and rational ones. The method of application of the fungicide on the diverse types of plants involved in row crops, orchards, etc., provides another set of problems: these will be dealt with in a later chapter. The fungicides themselves also pose questions: the choice of an effective material is perhaps fairly easy as the fungicides at present available are limited in number and are relatively specific. But this specificity can complicate matters particularly when more than one disease is, or may be, present on the same plant and two spray programmes with two different fungicides need to be integrated into one programme and into which insecticides also need to be incorporated. There are also considerations of cost, of possible phytotoxicity and almost certainly also other more local or personal aspects to be taken into account.

There can be no doubt that an appreciation of the biological background to each disease in relation to the cultivation of its host plant will provide the sound basis for the effective control measures to be described in the following chapters.

# CHAPTER 2

# THE CHEMICAL BACKGROUND

Although the use of chemicals for the control of the pests and diseases of crop plants is of long history, the greatest development, the advent of the synthetic pesticide, is but 30 years old. The range and number of compounds now available as pesticides is of surprising extent and their classification has to be on the basis of their chemistry though the grower will probably have no inclination to delve deeply into this subject. Nor will it be necessary for him to do so, for the chemicals can be classified on the basis of the particular pests or diseases against which they are used, a classification helpful in its indications of the method of usage but which suffers the defect that a given chemical may appear in more than one category.

## 2.1 INSECTICIDES

Compounds used mainly against insect pests are grouped as insecticides—an old term used in the days before it was found necessary to differentiate between the true insects and certain other groups of arthropods such as the red spider mites which have become troublesome pests. This wide, older definition is still of use and if a distinction is necessary, the 'insecticides' used against mites may be termed acaricides. Even among those compounds used against true insects, it is useful to classify according to the main group of insects against which they are effective, hence such terms as aphicide—effective against aphids—scalecide—effective against scale insects—etc.

Insecticides may also be sub-divided according to the way in which they penetrate into the insect body and thereby incapacitate it. Some, of necessity of greater or lesser volatility, enter the insect through its breathing pores and hence act as fumigants, though this term is seldom used outside of actual fumigation, a process not widely used by growers nowadays except for soil treatment. Nicotine, for example, the most effective of the older aphicides, acts mainly as a fumigant though applied in liquid or solid dilutions to its victims exposed on the leaf surface. In soil use, however, it is possible to use an insecticide of comparatively low volatility for the vapour is entrapped by the soil particles and retained for times long enough to be fatal to the soil pests against which it is applied.

Or the insecticide may be swallowed by the insect as it eats its way through the foliage, when it is called by the self-descriptive term of stomach poison. This term indicates that the poison must be applied to the foliage or plant tissue which the insect will eat and that, to be effective, it must pass from the gut into the body tissues of the insect, considerations to be borne in mind in deciding the time of application of the insecticide.

A third and the most important of the types of insecticide is the contact poison—a chemical which kills by contact with the exterior of the insect and which, therefore, must be capable of passing through the insect integument. Obviously such an insecticide must be applied at a time and in such a manner that the insect cannot avoid contact with it. With the older contact poisons such as nicotine and pyrethrum, insecticidal activity was soon lost after application—hence the term direct contact insecticide. But many modern contact insecticides retain their activity long after application and the insect may be killed even though only its feet come into contact with the insecticide deposit. The time of application is therefore less critical and the insecticide is able to protect the plant from infestation for some time after application; it is a protective contact insecticide.

The contact insecticides also include those compounds, such as the tar oils and DNOC applied when the tree is dormant. Such sprays were known as winter washes, though the term ovicide better indicates their purpose, for their main practical use is against the eggs of mites and insects.

Finally there are those insecticides which, when applied to the foliage or roots of growing plants, are taken up and translocated by the plant which thereby becomes systemically toxic to pests feeding on its sap or tissues. Such compounds are known as systemic insecticides. Their greatest role is the protection of young seedlings for which purpose they are applied in suitable form at the time of sowing or the emergence of new growth.

## 2.2 FUNGICIDES

Fungicides applied to foliage fall into two groups; those used against fungi the mycelium of which is exposed on the surface of the leaf, for example, the powdery mildews; in general, these fungicides kill by direct contact. But a larger group of fungi causing foliar diseases live within the plant tissue inaccessible to a surface-applied chemical. Against such fungi the primary object is the protection of the foliage from infection—such fungicides are protective and clearly should be applied at a time prior to the arrival of the fungal spores which otherwise would infect the leaves.

However if application is too late and the plant has become infected, it is still possible in certain cases to apply a fungicide which will kill the infected tissue, such a compound is usually called an eradicant fungicide.

Fungicides applied to 'seed'—using the word in its widest sense to include tubers, corms, etc.—are primarily direct in action against surface-borne spores but, in some cases, the fungicide is required to persist on the seed coat long enough to be effective against dormant mycelium contained within the seed. When applied to the 'seed' before sowing, the fungicide is called a seed disinfectant or seed dressing, though the latter term may include treatments not intended to counter seed-borne fungi or soil pests.

## 2.3 FORMULATION

At one time the grower intending to use pesticides would purchase the active component in more or less pure form, as with flowers of sulphur, or he would prepare the sprays he used from simple ingredients at the site of use. Bordeaux mixture, for instance, would be made in the spray tank from hydrated lime and a stock solution of copper sulphate; or he would dissolve soft soap in the spray tank and add to it the required amount of nicotine in the form of the 95% alkaloid. But with the modern pesticide it is rarely that he would handle the separate components for they are marketed only as formulated products ready for use.

### FORMULATION FOR SPRAYS

If for spraying, such formulated products need only to be added to water in the spray tank. As many of the active pesticides are insoluble in water, they must be formulated to a product which rapidly disperses to give a uniform dilution. If solid, it may be ground with, or treated with, other components to give a powder which, on addition to water, disperses automatically to give a suspension of fine particles. Such a preparation is called a wettable powder, usually abbreviated to w.p., the special properties of which rest in the surface-active component (or wetter) used to render the particles easily wetted by water. The content of active ingredient (a.i.) is usually adjusted to a suitable round number by the addition of inert carrier.

Alternatively if the a.i. is liquid or if it is to be used in a liquid such as oil which is not miscible with water, the usual method of formulation is the so-called emulsifiable concentrate (e.c.), at one time also called a miscible oil. When added to water the emulsifiable concentrate spontaneously disperses as fine droplets to form an emulsion. In this case, the special property of self-dispersion is imparted by the use of surface-active components (surfactants) which serve to prevent the re-coalescence of the oil droplets thereby giving an oil-in-water emulsion. Because by far the greater bulk of the emulsion when diluted to spray strength is water and because of the powerful emulsifiers now available, the so-called inverted emulsion, in which the water droplets are dispersed in oil, is no longer likely to trouble the grower. The older stock emulsion (s.e.) a concentrated emulsion prepared by milling oil, water and emulsifier and containing around 70% of oil, is often marketed these days.

Modern formulations have made it a simple matter for the grower to prepare his sprays. Rarely will he be required to do more than to add the requisite amount of wettable powder or emulsifiable concentrate to water in the spray tank. The need for the addition of supplementary materials to improve the wetting properties or the adherence of the spray deposit sometimes arises and care is still necessary when using mixtures of formulations to ensure that there is no unwanted interaction between the components of different formulations. In technical jargon the formulations must be compatible and the label will provide the necessary precautions. But a word on the reasons for care will be helpful. Wetting agents, emulsifying agents and other forms of surfactant fall into three broad groups according to their chemistry. Soaps and many of the modern surfactants used in wettable powders are the alkali salts of certain organic acids and, because the surface activity is due to the structure of these acids, they are called anionic surfactants. In a contrasted group, the special properties which confer surface activity reside in the basic part of the compound, they are cationic surfactants. If cationic and anionic wetters are mixed a reaction will occur resulting in the precipitation of the acidic and basic components as a water-insoluble grease with consequent loss of surface-active properties. Among the cationic surfactants are several compounds used as fungicides and care is therefore needed when their formulations are mixed with others.

In the third group of surfactants, these difficulties are avoided for, not being salts, they are not ionized in solution, hence they are called non-ionic surfactants. This particular type of incompatibility will not arise.

Wettable powders and emulsifiable concentrates are prepared for dilution with water to spray strength, the extent of dilution being determined by the amount of a.i. or of spray to be applied per acre. It requires energy to break up the spray into droplets and to impart to these droplets the velocity needed to carry them to the surface to be sprayed. At one time this energy was supplied, via the spray pumps, to the spray itself and, in this so-called conventional or H.V. spraying, large amounts of spray were applied per acre, naturally of a dilute spray. This method is still in wide use but machines are now available which use a strong current of air to convey the spray droplets to their target—a method which permits the use of smaller amounts per acre—naturally of a more concentrated spray—

hence the technique is called concentrate or L.V. spraying. These matters are discussed in Chapter 3 but a more recent development may be mentioned here though it is applicable only to the spraying of ground crops. In this method the spray is gravity fed and merely shaken on to the foliage by a to-and-fro movement of the nozzles.

The production of droplets by the use of a strong current of air is called atomization, though this term exaggerates the degree of reduction of droplet size. For this purpose, a solution of the a.i. in a volatile solvent an 'atomizing solution' is often used. A more accurate description of the spray so produced is 'aerosol', though this term is better limited to the spray mist produced by a device in which the liquid is dispersed from a pressurized container containing a solvent of low boiling point which, in fact, boils when the pressure is released. The formulation used is here a solution of the pesticide in an organic solvent containing a suitable propellant liquid, such as methyl chloride. The pressurized can has to be of small dimensions and the aerosol is a handy tool for the amateur when the target to be sprayed is small or for use in an enclosed space such as the frame or glasshouse. An alternative for use in an enclosed space is the 'smoke' in which the pesticide is dispensed by the use of heat usually generated from a pyrotechnic mixture, a method applicable only if the pesticide is heat-stable.


FORMULATION FOR DUSTS

If the pesticide is to be used as a dust, formulation is a comparatively simple matter, the compounding of the active ingredient into a form suitable for the purpose for which the dust is to be used. In most cases this process is the intimate grinding of the a.i. with an inert carrier or filler but certain qualities are required, such as freedom from caking tendencies when the product is on the shelf or in the store. If intended for application through a dusting machine, it obviously must flow readily and not 'ball' in the hopper or clog the air-ducts. Ground sulphur, for example, in notoriously prone to ball in the hopper of the dusting machine but this tendency, due probably to static electricity, is easily cured by the admixture of a small proportion of some other dust.

If to be used for the treatment of 'seed', the dust should adhere well to the dressed seed yet not so readily disperse that a dust hazard to the operator is created. This fault can be corrected by the addition of anti-dust agents.


FORMULATION AS GRANULES

For certain uses, there are advantages in the application of the a.i. in the form of small granules, a method which has become widely used for the treatment of seedling crops with systemic insecticides (see p. 78). The ease of handling and application greatly reduce hazards to operators.

## 2.4 THE ACTIVE COMPONENTS

It is now necessary to consider the chemicals employed as the active components of the formulated pesticide. As most of these chemicals now have common names agreed by the British Standards Institution, they may be simply arranged in alphabetical order. In this way, the reader may, at this stage, be spared the long, and to him probably unintelligible, chemical name though it is important that he should know something of the chemical affinities of the compound for, should he be unfortunate enough to encounter resistant strains of pest, he will have to select an alternative pesticide outside the particular group to which resistance has arisen. This chemical information has, for convenience, been placed in the Tables of the Appendix.

GENERAL NOTES ON PRECAUTIONS

It is self-evident that a compound used because it is toxic to an insect or fungus, may also be toxic to other forms of life. Precautions have to be taken to ensure the pesticide is used in a manner which will not expose other organisms, whether man, stock or wildlife, to danger. This topic is discussed in detail in Chapter 4 but there are general precautions to be taken even if the pesticide is not specified in the Agriculture (Poisonous Substances) Regulations and certain of these precautions may, for emphasis, be dealt with here for they will bear repetition.

For the protection of operators handling formulations for dilution before use, that is to say w.p., s.c. and e.c. formulations, the instructions may include: Wear rubber gloves when handling the concentrate; wash off with soap and water any concentrate on the skin; avoid inhaling mist from diluted material during application; wash the hands and other exposed parts of the body with soap and water on the completion of any operation and before eating, drinking or smoking. For those handling dust formulations, the instructions may include: do not handle the dust unnecessarily; avoid inhaling the dust; wash the hands and other exposed parts of the body on the completion of the operation and before eating, drinking or smoking.

Operators engaged in fumigation or the use of aerosols should take the following precautions: wear rubber gloves and face-shield and, if inhalation is unavoidable, wear an efficient respirator or suitable mask; if the skin is contaminated, wash thoroughly with soap and water and wash before eating, drinking or smoking; also wash rubber gloves inside and out after use and do not wear contaminated clothing longer than necessary.

For the protection of the consumer against possible contamination of his food, a time interval may be laid down under the Pesticides Safety Precautions Scheme between the last application of the pesticide and the harvesting of the treated crop. As this time is determined by the persistence of the pesticide deposit, it will be stated in the description of the compound concerned. Similarly for the protection of stock and poultry and for those engaged in the cultivation, thinning, pruning, etc., of the treated crop, access to the treated area should not be allowed until the lapse of a time interval specific for each chemical. Empty containers should be destroyed and disposed of out of reach of children. As many pesticides are powerful fish poisons, care is required to avoid the contamination of ponds, rivers or streams by the pesticide itself, by drainings from spray tanks, by wash water used to clean appliances or by empty containers.

Pesticides, and in particular insecticides, should not be applied to crops in flower, for bees and other pollinating insects would then be exposed; but these beneficial insects also visit flowering weeds, hence grass orchards should be mown before spraying and flowering weeds in plantations should be kept down.

Given effective pesticides, the responsibility for their safe use rests solely on the user. The discharge of this responsibility requires only common sense but, for the assistance of the user, his role has been defined in the 'Code of Conduct', a subject which belongs to Chapter 4, p. 95.

GENERAL NOTES ON USAGE

It is unfortunately impractical to give, in the description of a particular compound, the concentration to be used in pest control for, with each pest, the lowest effective concentration is recommended for reasons of both economy and safety. Such particulars must therefore be reserved for the disease or pest to be controlled. But even there the grower has to use a particular formulation so it might appear more convenient if the concentration of the formulation were given rather than the concentration of active ingredient. However with the older pesticides and those for which patent protection has expired, the grower is faced with a range of competitive formulations and it was to help him in his selection that the Agricultural Chemicals Approval Scheme was established. Also the concentration of a.i. in these formulations is seldom uniform and to phrase recommendations in terms of formulation would require lists of alternatives far longer than the somewhat tiresome lists of a.i. alternatives which he will meet in later pages.

**Aldrin** (36)*. This chlorinated hydrocarbon is of the cyclodiene group, so-called because cyclodiene derivatives are used in its manufacture. Its insecticidal properties, discovered in 1948, render the compound a potent

* The numerals refer to the Appendix, pp. 348–361.

stomach poison and contact insecticide, particularly effective against soil insects. It is of low solubility in water and somewhat volatile so that, of itself, it is not very persistent. However, aldrin is converted by biological oxidation to its epoxide, dieldrin, which is extremely persistent. It is compatible with almost all other pesticides, though certain minerals, used as diluents, induce a catalytic dehydrochlorination avoided by the incorporation of an inhibitor in w.p. and dust formulations. Dust formulations usually contain $1\frac{1}{4}$–2% aldrin and e.c. for spray preparation 30%. It is also used, at 30%, in seed dressings.

Because of its conversion to dieldrin and the uncertainty of the long-term effects of the presence of small amounts in the animal body, the uses of aldrin have been limited by agreement under the Pesticides Safety Precautions Scheme. The currently agreed uses of the dust and liquid formulations are: control of wireworm in potatoes at the rate up to 48 oz a.i./acre for sprays, 67 oz a.i./acre for dusts; of leatherjackets on DDT-susceptible varieties of spring barley at 24 oz a.i./acre for sprays, 45 oz a.i./acre for dusts; of cabbage root fly on brassicas as 42 oz a.i./acre; of narcissus bulb fly at 6 oz a.i./20 gal dip or at 7 lb a.i./acre as a dust, or at 3 lb a.i./acre as a spray, dust or spray being applied to the drills when planting out; of hop root weevil as spray or dust at 24 oz a.i./acre; of vine weevil at 2 oz $1\frac{1}{4}$% dust/bushel of potting compost; of strawberry seed beetle at up to 80 oz a.i./acre applied not less than 3 weeks before harvesting begins. The seed dressing may be used on winter-sown wheat up to the end of December where there is a real danger of attack from wheat bulb fly, at a rate of 0·8 oz a.i./bushel.

**Azinphos-methyl** (4) is an insecticidal and acaricidal organophosphorus compound first introduced in 1954 and since used for the control of a range of phytophagous insects and mites. Being relatively stable and of low vapour pressure, it is moderately persistent, and it is included in Part III of the Regulations, see p. 87. The least time interval permitted between application as a spray and the harvesting of edible crops is 3 weeks in the case of apples, pears, plums, blackberries, loganberries and brassicas; 2 weeks in the case of blackcurrants, gooseberries, strawberries, peas, mustard and brassica seed crops. Not more than five applications, each at 8 oz a.i./100 gal as e.c. or 10 oz a.i./100 gal as w.p., may be made in any one season to apples and pears; four applications each at 7 oz a.i./100 gal to strawberries; three, each at 6 oz a.i./100 gal, to plums; two, each at 7 oz a.i./100 gal, to blackcurrants and gooseberries; one at 4 oz a.i./100 gal to redcurrants after fruit-picking; one at 3 oz a.i./100 gal to blackberries and loganberries, applied at bud burst; two, each at 7 oz a.i./100 gal, to mustard and brassica crops grown for seed, applied before the flowers open; two, each at $4\frac{1}{2}$ oz a.i./100 gal/acre to peas. If applied L.V. the amount of a.i. used per acre should not exceed that which would have been applied

H.V. Stock and poultry should be kept from the sprayed area for 2 weeks.

Azinphos-methyl may also be used as a drench for brassicas, a single application at 8 oz a.i./100 gal applied one-eighth pint per plant at transplanting. As an aerosol it may be applied up to five applications each at $1\cdot5$ g a.i./1,000 ft$^3$ on cucumber and up to three at the same rate on tomatoes; the interval between last use and harvest must be at least 2 clear days.

The usual formulation is the 22% w/v e.c. but a w.p. formulation is available containing 25% azinphos-methyl plus $7\cdot5$% demeton-S-methyl sulphone for the control of the pre-blossom pest complex on top fruit.

**Azobenzene** (48). The acaricidal properties of this compound were first utilized in 1945 when, by virtue of its appreciable volatility, it was used for red spider mite control in glasshouses either as an aerosol or by painting on the steam pipes or by formulation in a pyrotechnic mixture for use as a smoke. Effective against the eggs and immature stages of mites; action is obtained at around $2\cdot5$ g/1,000 ft$^3$ but, at this concentration, there is a risk of damage to many plant species especially if dry at the root or in bright sunlight. Damage may also result if azobenzene is used within 7 days of an oil spray and it is well to cut all marketable blooms before treatment.

The reasons why azobenzene is toxic to mites are still unknown but they are probably associated with its molecular structure; reference to Appendix I shows a number of other acaricides of structure similar in that the molecule consists of two benzene rings linked by a 'bridge'. This bridged diphenyl structure is common to many insecticides of the DDT group, some devoid of useful acaricidal activity. Nevertheless if acaricide resistance to azobenzene develops, it would be wise to select an alternative acaricide outside the bridged diphenyl group.

**BHC**—*benzene hexachloride*. The insecticidal properties of this product were first put to use in the early 1940's. It is a mixture of several of the isomers of hexachloro*cyclo*hexane, a chlorinated hydrocarbon, of which the *gamma* isomer, see gamma-BHC, is the most toxic to insects.

The use of BHC is restricted by the readiness with which many treated crops, in particular potato and blackcurrant, acquire an unpleasant taint, a risk which is taken if susceptible crops are grown on land treated more than once in the current or the two preceding years. This period indicates that BHC is not rapidly decomposed in soil and a build-up may occur if it is applied annually on the same area. A period of at least 2 weeks should elapse between its use and harvest or the access of stock or poultry to the treated area. Both e.c. and w.p. formulations with various contents of gamma-BHC are available for spray preparation. The heat stability of the product permits its use in smokes, which may be enriched with gamma-BHC.

**Binapacryl** (65). This dinitro compound, introduced in 1960, has proved an effective acaricide and fungicide, particularly for use on apple against fruit tree red spider mite and powdery mildew. Up to seven applications, each at up to 8 oz a.i. (2 lb 25% w.p.)/100 gal H.V., or the equivalent/acre L.V., may be applied for this purpose; the interval between application and harvest should be at least 7 days and livestock should be kept from the treated area for at least 4 weeks.

**Bordeaux mixture.** The original copper fungicide for the protection of foliage and first devised in 1885. It is prepared by mixing a solution of copper sulphate with a lime suspension, a common recipe employs 10 lb copper sulphate crystals (bluestone) and 12·5 lb hydrated lime per 100 gal (for further details, see Chapter 6, page 121). The copper sulphate is specified in M.A.F.F. Technical Bulletin 1, as containing not less than 98% $CuSO_4.5H_2O$.

Fungicidal activity is associated with the slow formation, from the Bordeaux deposit, of water-soluble copper compounds, the ultimate toxicant being the cupric ion. The survival of the mixture for so long is largely due to the excellent adherence to foliage of the freshly prepared precipitate. The adherence of the precipitate deteriorates rapidly with time, so the spray should be used as soon as possible after preparation.

Bordeaux mixture is relatively safe to use, though care must be taken in its preparation as soluble copper salts are poisonous and the solution of copper sulphate should not be made in metal containers, otherwise the latter will receive a copper coating.

**Burgundy mixture** is a variant of Bordeaux mixture in which the lime is replaced by washing soda. The proportion of washing soda used is slightly greater, a typical recipe being 8 lb bluestone, 10 lb washing soda per 80 gal. The resultant spray is more likely to cause spray damage than is the similar Bordeaux.

**Calomel** (71) is the old name for mercurous chloride, long used both as an insecticide and as fungicide, though its strong phytotoxicity precludes its use on foliage except grass. Its action is thought to be due to its slow dissociation to metallic mercury, a reaction hastened by alkali, hence it is incompatible with lime and other alkalies. As an insecticide, its main use is against root maggots and, as a fungicide, it is used for the treatment of turf disease and against clubroot of brassicas and white rot of onion. For these purposes it is formulated as a 4% dust, though the undiluted chemical is often recommended as a dip or overall spray.

Although a salt of mercury, it presents, unlike the related mercuric chloride (corrosive sublimate), no serious toxicity hazard though its powerful purgative properties renders care in handling imperative.

**Captan** (72), a protective fungicide, introduced in 1949 and since widely used for foliage protection. Though of pungent odour, the compound is practically non-volatile and being stable and virtually insoluble in water, it is of long persistence and compatible with most other pesticides. The reasons for its fungicidal activity are unknown but it is somewhat specific, being ineffective against powdery mildews. The common formulation for spray preparation is the 50% w.p. and, for seed treatment, a 75% dressing. The compound is of low toxicity to warm-blooded animals and presents only the hazard of skin irritation. It is, however, harmful to fish hence care is necessary to avoid contamination of ponds, waterways and ditches with the chemical or used containers.

**Carbaryl** (73) is a contact and stomach insecticide, introduced in 1956. Being of low vapour pressure and but slightly soluble in water, it is of moderate persistence but it is incompatible with alkaline pesticides such as Bordeaux mixture or lime sulphur. The usual formulation for spray preparation is the 50% w.p.

Being a carbamate, its insecticidal action is probably due to the inhibition of the insect cholinesterases though the compound is not unduly toxic to warm-blooded animals. It is, however, toxic to fish and contamination of streams and ditches must be avoided. An interval of at least 1 week must elapse between application and the harvesting of peas, pears and apples.

**Carbon tetrachloride** (37), $CCl_4$, was first used in 1908 for the fumigation of nursery stock but, on the whole, its insecticidal potency is low. It has anaesthetic effects and is a potential liver and kidney poison. Any liquid spilt on the skin must be washed off at once and adequate protection must be provided to operators using the compound for fumigation. The vapour decomposes on hot surfaces to produce highly toxic gases, including phosgene, so smoking and the use of naked lights should be prohibited. At least a week's airing should be allowed before fumigated materials are used for human or animal consumption.

**Cheshunt compound.** A cuprammonium fungicide introduced in 1922 specifically for the control of damping off of seedlings. It is prepared by the intimate mixing of 2 parts by weight of bluestone (crystalline copper sulphate) and 11 parts of ammonium carbonate, the latter first being exposed to air to ensure the formation of the bicarbonate. Solutions of the mixture are used for the watering of soil and seed boxes.

**Chlorbenside** (49) is an acaricide of the bridged diphenyl type, first introduced in 1953. Being of low volatility and high stability, it is persistent, a property enhanced by its slow oxidation to the less volatile sulphone, a reaction which does not reduce its acaricidal activity to all stages of mites, including eggs. It is comparatively inert to true insects and to warm-blooded animals, for rats tolerated for 2 years a diet containing 1,000 ppm. Its high chemical stability renders it compatible with all other pesticides. It may be safely applied to all crop plants except cucurbits and, for spray purposes, it is formulated both as e.c. (20% a.i.) and as w.p. (20% a.i.).

D

**Chlordane** (38). This insecticide, first used in 1945, is included here because of its recommendation for the control of earthworms in turf and lawns. Neither dead worms nor worm-casts are found on treated turf for some time after its use. Label instructions must be observed.

**Chlorfenson** (50) is an acaricide of the bridged diphenyl group, first introduced in 1949. It is of high persistence because of its low vapour pressure and high chemical stability though, being an ester, it is broken down by alkalies, a reaction which is so slow at ordinary temperatures that the compound is compatible with all other pesticides. Though toxic to the eggs and immature stages of sap-feeding mites, it is virtually non-insecticidal and presents little hazard to warm-blooded animals. It is marketed as the 50% w.p.

**Chlorfenvinphos** (5). This organophosphorus compound, was introduced in 1963, particularly for use against soil insects. It is included in Part III of the Regulations (see p. 87) and is cleared, under the Pesticides Safety Precautions Scheme, for use on brassicas, turnips, swedes, carrots, onions, maize, beans, parsley, parsnips and cereals at rates up to 32 oz a.i./acre per application. On fen soils only, carrots may be treated at rates not more than 80 oz/acre per application. Its use as a liquid seed dressing for cereals sown in the autumn is also accepted. An interval of 3 weeks should elapse between application and harvest. Granule and e.c. formulations are marketed, as also is a seed dressing.

**Chlorobenzilate** (51). This acaricide, introduced in 1952, is so devoid of insecticidal properties that it has been recommended for the treatment of bees against tracheal mites. It has neither systemic nor fungicidal properties. Chemically it has the bridged diphenyl structure and bears some resemblance to DDT, but it is an ester and therefore susceptible to alkali. As experimental animals tolerated a diet containing 500 ppm for 2 years, its poison hazards appear to be negligible but a minimum interval of 3 weeks before application and harvest is advised. The principal formulations are the 25% w.p. and e.c.

**Chloropicrin** (74). An insecticidal fumigant first used as a pesticide in 1908. Because of its extreme toxicity, its use by inexperienced operators is seldom recommended but the intense irritation it causes to the eyes and mucous membranes enables it to be used as a warning agent in other fumigants.

**Copper fungicides.** Many formulations of copper compounds have been introduced for application as sprays or dusts and to avoid the troublesome preparation of Bordeaux and Burgundy mixtures. These products are prepared from a variety of copper compounds, usually from copper oxychloride, made by the aeration of scrap copper in cupric chloride–sodium chloride solutions. The resultant product approaches in composition to $3Cu(OH)_2.CuCl_2$, though, if calcium chloride is present, the precipitate will approximate to $3Cu(OH)_2.CaCl_2$. Alternatively, cuprous oxide may be used.

The copper dusts may have contents of 6–35% metallic copper, though the 6% dust is usual. W.p. formulations range from 20–50% metallic copper.

The so-called colloidal formulations consist of suspensions of copper oxychloride in a dispersing agent and subjected to an intense grinding by which the particles are reduced to a size such that they remain permanently in suspension both in the paste and when diluted to spray strength.

Copper compounds present few hazards though they may be injurious to sensitive varieties of plants. Livestock, and particularly sheep, should be kept from treated areas for at least 3 weeks.

**Corrosive sublimate** (76). At one time this compound, mercuric chloride, was used for soil treatment against clubroot of brassicas and pests such as root maggots, but it has been superseded by the less poisonous calomel. Though not included in the Agriculture (Poisonous Substances) Regulations, it is scheduled under the Pharmacy and Poisons Act 1933 and Poison Rules 1960. The only accepted use now is by groundsmen as a turf fungicide in formulations containing not more than 15% w/w mercuric chloride.

**Cresylic acid.** This name is applied to a coal tar derivative which consists mainly of the three isomeric cresols. The specified product contains 40% w/w of phenols when analyzed by the steam distillation of the acidified product. On treated soil, planting should be delayed for at least 3 weeks, 4 to 5 weeks on heavy soils. It should be kept off the skin and away from the eyes, food and food containers. It is hazardous to fish.

**Dazomet** (56) is a soil fumigant introduced in 1952 and found successful against soil-borne fungal pathogens, certain nematodes and weed seedlings. The compound breaks down in soil to a dithiocarbamate which, in turn, yields methyl isothiocyanate for which reason it is included in the dithiocarbamate group. It is available as a powder containing 98% a.i. which is used as a soil fungicide and nematicide. It is of low toxicity to mammals, but precautions should be taken to avoid inhalation or undue contact with the dust. Soil treatment should be carried out at least 8 weeks before planting.

**D-D mixture.** This by-product of plastics manufacture was first used as a soil fumigant in Hawaii in 1943 and has since proved an effective nematicide in cooler climates. It consists of a mixture, roughly one-half of 1,3-dichloropropene and about 30% of a mixture of 1,2-dichloropropane and 2,3-dichloropropene. Because of its phytotoxicity, a delay of at least 6 weeks is necessary before planting in treated soil. Its odour serves to warn operators against its harmful properties but care in handling is essential. On no account should the mixture come into contact with eyes, nose or mouth. Any liquid splashed on the skin must be washed off with paraffin and copious quantities of soap and water. Avoid breathing the concentrated fumes, do not open containers in confined spaces or apply in a closed house. All clothing, including rubber or leather boots and gloves, must be washed immediately with soap and water and must not be used as long as the odour remains. D-D mixture must be kept away from foods, feedingstuffs, and food containers and contamination of ponds, streams and water courses must be avoided. As the mixture is corrosive to many metals, equipment must be thoroughly washed out with paraffin immediately after use.

**DDT** (39). The first protective insecticide for outdoor use, the insecticidal properties of which were discovered in 1939 and which, by 1945, had attained a production of about 30 million lb. This success encouraged a search for insecticidal properties in related compounds giving rise to a series usually called the chlorinated hydrocarbon group. Strictly speaking this name would also be applicable to many other insecticides such as 'D-D mixture' or even chloroform which, however, are volatile and are used as fumigants.

The name DDT is an abbreviation for *d*ichloro*d*iphenyl*t*richloroethane though the accepted chemical name for the compound is 1,1,1-trichloro-2,2-di-(4-chlorophenyl)ethane. The technical product is, however, a mixture of compounds and may contain up to 30% of the *op'* isomer. The specification universally used for DDT and approved by W.H.O. calls for a content of at least 70% of the *pp'* isomer, which is the more effective insecticide.

The persistence of DDT is due to its low volatility and water solubility and to its good stability though, when in solution, it is readily decomposed by alkalies with loss of hydrochloric acid and insecticidal activity. A similar dehydrochlorination by an enzyme is thought to occur in those insects from which the continued use of DDT has resulted in the selection of resistant strains.

DDT is both a stomach poison and a contact insecticide; although without action as an ovicide and with little effect on pupal stages, its persistence ensures that the newly-hatched larvae and emerging adults acquire a lethal dose. No satisfactory explanation of its lethal action is yet known but it causes violent tremors and nervous and muscular activity become unco-ordinated.

DDT is harmless to plants, except to cucurbits and many varieties of barley. To man it is innocuous enough to be used for the control of body lice though it is not rapidly degraded in the body and may persist in the body fat. The ubiquitous contamination of the environment by DDT and by other persistent organochlorine pesticides which has resulted from their widespread use is deplorable and it should be used only where a persistent insecticide is really necessary. Livestock should be kept from treated areas for at least 14 days and the same interval should elapse between application and the harvesting of edible crops, except when used as an aerosol or smoke, when 2 days is long enough.

Dust formulations for agricultural use range in DDT content from 1–10%, the usual being 5%. W.p. formulations range from 20–50%, the usual being 50%. Solutions for e.c. or for aerosol use likewise have various DDT contents. DDT is sufficiently heat-stable to be used in smokes either of itself or in combination with BHC or gamma-BHC.

**Demephion** (6). This systemic organophosphorus insecticide was introduced in 1968, mainly as an aphicide and acaricide for use on fruit and vegetable crops. It is placed in Part III of the Regulations (see p. 87) and an interval of 3 weeks should elapse between application and harvest. Precautions to be taken for the safety of the user and consumer are similar to those given for demeton-S-methyl. The usual formulation is the 30% e.c.

**Demeton-S-methyl** (7). This systemic organophosphorus insecticide is an improved form of demeton-methyl, introduced in 1950 as an aphicide and acaricide for fruit, vegetable and ornamental crops. Demeton-methyl consisted of a mixture of $OO$-dimethyl $O$-2-(ethylthio)ethyl phosphoro-thioate and the related thiolate $OO$-dimethyl $S$-2-(ethylthio)ethyl phos-phorothioate but was, in 1963, replaced by preparations containing mainly the latter compound; the usual formulation is the 58% w/v e.c. Its insecti-cidal activity is due to its inhibitory action on the enzyme, cholinesterase, and a similar action in the animal body renders it poisonous to mammals. It is placed in Part III of the Regulations (see p. 87). Hence a strict adherence to the requirements of the Regulations and to instructions on the label is imperative. At least 3 weeks should elapse between application and harvest (2 weeks in the case of wheat and barley) and 2 weeks before allow-ing access of stock or poultry to treated areas; mangolds should not be clamped within 10 days of treatment to avoid risks to workers. No part of the treated crop should be used as fodder until 3 weeks have elapsed. Brassica crops should not be treated after the end of October. Some ornamentals are sensitive to sprays.

**Demeton-S-methyl sulphone** (8). Oxidation of demeton-s-methyl leads first to the sulphoxide, given the common name of oxydemeton-methyl, and then to the sulphone, which was introduced in 1967 as a component of a formulation of azinphos–methyl. Safety precautions to be taken in the use of this formulation are accordingly those given for azinphos-methyl.

**Derris** (1). The use, as an insecticide, of the ground root of certain species of Derris was patented in 1912, since when it has been established that the active components are a series of compounds called rotenoids, of which the main insecticide is rotenone. The rotenoids have also been found in other species of tropical plants used by the natives as fish poisons, including *Lonchocarpus* spp. from South and Central America. The reasons for the insecticidal activity of derris remain obscure but the insect respiratory system and heart are slowly paralysed. The rotenoids are of low toxicity to most warm-blooded animals but, on edible crops, at least 1 day should elapse between use and harvest. Their extreme toxicity to fish, however, makes care necessary especially in the disposal of empty containers. Although extracts are used in some formulations, it is not usual to extract the rotenoids from the root, which is used for dust formulations, generally containing 0·2–0·5% rotenone. W.p. and e.c. formulations are available with rotenone contents ranging from 1–6%.

**Diazinon** (9). This non-systemic organophosphorus insecticide and acaricide was first introduced in 1952 and has found particular use against aphids, thrips, flies and mites. Recent experience has shown it to be useful against some species of soil pests. It is of low water solubility and but slightly volatile. It is compatible with other pesticides but should not be combined with copper fungicides. Although an inhibitor of the enzyme cholinesterase, the compound does not present serious poison hazards to man and it is not included in Regulations though care in its handling is important. Two weeks should elapse between use and harvest, though with aerosols 2 clear days is long enough. Livestock and poultry should be kept from treated areas for at least 2 weeks. Cucumbers and tomatoes should not be treated before mid-May as there is risk of damage; maidenhair fern is sensitive. The usual formulations are the 20% e.c. and 20% atomizing solution; 5% granules and a 40% w.p. are available for soil use; the latter as a drench.

**Dichlofluanid** (77). This substituted sulphenamide was introduced in 1966 as a wide-range fungicide, especially for the control of powdery mildews and of Botrytis on soft fruit, vegetable crops and ornamentals. It is not included in the Regulations but an interval of at least 2 weeks should elapse between its application and the harvesting of strawberries and raspberries, and of 3 weeks in the case of blackberries, blackcurrants, gooseberries, loganberries, autumn-sown salad onions, cauliflower seedlings grown under glass, lettuce grown under glass and treated with a dust. It should not be used on strawberries grown under glass or polythene because of possible phytotoxicity. It is harmless to bees. The usual formulation is the 50% w.p.

**Dichlorvos** (10). Although this organophosphate was introduced in 1955 mainly for use against houseflies, it has recently been adapted for horticultural use in the form of e.c. and treated plastic strips from which the compound is slowly volatilized. It differs from other organophosphorus insecticides in its comparatively high vapour pressure and its ready decomposition. These properties reduce hazards to operators to the extent that, although the compound is placed in Part III of the Regulations, it is safe to enter treated areas unprotected after 12 hours. Edible crops should not be harvested within 24 hours of use. It should not be applied to cucumbers or roses as damage may occur; certain chrysanthe mum varieties are also sensitive. On mushrooms it should not be used at rates more than 5 g a.i./1,000 ft$^3$; on other edible crops the maximum rates are 2 g a.i./1,000 ft$^3$ for indoor crops and 16 oz a.i./acre for outdoor crops.

**Dicloran** (78). This fungicide, introduced in 1959, has been successfully used for the control of Botrytis, especially on lettuce. Its mode of action differs from that of most other protective fungicides in that it has little effect on spore germination but appears to intervene in cell division. Being practically insoluble in water, stable and of low volatility, it is persistent and compatible with other pesticides. Its mammalian toxicity is negligible though commonsense precautions should be taken in handling and the disposal of containers. The usual formulation is the 4% dust and crops should not be harvested within 3 weeks of the last postplanting application. On lettuce the maximum rate and frequency of application, in terms of the 4% dust, are: pre-planting 1 oz 4% dust/yd$^2$; post-planting $\frac{1}{4}$ oz 4% dust/yd$^2$, except for crops to be cut before Christmas when the rates are: pre-planting $\frac{1}{2}$ oz 4% dust/yd$^2$; post-planting $\frac{1}{4}$ oz 4% dust/yd$^2$ applied once only at 2 to 3 weeks after planting.

**Dicofol** (52). This acaricide was introduced in 1955. If its chemical structure is compared to that of DDT, it will be seen that the only difference is that the hydrogen (H) of the bridge of DDT is replaced in dicofol by the hydroxy (OH) group, a change which converts the highly insecticidal but non-acaricidal DDT to an acaricidal but poorly insecticidal dicofol. The latter is effective against all stages, including eggs, of phytophagous mites but the reasons for this activity are not known. Experimental animals fed for a year on a diet containing 300 ppm suffered no ill effect, so there seems to be little hazard to users or consumers. The usual formulations are the 20% c.c. and a 20% atomizing solution. To the latter, diazinon may be added to secure a more rapid effect on mites.

Dicofol should not be used on seedlings or young plants under glass before mid-May or in bright sunshine as damage may occur. On apples, pears and strawberries at least 7 days should elapse between application and harvest but, on edible indoor crops, this time may be reduced to 2 days.

**Dieldrin** (40). This cyclodiene derivative was first employed against insects in 1948, since when it has been widely used mainly as a soil insecticide. Though its molecule contains oxygen it is grouped with the chlorinated hydrocarbons for it is the epoxide produced from aldrin; indeed it is manufactured by the oxidation of the latter. The name is reserved for a product containing at least 85% of the epoxide. It is of high chemical stability and is compatible with all other pesticides.

The reasons for its insecticidal activity are still obscure and, like DDT, it disrupts nervous activity in the insect which does not suffer the tremors so characteristic of DDT poisoning. It is not included in the Agriculture (Poisonous Substances) Regulations, but great care should be taken to avoid skin contact for the compound is suspected of a hazardous dermal toxicity. For the reasons given under aldrin (see p. 29), the uses of dieldrin have been limited by agreement under the Pesticides Safety Precautions Scheme. Liquid and w.p. formulations may be used for the control of wireworm on potatoes at a rate up to 44 oz a.i./ acre; for the control of cabbage root fly on brassicas to a rate of 21 oz a.i./acre; as a dip against narcissus bulb fly to 6 oz a.i./20 gal. Seed dressings should not be used on poor seed or seed with a moisture content of more than 16% because of phytotoxic risks; their uses are on winter-sown wheat up to the end of December and only when there is a real danger of attack by the wheat bulb fly at rates up to 1·2 oz a.i./bushel; on rubbed or graded sugar beet seed for precision drilling up to 3·6 oz a.i./cwt; on onion seed for the control of onion fly at a rate up to 0·6 oz a.i./lb; on seed of french or kidney beans and runner beans up to $\frac{3}{4}$ oz a.i./28 lb; on spinach seed up to 3·6 oz a.i./cwt; as a dust on compost used for the potting of ornamentals against vine weevil at rates up to 2 oz $1\frac{1}{4}$% dust/bushel.

Formulations as w.p. usually contain 50% dieldrin and e.c. formulations, 15%. It is used, in combination with organomercury compounds, in cereal seed dressings, the usual content being 40–60%, and with thiram for vegetable seed treatment at 75–80%.

**Dimefox** (12) is one of the earliest of the systemic organophosphorus compounds discovered in Germany during the 1939–45 war. It was introduced in the U.K. in 1949 but its range of usefulness is limited by its high toxicity to warm-blooded animals, for which reason it is included in Part I of the Agriculture (Poisonous Substances) Regulations. The common formulation is a 50% e.c. for use on hops, at least 4 weeks before picking, at up to two applications each at $1\frac{1}{2}$ lb a.i./acre. Other accepted uses are on strawberries and Brussels sprouts, one application at 2 lb a.i./acre, on strawberries before flowering begins; on mangolds, fodder and sugar beet, up to two applications each at 3 lb a.i./acre.

**Dimethoate** (12). This systemic phosphorus compound was introduced in 1956, simultaneously in U.S.A. and in Italy, for the control of certain fruit flies but its use has now extended to many other sap-feeding insects. It is rather soluble in water and has a comparatively short life in plant tissues. Moreover it is of low mammalian toxicity permitting it to be tested for the systemic control of cattle grubs. As it is readily degraded by alkalies, it is incompatible with pesticides of an alkaline reaction, such as lime sulphur or Bordeaux mixture.

It is not sufficiently toxic to man to warrant regulation. A minimum period of 1 week should be allowed between the last application and the harvesting of edible crops or of allowing stock or poultry access to treated areas. The usual formulation is a 40% e.c. It should not be used on chrysanthemums, hops (except the variety Fuggles) or on ornamental *Prunus* spp.

**Dinobuton** (66). This dinitro compound was introduced in 1963 as a non-systemic acaricide and direct fungicide of value for field and glasshouse use for the control of red spider mites and powdery mildews. Care is necessary for it may be phytotoxic to tomatoes and to some cultivars of chrysanthemum and rose. It should not be mixed with alkaline pesticides nor with carbaryl. It is not included in the Agriculture (Poisonous Substances) Regulations but its use on edible crops should be restricted, on apples and pears, up to three applications each at 16 oz a.i./100 gal H.V. *or* up to ten applications each at 8 oz a.i./100 gal H.V. at 10–14 day intervals; on cucumbers, up to five applications each at 8 oz a.i./100 gal as an aqueous suspension, *or* up to five applications each at 4 g/1,000 ft$^3$ as an aerosol. If applied L.V., the quantity of a.i./acre should not exceed that which would have been applied H.V. At least 3 weeks should elapse between application and allowing the access of livestock or the picking of apples and pears, of 3 days between application and the picking of cucumbers.

**Dinocap** (67). This compound arose from a search among dinitro derivatives for an acaricide less phytotoxic than DNOC. In 1949 it was found to be fungicidal, specifically against powdery mildews, for which purpose it finds its main use today. Originally thought to be a simple compound as given in the Appendix, it is now known to be a mixture of 2,4- and 2,6-dinitro isomers. It is an ester and therefore unstable in the presence of alkalies so it should not be mixed with lime sulphur or other alkaline pesticides; nor should it be used with oil-containing sprays on account of its oil solubility.

Although closely related chemically to the poisonous DNOC, dinocap is less toxic to man though care should be taken to avoid inhalation and the skin and hair should be protected from its staining properties. Dinocap is toxic to fish and care is therefore required in the disposal of empty containers. A minimum interval of 1 week must be observed between last treatment and the harvest of outdoor crops and 2 days in the case of crops under glass. The usual formulations are the 50% e.c. and the 25% w.p. It should not be used on some varieties of chrysanthemum.

**Disulfoton** (10). This organophosphorus systemic, introduced in 1956, has found its greatest use for the protection of young plants against aphids. It is almost insoluble in water and stable except to strong alkalies. Being a toxic chemical, it is placed in Part II of the Regulations (see p. 87) and the clearance maxima under the Pesticides Safety Precautions Scheme are as follows: on leaf brassicas 48 oz a.i./acre; on carrots and parsnips 32 oz a.i./acre; on potatoes, mangolds, fodder and sugar beet, marrows and strawberries 24 oz a.i./acre; on broad, horse, field and tick beans, peas and red beet 17 oz a.i./acre. If a single application is made within 7 days of planting, granules to a rate of 17 oz a.i./acre to french and runner beans, and to a rate of 32 oz a.i./acre to celery are approved. There should be an interval of at least 6 weeks between last application and harvest of an edible crop. The usual formulation is the 7·5% granules.

**Dithianon** (81). This compound was introduced in 1962 as a protective fungicide of promise for the control of many diseases of pome and stone fruit, though not effective against powdery mildews. In the U.K. it is used for the control of apple and pear scab, being applied at least 8 weeks before picking. A colloidal formulation is available.

**DNOC** (68), dinitro-*o*-cresol also known as DNC, was the active component of a product introduced in 1892 for the control of the Nun moth. In a systematic examination of substituted aromatic compounds, carried out at the Rothamsted Experimental Station in 1925, it was found to be an excellent ovicide. Unfortunately its action on foliage is sufficiently drastic for it to be used as a herbicide, hence its use as an insecticide is limited to dormant trees. For this purpose it proved useful, for its toxicity to aphid and sucker eggs enabled its use as an alternative to tar oils in the winter wash. DNOC-petroleum oil washes are specified, the M.A.F.F. specification requiring a content of not less than 60% neutral oil and a content of DNOC not less than 1/45th of the weight of neutral oil in the product. The usual formulation, generally of the stock emulsion type, contains 65–75% oil and 1·35–3·5% DNOC. The range of pests controlled can be extended by the addition of DDT which is incorporated in some DNOC-petroleum oil concentrates in amounts of 1·5–2%.

DNOC-petroleum washes should not be applied to the plum variety Myrobalan nor to the red currant variety Raby Castle.

The reasons for the ovicidal action of DNOC are unknown but the compound is also toxic to man and Statutory Regulations apply to the use of products containing more than 5% DNOC. Although the winter wash is thereby excluded, protection of the skin and hair is advisable to avoid the staining properties of the cresol.

DNOC is itself specified in M.A.F.F. Tech. Bull. 1, but is mainly used as a weedkiller and its sodium salt, DNOC-sodium, is sometimes used for the disinfection of mushroom houses. This compound is subject to Part II of the Regulations and an added hazard is its inflammability for which reason DNOC-sodium is marketed with water as a paste. The M.A.F.F. specification requires that the product should contain between 97 and 103% of the amount of DNOC stated on the label, but is applicable only to herbicides.

**Dodine** (82). This protectant fungicide, introduced in 1956, has proved of special value against apple scab because of its ability to eradicate recent infections. Chemically, the compound has the features of a cationic wetter (see p. 26) and should therefore not be mixed with formulations of pesticides containing an anionic wetter except with the approval of the manufacturer. Its toxicity to warm-blooded animals is low but it may produce severe skin irritation, hence care is necessary in handling the concentrate which is normally a 65% w.p.; a 20% liquid formulation is also available.

**Endosulfan** (41). This cyclodiene insecticide, introduced in 1956, is sometimes included in the chlorinated hydrocarbon group though its molecule includes a sulphite grouping. Chemically the product is a mixture of two isomers which do not, however, appear to differ greatly in insecticidal activity. Being practically insoluble in water, non-volatile and stable, it is of long persistence, but its use is limited by a high mammalian toxicity for which reason it is included in Part II of the Regulations. The usual formuation is the 20% e.c. Its use on edible crops is limited to black-currants and strawberries. On fruiting blackcurrants to which up to two applications are permitted, one at first open blossom stage and one 3 weeks later, each at 8 fl. oz (40 fl. oz 20% e.c.)/100 gal H.V. or the equivalent L.V. At least 6 weeks should elapse between the last application and picking; stock and poultry should be kept from sprayed areas for at least 3 weeks. Its use on strawberries should be limited to the period between the completion of picking and the next season's flowering. Apart from the limited uses on edible crops specified above, uses on non-edible crops are currently restricted by agreement to the control of bulb scale mite on narcissus, and the control of big bud mite on non-fruiting blackcurrant bushes for propagation.

**Endrin** (42). This insecticide, introduced in 1951, is an isomer of dieldrin and, like the latter, is of long persistence. But as it finds use for foliage rather than soil application, the danger that its repeated use will encourage the selection of resistant strains is less significant than with dieldrin. Its mammalian toxicity is, unfortunately, higher than that of dieldrin and it is scheduled in Part II of the Regulations (see p. 87). On apples, endrin must not be applied later than 1 week after 80% petal fall. On fruiting blackcurrants or blackberries the last application must not be made later than immediately before flowering; and on strawberries use should be limited to the period between the completion of picking and the next season's flowering. Stock and poultry should be kept from sprayed areas for at least 3 weeks. Endrin is compatible with other pesticides and the usual formulation is the 20% e.c.

Apart from the limited uses on edible crops described above, uses on non-edible crops are currently restricted to the control of Tarsonemid mite on narcissus and ornamentals grown under glass.

**Ethion** (14). This organophosphorus compound was introduced, in

1955, as an insecticide and acaricide but its main use, in Great Britain, is now in seed dressings for the control of wheat bulb fly. It is placed in Part III of the Regulations (see p. 87).

**Ethoate-methyl** (15). This systemic organophosphorus insecticide, introduced in 1967, is employed as an aphicide and acaricide on fruit and vegetable crops. It is not included in the Regulations (see p. 87) but an interval of at least 1 week should elapse between application and harvest or permitting the access of livestock and poultry to treated areas.

**Ethylene dichloride** (43). This insecticidal fumigant has been in use since 1927, mainly for the treatment of stored products. To reduce fire hazards it is mixed with carbon tetrachloride, usually in the ratio of 3 : 1. Operators are warned not to inhale the vapour, to keep the liquid off skin and clothes and not to smoke or use naked lights for the mixture will decompose in contact with heated surfaces to produce the highly toxic phosgene. At least 1 week's airing should be allowed before treated materials are used for human consumption or as fodder.

**Fenchlorphos** (16). This systemic organophosphorus insecticide, introduced in 1954, is so non-poisonous that it has been proposed for the treatment of cattle against internal and external pests. As an insecticide it is not of much use against lepidopterous larvae but is of promise against flies and sap-feeding insects. Its low mammalian toxicity is thought to be associated with a rapid break-down in the body, a break-down not accomplished by the insect before oxidation to an active anti-cholinesterase. Normal precautions will protect the operator. On sugar beet, mangold and fodder beet an interval of at least 6 weeks between application and harvest is stipulated.

**Fenson** (53). This acaricide, introduced in 1952, is chemically related to chlorfenson which, as indicated by the common name, has the molecular structure of fenson but has an additional chlorine in its build-up. Fenson has much the same chemical, physical and biological properties as chlorfenson but fell from favour when it was suspected of russeting certain varieties of apple; it also damages cucurbits. Its formulations are a 20% w.p. and e.c. and an aerosol for glasshouse use.

**Fentin acetate** (83). This organotin compound was introduced in 1954 particularly for the control of potato blight. It is included in Part III of the Regulations (see p. 87) and, provided the rate of each application does not exceed 4 oz a.i. (7 oz 60% w.p.)/acre, should present no hazard to consumers. Livestock should be kept for at least 1 week from treated areas. In the U.K. it is also available in combination with maneb as a w.p. containing 60% fentin actate and 20% maneb. It should not be mixed with oil-based formulations.

**Fentin hydroxide** (84). Like fentin acetate, this fungicide is used mainly for the control of potato blight and is included in Part III of the

Regulations. The rate of each application should not exceed 4 oz a.i. (20 oz 20% w.p.)/acre. It is also used as a canker paint on apples and pears. Livestock should be kept from treated areas for at least 1 week. It is formulated as a 20% w.p., with or without maneb.

**Folpet** (85) is a recently introduced fungicide closely related, chemically, to captan which it resembles in physical and biological properties but it may be found better than captan for certain purposes. It is currently available only in w.p. formulations in combination with dinocap for use on roses.

**Formaldehyde.** The commercial solution of this compound is known as formalin and was, at one time, used for the treatment of cereal seed but it now finds a wider use as a soil disinfectant. It is specified by the M.A.F.F. as a clear solution containing not less than 37·5 gram and not more than 40·5 gram formaldehyde (H.CHO) per 100 ml at 20°C. The solution may become cloudy through a polymerization of the formaldehyde, a process hastened by low temperatures; hence long storage in the cold should be avoided. Treated soil should not be sown or planted until all smell of the chemical has gone which takes about 3 weeks on light soils but up to 6 weeks on heavy clay soils.

**Formothion** (17). This systemic organophosphorus aphicide and acaricide was introduced in 1964 and is suitable for application to most crops except chrysanthemums and hops. It is not included in the Regulations but an interval of at least 1 week should elapse between application and harvesting or allowing access of stock or poultry to the treated area.

**Gamma-BHC** (44), a compound for which the common name, outside the U.K., is lindane, is the most active insecticide present in BHC (*q.v.*) from which it is prepared by crystallization. Although practically insoluble in water and of low volatility, it is sufficiently insecticidal to act in a pseudo-systemic manner on insects within the plant tissue adjacent to the deposit. It is compatible with other pesticides though, like DDT, it is dehydrochlorinated by alkali when in solution. An interval of 2 weeks should elapse between application and harvest or permitting the access of stock or poultry to the treated area. Though not of the strong tainting propensities of BHC, the hazard is there if applied to susceptible crops or if potatoes or carrots are planted where gamma-BHC has been used during the previous 18 months.

Dust formulations range from 0·2–0·65%; e.c. products range from 12–20%. Solutions for aerosol use are available and the compound is sufficiently heat-stable to be used pelletted with a pyrotechnic mixture or in smoke generators. When so used the interval before harvesting should be at least 2 days. Gamma-BHC is also incorporated, in amounts of 20–40%, in certain organomercury seed dressings to extend activity to soil insects such as wireworm and the wheat bulb fly. The precautions to be taken are given under organomercury seed dressings.

**Hexachlorobenzene** (40). This chlorinated hydrocarbon, introduced in 1945, has been used in North America for the seed treatment of wheat against bunt. It is not effective against other seed-borne diseases but its slight volatility enhances its activity against dwarf bunt.

**Lead arsenate** (86) (diplumbic hydrogen arsenate, $PbHAsO_4$) is the classic stomach poison first used in 1892. It is specified by the M.A.F.F. both as powder and as paste. The powder should contain not less than 62% lead oxide (PbO) and not less than 32% arsenic, calculated as arsenic oxide ($As_2O_5$). As water-soluble compounds of arsenic are generally phytotoxic, their content must not exceed 0·25% $As_2O_5$. The corresponding figures for the paste are 28·4% PbO, 14% $As_2O_5$ and 0·25% $As_2O_5$ in water-soluble form, with a content of water not more than 52%. Hazards to operators are not serious if commonsense precautions are taken, but those to consumers require an interval of at least 6 weeks between application and harvest. Stock and poultry should be kept from treated areas for at least 6 weeks though, if heavy rain intervenes, this period may be reduced to not less than 3 weeks. It should be noted that, under the Food and Drugs Act 1955(a), it is illegal to sell, deliver or import any food containing arsenic in amounts by weight, exceeding 1 ppm expressed as elementary arsenic (As) or lead (Pb) in amounts exceeding 2 ppm.

**Lime sulphur** (87) has been used as a fungicide for over a century. It is made by boiling sulphur and lime water and, chemically, is a solution of calcium polysulphides ($CaS.S_x$) with a smaller content of calcium thiosulphate. It is specified by the M.A.F.F., the specification requiring that the solution should be clear and free from sludge, should have an apparent density at 20°C of 1·30 ± 0·1, expressed as gram/ml, and should contain not less than 24% w/v of polysulphide sulphur (the $S_x$ of the above formula). On dilution with water the solution acquires an alkaline reaction which limits its compatibility with other pesticides. On exposure to air, the solution decomposes and the polysulphide sulphur is liberated as elementary sulphur. Lime sulphur may therefore be regarded as a formulation of sulphur and its biological properties are accordingly discussed under that title. Lime sulphur is, however, more effective than sulphur against scale insects, no doubt because of the alkalinity of the undecomposed solution.

**Liver of sulphur** is prepared by the fusion of caustic potash and sulphur. It is a water-soluble solid yielding a solution containing potassium polysulphides which, after application, decomposes to a deposit of sulphur; its fungicidal and acaricidal properties are therefore those of sulphur (*q.v*) but the alkalinity of the solution tends to phytotoxicity and it is rarely used nowadays.

**Malathion** (18). This non-systemic organophosphate was introduced in 1950 and is used for the control of sap-feeding insects and mites. It is slightly soluble in water and is of short to moderate persistence. Being susceptible to decomposition by alkalies it should not be used with alkaline spray materials. It is apt to injure certain ornamentals. Consumer hazards are low but an interval of at least 1 day should elapse between application and harvest though, to avoid taint, it is advisable to allow 4 days. Formulations include the 4% dust, 60% e.c. and an atomizing solution, usually of 25% a.i.

**Mancozeb** (57). This complex of zinc ion and maneb containing 20% manganese and 2·5% zinc was introduced in 1961 as a dithiocarbamate fungicide particularly for use against potato blight. The following intervals should elapse between application and harvest: 7 days on apples, blackcurrants, celery, gooseberries, hops, lettuce and pears; 2 days on mushrooms and tomatoes. It is formulated as an 80% w.p. and also in combination with zineb.

**Maneb** (58). This dithiocarbamate fungicide was introduced in 1950 and has been found useful, particularly for the control of downy mildews. There is evidence that the manganese present accelerates decomposition to the active fungicide, thought to be ethylene thiuram monosulphide. Maneb is of low toxicity to animals, whether man or insect, but care should be taken to avoid inhalation or skin contact. Treated outdoor edible crops should not be harvested within 7 days of application, indoor edible crops 2 days. The usual formulation is an 80% w.p.

**Manganese, Zinc, Iron Dithiocarbamate Complex.** This dithiocarbamate fungicide was introduced, in 1966, for the control of potato blight and of tulip fire. It can be irritating to the skin, eyes and nose.

**Mecarbam** (19). This organophosphorus compound was introduced in 1961, and has already proved useful as an aphicide; it also has some acaricidal activity. It exerts a degree of systemic effect, is of moderate persistence and is compatible with all but highly alkaline pesticides. It is sufficiently toxic to man to be placed in Part III of the Regulations (see p. 89) and its use on edible crops is, at present, limited to apples and pears, on which up to four applications, each at 10 oz a.i. ($12\frac{1}{2}$ fl. oz 80% e.c.)/100 gal H.V., a quantity not to be exceeded if applied L.V. and to celery. An interval of at least 2 weeks should elapse between application and harvest or permitting stock and poultry access to treated areas. It may also be used as a drench on celery within 7 days of transplanting. For these purposes it is available as an 80% s.c. It is also available as 4% granules for the control of cabbage root fly, carrot fly and celery leaf miner.

**Menazon** (20) was introduced in 1961 as a systemic organophosphate with selective insecticidal properties, being a potent aphicide yet without marked toxicity to other animals. It is sparingly soluble in water and with a negligible vapour pressure yet is taken up by the plant which is kept free from aphids for long periods. Its heat stability is poor but it is compatible with all but strongly alkaline pesticides. It is not included in the Regulations but livestock should be kept from treated areas for at least 3 weeks, and edible crops should not be harvested until at least 3 weeks after treatment. It is marketed as a 40% liquid formulation and as a 70% w.p., the latter being used for dips and soil drenches. Granules and dry seed dressings are available for the protection of sugar beet from early aphid attacks.

**Metaldehyde.** This compound, the slug-killing properties of which were discovered about 1936, is specified by the M.A.F.F., the specification requiring a content of not less than 95% by weight of metaldehyde calculated as acetaldehyde ($CH_3.CHO$). Its molluscicidal properties are usually attributed to an anaesthetic action which prevents the slug from regaining shelter with consequent desiccation; it is also toxic to warm-blooded animals. The compound has a domestic use as a solid fuel and must be kept out of reach of children. Poultry should be kept from treated areas for at least 7 days. For use against snails and slugs the crushed tablets are mixed with a protein-rich milling offal such as bran, but a number of approved slug baits are available with contents of 2·5–4% a.i. and liquids with a higher a.i. content.

**Metham-sodium** (59). This soil fumigant was introduced in 1954 and has been found useful for the partial sterilization of soil and for the control of soil nematodes and fungi. Chemically it is a dithiocarbamate readily soluble in water and, though stable in concentrated solutions, is decomposed in the soil to form methyl isothiocyanate which is thought to be the active agent. Users should follow label instructions carefully for the compound is irritant to the eyes and mucous membranes. Metham-sodium is also so strongly phytotoxic that it has been used as a weedkiller. Treated soil should not be planted until decomposition is complete and all smell of the chemical gone. A cress germination test (see p. 229) is advisable and care must be taken to ensure that the fumes do not reach other plants. The compound is sold as a stock solution containing 33% of the anhydrous salt.

**Methiocarb** (88). This carbamate, introduced in 1965 as a non-systemic insecticide and acaricide, has found a special use as a molluscicide. It is not included in the Regulations but it should not be used on land growing edible crops within 7 days of harvest. Poultry should be kept from treated areas for at least 7 days; containers and compound should be kept from children and birds; operators should wash hands and exposed skin before meals and after work. For the control of slugs and snails, the usual formulation is as pellets containing 4% w/w, broadcast at 3·2 oz a.i. (5 lb pellets)/acre.

**Methyl bromide** (89), was first used as an insecticide in 1938 and has since proved of great value in the fumigation of stored products, as a soil disinfectant and nematicide, and as a herbicide. It is a colourless gas, almost insoluble in water and with a chloroform-like odour which is usually masked by the chloropicrin added as warning agent. Methyl bromide is an insidious poison, quite unsuitable for use by untrained personnel.

**Dimethirimol** (79). This name has been suggested by B.S.I. for a substituted pyrimidine introduced, in 1968, as a systemic fungicide effective against powdery mildews. It has been cleared for certain experimental uses and is available as stock solutions of the hydrochloride intended for use as soil drenches or the spraying of woody plants, or in granular formulations on a fertilizer base.

**Metiram** (60). This dithiocarbamate fungicide is a complex of zineb and polyethylenethiuram disulphide, also known as Zinc PETD. It was introduced in 1961, mainly for the control of potato blight, as a 70% w.p. Inhalation and skin contact should be avoided. An interval of at least 1 week should elapse between application and the harvesting of apples, black-currants, hops, lettuce, potatoes and tomatoes.

**Mevinphos** (21). This systemic organophosphorus insecticide was introduced in 1953. It is of brief persistence for it is water-soluble and is decomposed by water. It has a high mammalian toxicity which places it in Part II of the Regulations (see p. 89) but its rapid decomposition permits its use, when required, shortly before harvest and a period of 3 days suffices to render produce free from harmful residues. The maximum rate of application per season is 6 oz a.i./acre on mangolds and fodder beet, 4 oz a.i./acre on apples, cherries, pears, plum and brassicas, 3 oz a.i./acre on hops, lettuce, peas and sugar beet, 2 oz a.i./acre on beans and straw-berries. Unprotected personnel, livestock and poultry may be admitted to treated areas after 1 day but wild life should be flushed from the area before mevinphos is applied. Being readily water-soluble the compound does not require formulation.

**Morphothion** (22) was introduced in 1957 as a systemic organophosphate effective against sap-feeding insects and mites. It is slightly soluble in water and is stable in solution and hence of moderate persistence. As its mammalian toxicity is low, it is not included in the Regulations, but its use on edible crops is mainly limited to hops, potatoes and sugar beet; also on chrysanthemums. An interval of at least 2 weeks should elapse between application and the harvesting of potatoes or sugar beet; 3 weeks in the case of hops. Livestock should be kept from the treated area for at least 2 weeks. The usual formulation is the 20% e.c.

**Nabam** (61). When the fungicidal activity of this compound was discovered in 1943, it aroused interest as the first example of a water-soluble compound exerting protective action. The explanation was later found to be a decomposition on aeration to form the insoluble and fungicidal ethylene thiuram monosulphide. But in the field, nabam proved unsatisfactory partly because of phytotoxicity and it was found necessary to

E

add zinc sulphate to precipitate the nabam as the zinc salt, since called zineb. Later the manganese salt, maneb, was found better for some uses. Accordingly, nabam is mainly used apart from soil application, for the tank-mix preparation of these two fungicides. Nabam is available both as as powder containing at least 93% of the anhydrous salt and as a stock solution containing at least 22% of the anhydrous salt.

It is of low toxicity but when used on edible crops, either alone or in conjunction with zinc or manganese salts, an interval of at least 7 days should elapse between use and harvest of outdoor crops, 2 days for crops under glass.

**Naled** (23). This organophosphate, introduced in 1956 in the United States, has useful fumigant properties as an insecticide and is of comparatively low mammalian toxicity. In Great Britain, it is still in the experimental stages.

**Nicotine** (2) is the insecticide of the tobacco steep which, 50 years ago, was about the only available general aphicide. Nowadays nicotine is extracted from tobacco waste and is specified by the M.A.F.F. as the mixed alkaloids of tobacco containing not less than 95% by weight of nicotine. It is readily soluble in water and is sufficiently volatile to act as a fumigant, for which reasons it is non-persistent. Its toxicity to mammals warrants its inclusion in the Regulations as a Part III substance though formulations containing less than 7.5% w/w are exempt. A particular hazard is due to its rapid absorption through the skin; splashes of the concentrate must be at once washed off with cold water. Edible crops should not be harvested for at least 2 days after treatment, except as smokes when the minimum interval is 1 day. Unprotected personnel should not enter treated glasshouses for at least 12 hours; and livestock should be kept from treated areas also for 12 hours.

Formulations include dusts, usually containing 3% nicotine, and a number of liquid formulations of various nicotine contents. The concentrate generally used is the 95% material though it may be noted that the sale of this material is not practised in many other countries where the solution of nicotine sulphate containing 40% nicotine is the rule.

**Organomercury seed dressings.** The fungicidal components of these pesticides will not be dealt with alphabetically for, nowadays, the dressings are usually applied by the seed merchant, a practice to be encouraged. A number of compounds are used and among those of cereal seed treatment for the control of seed-borne fungi may be mentioned methoxyethylmercuric silicate and phenylmercuric urea, used in dry treatment, and methylmercuric dicyandiamide, used in liquid treatment. The dry dressings contain from 0.6–1.5% mercury, expressed as metal, and the liquid dressings 0.6–2.0% mercury. The latter are supplied only to authorized seed merchants and are subject to The Factory Acts 1937–59.

Growers using dry dressings must observe the following precautions: wear rubber gloves and protect the wrists and forearms, for some organomercury compounds are strong vesicants; protect the nose and mouth from any dust; wash with soap and water, immediately if the skin is contaminated, after dressing the seed and before eating, drinking or smoking. Similar precautions should be taken when drilling treated seed. Treated seed must never be used as food or fodder and sacks used for treated grain should never be used for millable grain and should be well shaken and washed before use for feedingstuffs. The practice of reinforcing organomercury seed dressings with insecticides such as dieldrin or gamma-BHC is referred to under the heading of the insecticide used.

**Oxydemeton–methyl** (24). As indicated by its common name, this organophosphate is a derivative of demeton-methyl which, it is thought, is oxidized in the animal body to the sulphone with oxydemeton-methyl as an intermediate product. Accordingly the latter, which is also known as demeton-S-methyl sulphoxide, was introduced in 1960 as a systemic insecticide and acaricide. Because of its mammalian toxicity, it is placed in Part III of the Regulations (see p. 89). A minimum interval of 3 weeks should elapse between application and harvest, except in the case of wheat and barley when 2 weeks is stipulated. When used on mangolds or fodder beet there should be an interval of at least 10 days before clamping and the treated crop should not be used as fodder for at least 3 weeks. Livestock should be kept from treated areas for at least 2 weeks. It is available as a 57% w/v e.c.

**Paradichlorobenzene** (46). This chlorinated hydrocarbon insecticide, though introduced in 1915, still finds its main use as a domestic fumigant against clothes moth, for which purpose it is marketed as crystals of high purity.

**Parathion** (25), the first organophosphorous compound to find insecticidal use, was introduced in 1944. It is of slight water solubility and is sufficiently volatile to exert a pseudo-systemic action on insects, such as leaf miners, feeding near the plant surface. It is therefore of moderate persistence. As it is rapidly decomposed by alkalies, it is not compatible with alkaline pesticides. Its insecticidal action is due to an inhibition of the cholinesterases causing a failure of nervous function though, to accomplish this, parathion must first be oxidized to the corresponding phosphate. As this oxidation proceeds in the animal body faster than detoxication, parathion is extremely toxic to man and is therefore placed in Part II of the Regulations (see p. 89). When used as a spray the maximum rate and frequency of application per season are: for peas, one application at 3 oz a.i./100 gal; for fodder and sugar beet and mangolds, two applications each at 3 oz a.i./100 gal; on tomatoes and cucumbers under glass, up to three applications each at 2 oz a.i./100 gal. For soil application as a spray on cucumbers, one application at 1 oz a.i./100 gal; on tomatoes, one application at 6 oz a.i./100 gal *or* up to three applications each at 2½ oz a.i./100 gal are specified. If sprayed L.V. the amount applied per acre should not exceed that which would have been applied H.V. For soil application as granules on barley, brassicas and potatoes, one application at 5 lb a.i./acre is accepted before or soon after planting. In glasshouse use as an aerosol, up to five applications each at 1·4 g a.i./1,000 ft³; as a smoke, up to five applications each at 4 g a.i./1,000 ft³ are accepted. There should be an interval of at least 4 weeks when used as granules or spray, 2 days when used as an aerosol and 24 hours when used as a smoke, between use and the harvesting of an edible crop. Unprotected personnel should not enter treated areas for at least 1 day; livestock should be kept out for at least 10 days.

The usual formulation is the 20% e.c.; granules and smokes are available.

**Paris green** (90). This pigment was used, a century ago, against Colorado beetle but, for foliage application, it was soon replaced by lead arsenate. But it still finds use as the toxic component of baits for the control of slugs and soil insects for which purpose it is mixed with dried blood or a protein-rich milling offal. It is specified by the M.A.F.F. as a fine green powder containing not less than 30% copper, expressed as cupric oxide; not less than 55% arsenic, calculated as arsenious oxide; not more than 1·5% arsenic, as arsenious oxide, soluble in water; not less than 10% acetate, expressed as acetic acid. The powder should pass completely through an 80 mesh B.S. sieve and not more than 5% should be retained on a 300 mesh B.S. sieve. The compound is a powerful poison and domestic animals should be kept from the bait; moreover it causes suppuration of open wounds which should be protected from contact.

**Petroleum oils.** Distillates from petroleum, often called mineral oils, have in one form or another been used for pest control since the days when paraffin or kerosene was first used as an illuminant. Nowadays their use for crop protection is limited to certain high-boiling fractions from the lubricating oil range. Chemically such oils are predominantly hydrocarbons but, to ensure safety to foliage, a high degree of refinement is necessary; in technical terms the percentage of unsulphonated residue must be high. Their main use is acaricidal either on the dormant tree or on foliage. For these purposes, four types are specified by the M.A.F.F.

(1) Winter washes of the e.c. (miscible oil) type should contain not less than 75% by weight of neutral oil with the following characteristics: a specific gravity of between 0·86 and 0·93 at 15·5°C/15·5°, a distillation range such that less than 5% by volume will distil over at an oil temperature of 350°C and an unsulphonated residue not less than 65% by volume. As phenols are sometimes used as mutual solvents in this method of

formulation and tend to be phytotoxic, their content must not be greater than 6% by weight.

(2) Winter washes of the s.e. type should contain not less than 60% by weight of neutral oil of the same characteristics as under (i).

(3) Summer washes of the s.e. type for orchard use should contain not less than 60% by weight of neutral oil of the following characteristics: a specific gravity of between 0·84 and 0·92 at 15·5°C/15·5°, a distillation range such that 10% of the oil distils between 310 and 340°C, 50% between 350 and 375°C and 80% between 380 and 400°C, a viscosity between 100 and 200 sec. Redwood at 70°F and an unsulphonated residue not less than 92% by volume.

(4) Summer washes for glasshouse use should be of the s.e. type and contain not less than 60% by volume of neutral oil of the same characteristics as (3), except that the specific gravity should be between 0·84 and 0·89 at 15·5°C/15·5°, and an unsulphonated residue not less than 95% by volume. In all four types, the specifications require that the concentrated and diluted emulsions shall pass prescribed stability tests.

The action of the oil on the mite is presumably physical for oils of these characteristics could hardly be toxic in the usual sense, indeed oils of the glasshouse type approach medicinal paraffin in blandness. To extend their usefulness the incorporation of other pesticides is commonly practised, either to the formulation (see DNOC) or to the diluted emulsion.

**Phenkapton** (26). This organophosphate, introduced in 1957, is of interest because its insecticidal properties are slight yet it is toxic to all stages, including eggs, of phytophagous mites. It is included in Part III of the Regulations and its use on edible crops has so far been defined, under the Pesticides Safety Precautions Scheme, only on top fruit, when the recommended interval between application and harvest is 2 weeks, and on blackcurrants with an interval of at least 4 weeks. Stock and poultry should be excluded from treated areas for at least 2 weeks. Although phenkapton appears to be tolerated by bees, full precautions should be taken. The usual formulation is the 20% e.c.

**Phenylmercury compounds.** Phenylmercury chloride has found use as a foliage fungicide because of its ability to eradicate recent infections of apple and pear scab, for which purpose it was introduced in 1942. It is virtually insoluble in water and, unlike the related phenyl-mercury acetate which is used for seed treatment, is non-volatile. Certain apple and pear varieties are liable to damage especially in wet weather and the label instructions on varieties must be obeyed. Apples and pears should not be sprayed after the end of July and an interval of at least 6 weeks is advised between application and picking whereas stock and poultry should be kept out for at least 2 weeks. Formulations may include lead arsenate to extend activity to codling moth and leaf-eating caterpillars or sulphur to add fungicidal properties against apple mildew. Phenylmercury chloride is available as a $2\frac{1}{2}\%$ ($\equiv$1·5% Hg), w.p. used at 2 lb/100 gal H.V.; other organomercury scab fungicides are phenylmercury dimethyldithiocarbamate in 3% ($\equiv$1·5% Hg), w.p. used at 1 lb/100 gal H.V.; phenylmercury nitrate in $2\frac{1}{2}\%$ ($\equiv$1·6% Hg), w.p. used at 2 lb/100 gal H.V.; and a liquid formulation of phenylmercury salicylanilide at 6% ($\equiv$4% Hg) used at 1 pint/acre, H.V. or L.V. Phenylmercury salicylate is used in greenhouse aerosols, the concentrate containing 0·3% Hg. This product comes within Part IV of the Regulations and requires an interval of at least 12 hours between use and the harvesting of an edible crop, of which only tomato is recommended. The number of applications is limited to five at intervals of at least 7 days, the dosages at each application should not exceed 40 mg organically-combined $Hg/1{,}000\ ft^3$ (i.e., 1 fl. oz 0·3%/ 1,700 $ft^3$).

**Phorate** (27). This systemic organophosphate, introduced in 1954, has found its main use for seed or soil treatment to protect young plants from attack. For this purpose it is formulated as granules containing 10% a.i., whereby the hazards arising from its mammalian toxicity are reduced. Phorate is practically insoluble in water and is unstable in solution; in the plant sap it is thought to be oxidized, ultimately to a stable sulphone, but the intermediate products are unstable. It is placed in Part II of the Regulations (see p. 89) and the greatest amounts which should be applied in any one season are: on brassicas, except root crops, 64 oz a.i. (40 lb 10% granules)/acre; on carrots, mangolds, fodder beet, red beet and sugar beet, parsnips and potatoes, 48 oz a.i. (30 lb 10% granules)/acre; on brassica root crops, dwarf, french and runner beans, peas, 32 oz a.i. (20 lb 10% granules)/acre; on broad, horse, field and tick beans, strawberries and sweet corn, 24 oz a.i. (15 lb 10% granules)/acre. There should be an interval of at least 6 weeks between application and harvest or before allowing livestock to enter treated areas. The usual formulation is as 10% granules.

**Phosalone** (28). This non-systemic organophosphorus compound was introduced in 1963. It is not included in the Regulations but its recommended use is restricted, on apple and pear, to not more than five applications per season at 10 oz a.i./acre, not more than four postblossom; on plums, to three at 10 oz a.i./acre of which not more than two should be postblossom; on brassica seed crops, three applications at 6·6 oz a.i./acre at green bud, early yellow bud (if weevil attack is severe) and yellow bud. An interval of at least 3 weeks should elapse between use and harvest and livestock should be kept from treated areas for at least 4 weeks. The usual formulation is the 33% e.c.

**Phosphamidon** (29). This systemic organophosphate, introduced in 1957, is readily soluble in water and, in solution, is rapidly decomposed;

its half-life in plant sap is estimated to be 2 days. It is effective against sap-feeding insects and mites. It is sufficiently toxic to be included in Part III of the Regulations (see p. 89). Its use on non-edible crops is not restricted but, on edible crops, the number of applications and amounts applied are defined, under the Pesticides Safety Precautions Scheme: on apples and pears, up to five applications at 5 oz a.i. (25 fl. oz 20% s.c.)/100 gal; on hops, up to a total of 32 oz a.i./acre, the last application not being greater than 8 oz a.i./acre; on sugar beet, fodder beet and mangolds, up to three applications at 5 oz a.i./acre; on brassica crops, up to four applications at 4 oz a.i./acre; on broad and field beans, up to two applications at $3\frac{1}{4}$ oz a.i./acre; on potatoes, up to 4 applications at $3\frac{1}{4}$ oz a.i./acre, on blackcurrants, one application at $3\frac{1}{4}$ oz a.i./100 gal; and on strawberries, up to two applications at $3\frac{1}{4}$ oz a.i./100 gal. A minimum interval of 3 weeks between application and harvest is recommended. Livestock should be kept from treated areas for at least 2 weeks. The usual formulation is a 20% s.c.

**Propineb** (62). This dithiocarbamate fungicide, once named mezineb, was introduced, in 1966, mainly for the control of downy mildews and potato blight. It is not included in the Regulations (see p. 89) and is formulated as a 70% w.p.

**Pyrethrum** (3). The ground flower heads of certain species of pyrethrum, in particular *Chrysanthemum cinerariaefolium*, have been used as a domestic insecticide for over a century. The active components are a group of esters known as pyrethrins, present in amount from 0·7 to 3%, which are potent contact insecticides with a rapid paralytic effect ('knockdown'). But these esters are susceptible to decomposition especially in sunlight and their persistence in the open is brief. Moreover they are rapidly decomposed in the insect body though a group of compounds, pyrethrum synergists, are known which intervene in the detoxication. Both pyrethrum and its synergists are virtually harmless to warm-blooded animals though the flower heads contain an oil which may cause dermatitis on sensitive persons. For dust use, the ground flower heads are diluted with a non-alkaline carrier; for spray or aerosol use, a synergist is usually added and often derris or DDT to ensure that the insects knocked down do not recover.

**Quinomethionate** (94). This compound, once named oxythioquinox, was introduced in 1965. It is a cyclic carbonate and has both fungicidal and acaricidal properties, the latter of use against mites resistant to organophosphorus compounds. Its main uses are as an acaricide on apples and against powdery mildews and leafspot on currants and gooseberries. It is not included in the Regulations (see p. 89) but skin contact and inhalation should be avoided. An interval of 3 weeks should elapse between application and the harvesting of apples and pears, of 2 weeks for currants, gooseberries, and strawberries. On cucumbers under glass, 2 days is long enough but 1 week for outdoor marrows.

**Quintozene** (69). Although this fungicide, also known as PCNB, was first introduced in the late 1930's, it has but recently become widely used as a soil fungicide and for the control of Rhizoctonia. As it is practically insoluble in water, of low volatility and stable in soil, it is of long persistence; it is of low toxicity to man and other animals. Certain crops are sensitive to it and cucurbits and tomatoes should not be planted on treated soil, though potatoes are tolerant. The usual formulation is the 20% dust.

**Salicylanilide** (30). This compound was applied, in 1930, for the mildew-proofing of textiles and has found a limited application in crop protection, mainly because of the phytotoxicity of its alkali salts. The anilide is itself almost insoluble in water and non-volatile but is soluble in alkali. Its main use is for the control of tomato leaf mould and the hazards to man and stock appear to be slight.

**Schradan** (30) was one of the first systemic organophosphorus compounds, a property discovered in 1941–2. The name was applied to a mixture of phosphoramides, which are soluble in water yet stable in solution, and in which octamethylpyrophosphoramide predominates. Schradan is used for the control of sap-feeding insects and mites but its use is restricted by its high mammalian toxicity for which reason it is placed in Part II of the Regulations (see p. 89). Its use on edible crops are restricted both to rates and times of application: on brassica crops to one application at 24 fl. oz a.i./acre, except on Brussels sprouts when up to 40 fl. oz a.i./acre may be used; on maincrop cucumbers, one application at 16 fl. oz a.i./acre; on field beans, one application at 12 fl. oz a.i./acre; on hops, two applications totalling 32 fl. oz a.i./acre; on mangolds, fodder beet and sugar beet, one application at 32 fl. oz a.i./acre; on strawberries, two applications, the first before flowering, the second after fruit-picking, totalling 32 fl. oz a.i./acre and on apples and peaches, one application at 12 fl. oz a.i./100 gal H.V. or proportionately L.V. Schradan should not be applied to edible crops after the 15th September. The minimum intervals between application and the harvesting of an outdoor crop should be 4 weeks if schradan is applied from April to July and 6 weeks if applied from the beginning of August to mid-September. Cucumbers grown under glass should not be treated within 4 weeks of harvesting. Stock and poultry should be kept from treated areas for at least 4 weeks. Schradan is usually marketed as a solution containing 40–55% a.i.

**Sodium nitrite** (94). This fungicide is used specifically for the treatment of freshly cut stump surfaces to protect them from infection by *Fomes annosus*. It is not included in the Agriculture (Poisonous Substances) Regulations but care must be taken to ensure that water supplies are not contaminated for it is a poison, especially to fish. Livestock should be excluded from treatment areas for at least 2 weeks. It is applied as a 10% solution with an added marker dye.

**Sodium pentachlorophenate.** Pentachlorophenol has been widely used for timber preservation since 1936 and its sodium salt, being water-soluble, is a handy form in which to employ its potent fungicidal properties. Unfortunately it is rather toxic to mammals, the acute oral LD50 for rats being 210 mg/kg; solutions stronger than 1% may cause intense skin irritation and its dust and vapour provokes intense sneezing. Hence the greatest caution is needed in its use for the disinfection of wooden structures such as mushroom houses.

**Streptomycin.** This antibiotic, isolated in 1944, is of wide medicinal use but has been applied for the control of certain bacterial diseases of stone fruit. It is sometimes mixed with another antibiotic, oxytetracycline, and such products for crop protection uses containing one or both compounds must, by law, be rendered unpalatable and unfit for medical or veterinary use. Uses on edible crops are as a spray on cherries at a rate not in excess of $3\frac{1}{2}$ oz streptomycin and/or $\frac{1}{3}$ oz oxytetracycline per 100 gal H.V., or the equivalent L.V., applied at a period not between petal fall and harvest. Streptomycin by itself may be used on hops, two applications each at 17 oz a.i./100 gal but not within 2 months of picking. Protective gloves should be worn when handling or stringing hops within 10 days of treatment. The use of an emulsion paint containing 1% streptomycin on stone fruit after harvesting should not lead to a consumer hazard but the operator should wear rubber gloves, rubber boots, face-shield and a mackintosh or overalls with sleeves rolled down and collar fastened when using the paint. If extensive use is to be made of these antibiotics, general practitioners and hospitals in the area should be asked to look out for signs of untoward effect.

**Sulfotep** (31). This organophosphorus compound was introduced in 1944 as a contact insecticide of low persistence but its main use today is as a fumigant in closed spaces. Chemically it is a sulphur analogue of TEPP but is less soluble in and more stable to water. Its strong mammalian toxicity places it in Part II of the Regulations (see p. 89). Although these regulations do not require operators to wear protective clothing when lighting a smoke generator, protective clothing must be worn if it is necessary to re-enter the house to relight a generator. Any person entering a treated glasshouse during the 12 hours following treatment is required to wear an overall, hood, rubber gloves and respirator. Its use as a smoke on any edible indoor crop should be at a rate not greater than 1 gram a.i./1,000 ft³ and an interval of at least 24 hours should elapse between application and havest.

**Sulphur** has been in long use as a fungicide and acaricide but the reasons for its toxicity to these organisms is still little understood. It is insoluble in water and almost non-volatile, though, with suitable equipment, it is used for the fumigation of glasshouses. Chemically it is stable and is compatible with other pesticides. It is non-toxic to man and warm-blooded animals and to plants, though certain varieties are peculiarly sensitive to sulphur; the label instructions on 'sulphur shy' varieties must be carefully followed. Dusting sulphurs usually contain 95% sulphur. W.p. formulations are specified by the M.A.F.F. as dispersible sulphurs and the specification requires a content of not less than 70% sulphur with a particle size such that less than 40% of the sulphur shall be present in particles of $6\mu$ or less diameter and not less than 9% in particles of $2\mu$ or less diameter. Paste formulations are specified as colloidal sulphurs of which the sulphur content should not be less than 40% and the particle size of this sulphur should be such that not less than 90% should be of particles of $6\mu$ or less diameter and not less than 55% in particles of $2\mu$ or less diameter.

**Tar oils,** derived by the distillation of coal tar, have for years been used as fungicides for wood preservation. But tar distillate washes for dormant use on fruit were first introduced in the U.K. in 1921, and at once became popular for the control of aphids, scales and suckers and for the removal of lichens. The tar distillates used are those from which anthracene has been recovered, hence the anthracene oils, which consist mainly of aromatic hydrocarbons with small amounts of derivatives soluble in alkali, tar acids or phenols, or soluble in acids, tar bases. The ovicidal compounds, specifically against aphid and psyllid eggs, reside in the fraction insoluble in acids or alkalies, the neutral fraction. Strong phytotoxicity limits the use of tar oils to fully dormant plants but they should not be applied to the plum variety Myrobalan nor to the currant variety Raby Castle. They are free of poison hazards though contact may cause dermatitis.

Tar oil washes are specified by the M.A.F.F.; the specification for the e.c. (miscible oil) type calls for a content of not less than 80% by weight total oil, not less than 52% by weight of the wash should be soluble in dimethyl sulphate and not more than 10% by weight of the wash shall be phenols. The neutral oil soluble in dimethyl sulphate (i.e., aromatic hydrocarbons) shall have a distillation range such that at least 40% by weight of the wash should distil above 230°C, not less than 22% above 290°C, and not less than 10% above 335°C. The specification of the s.e. type requires a content of not less than 60% by weight of neutral oil; not less than 54% by weight of the wash should be neutral oil soluble in dimethyl sulphate and not more than 5% by weight of the wash shall be phenols. The neutral oil soluble in dimethyl sulphate should have a distillation range such that the fraction distilling above 230°C should constitute at least 48% by weight of the wash, that distilling above 290°C at least 27% by weight of the wash and that distilling above 335°C, at least 11% by weight of the wash. In both types, the undiluted and the diluted wash should pass certain stability tests.

To extend the ovicidal properties of the wash to such insects as capsid bugs and to mites, the addition of petroleum oils to the concentrate is possible but the amount of petroleum oil needed is roughly twice that of the neutral tar oil. But the combined tar-petroleum wash, with its 10% oil content, is seldom used today and it seems unnecessary to cite the specifications which are to be found in the M.A.F.F. Tech. Bull. 1.

**TDE** (47). This chlorinated hydrocarbon, tetrachlorodiphenyl ethane, is closely related to DDT, as is indicated by its usual chemical name. 1,1-dichloro-2,2-di-(4-chlorophenyl)ethane. Its insecticidal properties were discovered in 1944 but, as a general insecticide, it is not as useful as DDT though better for the control of some leaf-eating caterpillars, particularly those of the tortrix moths. Being highly persistent, its uses should be confined as far as possible to those cases where no alternative insecticide of less persistence is available. It is not included in the Regulations (see p. 89) but an interval of at least 14 days should elapse between application and the harvesting of apples, pears, loganberries, raspberries and strawberries; livestock should be kept from treated areas for the same period. The usual formulation is the 50% w.p.

**Tecnazene** (70). This fungicide, also called TCNB, was introduced in 1947, and is, nowadays, mainly used for the control of the dry rot of potatoes caused by *Fusarium* spp. Its fungicidal properties are similar to those of the related quintozene (PCNB) but it more volatile and hence of lower persistence in the open. Moreover its ability to inhibit the sprouting of potato tubers enhances its usefulness for the storage of ware potatoes. It is marketed either as a dust of 3–6% a.i., as a 6% e.c., or for use as a smoke. The latter are used for the control of Botrytis in glasshouses and gamma-BHC may be added to extend activity; but the mixture should not be used on roses. At least 2 days should elapse before smoked edible crops are harvested.

**TEPP** (32). The aphicidal properties of this organophosphorous compound were discovered in 1939 and it was introduced as a nicotine substitute in Germany during the 1939–45 war. The original German material was at first thought to be a mixture of higher anhydrides of diethylphosphoric acid but it was later found that the simple anhydride, tetraethyl pyrophosphate, was the main component. TEPP is readily soluble in water but is rapidly decomposed by water to the non-toxic diethylphosphoric acid. Hence, whereas the brief persistence of nicotine is due to volatility, that of TEPP is due to instability. Nevertheless the high mammalian toxicity of TEPP places it in Part II of the Regulations (see p. 89). Its use on non-edible crops is acceptable under the Pesticides Safety Precautions Scheme and, provided there is an interval of at least 2 days between application and harvest, is use on growing edible crops at a rate not more than 2 fl. oz a.i./acre per application should not present a hazard to consumers. Unprotected persons must be kept from treated areas for at least 1 day, livestock for at least 2 days, after use. For pesticide use, TEPP is a mixture of polyphosphates containing at least 40% of the pyrophosphate, and is marketed either in solution in a non-water solvent (20–40% TEPP) or for use as an aerosol.

**Tetradifon** (54). This acaricide was introduced in 1954 and is effective against all stages and eggs of phytophagous mites. It is virtually insoluble in water, non-volatile and of a stability which renders it of good persistence and compatible with other pesticides. Yet it is capable of a pseudo-systemic effect and is toxic to mites on the side of a leaf opposite to that sprayed. It is of low toxicity to insects and mammals and harmless to plants except young cucumbers or plants that are wet, and is available as a concentrate of 8% a.i. and in smoke formulations.

**Tetrasul** (55). The acaricidal properties of this compound were first reported in 1957 and it is effective against the eggs and immature stages of phytophagous mites. Its sulphone, tetradifon, is more widely used.

**Thionazin** (33). This systemic organophosphorus compound, introduced in 1959, has found its main uses as a nematicide for the control of stem and bulb eelworm of narcissus and tulip and of leaf and bud eelworm of chrysanthemum; as an insecticide against cabbage root fly and as an acaricide against bulb scale mite. It is included in the Regulations (see p. 89) as a Part II substance and its uses, apart from bulb treatment, are restricted to use as a soil drench on leaf brassicas to which it should be applied within a week of transplanting, and to tomatoes which should not be harvested within 11 weeks of treatment. Livestock should be kept from treated areas and from dipping premises for at least 8 weeks. The available formulations are a 46% e.c. and 10% granules.

**Thiram** (63). This dithiocarbamate was introduced as a fungicide in 1934 and has since been used both on foliage and seed. It is practically insoluble in water and non-volatile and is compatible with most other pesticides. Its action on fungi is not fully understood but it is different from the other dithiocarbamates listed here for, being the derivative of a secondary amine, it is therefore not susceptible to the reactions thought to be associated with the fungitoxicity of these primary amine derivatives. It is harmless to plants at the concentrations used and is of low toxicity to warm-blooded animals, except perhaps poultry. But it should not be applied to fruit intended for canning or deep freezing because of the possibility of taint and a discoloration of the tins. Care in handling is necessary because of irritation to the skin and mucous membranes. The usual formulations for spray use are an 80% w.p. and a 50% colloidal suspension; for seed treatment a 50% powder is available but the usual method of treating seed is the thiram soak treatment described on p. 277. Edible crops should not be harvested within 7 days of treatment.

**Trichlorphon** (34). This organophosphonate was introduced in 1952 mainly for the control of DDT-resistant house flies but it has found some use in crop protection. It is moderately soluble in water but is decomposed by alkalies and hence incompatible with alkaline pesticides. Its mammalian toxicity is low enough for it to be excluded from the Agriculture (Poisonous Substances) Regulations but the usual precautions must be taken especially by those who have to handle other organophosphates. It exerts a pseudo-systemic action which renders it effective against leaf-miners but its main current use is against cabbage root fly larvae attacking Brussels sprouts in the month before harvest for only 2 days need elapse between application and harvest. It is marketed as an 80% s.c.

**Vamidothion** (35). This systemic insecticide and acaricide was intro-duced in 1962 and is included in Part III of the Regulations. Under the Pesticides Safety Precautions Scheme, the maximum rates and frequency of application are given as: on apples and pears, one application at 3·2 oz a.i. (8 fl. oz 40% s.c.)/100 gal between green cluster and pink bud and up to two applications later at 6·5 oz a.i. (16 fl. oz 40% s.c.)/100 gal H.V. or the equivalent L.V.; on cherries and plums, one application at 6·5 oz a.i. (16 fl. oz 40% s.c.)/100 gal H.V. or the equivalent L.V.; on hops, up to three applications each at 10 oz a.i. (24 fl. oz 40% s.c.)/acre; on brassicas, fodder beet, sugar beet, mangolds and potatoes, up to three applications each at 8 oz a.i. (20 fl. oz 40% s.c.)/acre. Brassicas should not be treated after the end of September. There should be an interval of at least 6 weeks between application and the picking of apples, pears, cherries or plums; and of 4 weeks in the cases of brassicas, fodder and sugar beet, mangolds, hops and potatoes. Livestock should be kept from treated areas for at least 4 weeks. The usual formulation is the 40% s.c.

**Zinc, manganese, copper, iron dithiocarbamate complex.** This dithiocarbamate fungicide was introduced in 1967 for the control of potato blight, downy mildews and other foliage diseases. It is marketed as an 80% w.p. and is compatible with most insecticides and with copper oxychloride, but should not be mixed with lead arsenate, lime or lime sulphur.

**Zineb** (64). When, in 1943, it was found that the addition of zinc sulphate improved the fungicidal efficiency of nabam, zineb was prepared in the tank in the field (see p. 49). But nowadays the ready-prepared compound is available and finds use against the downy mildews. It is practically insoluble in water and non-volatile and is compatible with other pesticides. Its fungicidal action is thought to be that of nabam and it is harmless to all, except zinc-sensitive, plants. Its toxicity hazards are low but inhalation and skin contact should be avoided. On edible crops at least 7 days should elapse between application and harvest except for crops under glass and mushrooms, 2 clear days is adequate. The usual formula-tions are dusts of 8 and 15%; w.p. usually of 65% a.i.

# CHAPTER 3

# THE APPLICATION OF PESTICIDES

In protecting crops against pest and disease it is necessary to distribute a small amount of pesticide over the surface of the host plant, a task much more difficult than the treatment of seed or soil. On the crop, the pesticide may be applied as a dust or, more usually, diluted with water and sprayed on. A variety of methods are available for making the application and these must be selected to suit the particular requirements of the crop or pest.

## 3.1 SPRAYING

### 3.1A METHODS AND TERMINOLOGY

**Volume.** In almost all cases in Great Britain, water is used as the diluent of the spray. The chemical is supplied as a concentrate, the several types of which have been described in Chapter II. Various amounts of spray may be applied to each acre of crop and the volume used may affect the distribution of the chemical. A large volume increases the chances of full cover and may assist the wetting of difficult targets. When the volume is reduced and the same level of pesticide is required, it is necessary to increase the concentration of the spray and to use a finer drop 'spectrum' (a term for the distribution of the drops of a spray into size groups). The cover may vary from a complete film of spray on the target to a distribution of discrete droplets.

The volume applied per acre will give different degrees of cover depending on the area of foliage of the crop and the following terms are commonly applied:

|  | Gal/acre on bushes and trees | Gal/acre on ground crops |
| --- | --- | --- |
| Very low volume | under 20 | under 5 |
| Low volume | 20–50 | 5–20 |
| Medium volume | 50–100 | 20–60 |
| High volume | over 100 | over 60 |

Volumes as little as a few pints per acre have been used in certain oversea insect control measures for many years. The insecticide in this case is in a diluent which is much less volatile than water and this is formed into a very fine spray of drops just large enough to settle on the target. The technique, which has become known as ultra low volume (ULV) application, is now being tried on a variety of crops in Britain with specially designed equipment and formulations.

**Concentration.** The amount of chemical to be applied to ground crops is usually recommended as amount per acre and recommendations for fruit tree spraying may be quoted as amount per 100 gal for high volume application or, for all types of application but usually for low volume, as amount per acre. The volume of spray may be adjusted to suit the density of the crop and, in all cases, label instructions must be followed.

**Dose and Cover.** Dose and cover in a spraying context are usually taken to mean respectively the amount of chemical on a unit area of the target and the proportion of that area covered. Fortunately for the grower it is unnecessary to pursue these concepts and it may be accepted that dose is roughly governed by the amount of active chemical applied per acre of ground and that cover is often dependent on the amount of water used. Where high or medium volume is used, the addition to the spray of surface-active 'wetters' or 'spreaders' may be recommended in order to increase the cover obtained. Such materials can markedly affect the spread and retention of spray on the plant and they should only be used as recommended. The distribution of the spray, i.e. the uniformity of dose, can obviously be influenced by the type of the spraying equipment and its operation.

**Requirements.** It is difficult to generalize since some pests and diseases can be readily controlled by chemicals even when the latter are indifferently applied, whereas, with other pests and diseases, it is essential to obtain as complete a cover as possible. Generally, the application of insecticides and fungicides must be more carefully carried out than with herbicides and foliar nutrients. Systemic compounds, which are translocated by the plant, do not require complete cover to be effective but this property does not mean that they may be carelessly applied.

## 3.1B EQUIPMENT

### Types of machine

**The ground crop sprayer.** The ground crop sprayer most widely used by the farmer and smallholder is tractor mounted and comprises a tank of 40–100 gal capacity, a pump driven directly from the tractor power take-off and a horizontal boom with a span up to about 20 ft. The boom is adjustable for height above the crop and is fitted with nozzles so placed to ensure a complete swath of uniform throughput. This type of machine is widely used at low and medium volumes for weed control.

Tank contents may be agitated either by fluid returned to the tank from the pump or by a mechanical stirrer. The working pressure of the pump is about 40 p.s.i. though it can be regulated by adjustment of a relief valve. In order to reduce spray drift, this pressure is often reduced to the lowest consistent with a satisfactory nozzle performance, but the precaution is of more importance in the application of herbicides than in the application of insecticides and fungicides for which higher pressures are recommended. The pump capacity is selected to suit the volume of spray to be applied per acre, the swath width and the speed of the tractor. The tractor driver of this type of machine can be readily drenched by spray drift and the special precautions necessary when using hazardous pesticides are given in Chapter 4 (pp. 83–96).

The spray contractor or the farmer with a large acreage uses a similar but larger machine. The boom may be up to 50 ft in length and these machines often employ a trailed tank or have saddle tanks mounted on a large tractor. The tank capacity and the size of the pump must be matched to the requirements of the machine, and with extensive use and particularly with toxic materials, it is well to fit an air-conditioned cab to the tractor.

Aircraft spraying in the United Kingdom is restricted by the small acreages of the crops and the hazards presented to the pilot. Its most successful use is on potatoes.

**Tree sprayers.** Ground crop sprayers, when fitted with vertical booms, can be used for the automatic spraying of small trees and bushes or the pump of these machines may be used to supply spray to hand-held lances. But the spray so produced can be projected for only a limited distance. High pressures and throughput are needed to spray large trees and, at one time, hydraulic nozzles working up to 600 p.s.i. were used. This method of application has been superseded by the use of air to blow the spray into the tree. These machines may have hydraulic nozzles or they may use the air stream to shear the spray liquid into droplets and to carry these to the target.

The machine fitted with hydraulic nozzles is used for high volume application but, if fitted with well-designed nozzles of low throughput, it is suitable for low volume work. Airblast machines fitted with air-shear nozzles require a high pressure fan and will give a spray fine enough for very low volume work, but can be used at all volumes.

Airblast orchard spray machines can be adapted for smaller crops and are used on other types of fruit. They may there do a slightly better job than the boom-and-nozzle machine, but their high capital cost is not usually justified for such work and their use on small fruit occurs mainly where there is top fruit to be sprayed on the same holding.

Such machines are designed to take all the power available from an orchard tractor, with due allowance for traction. The fan and pump unit is often mounted on the tractor linkage. The tank capacity may be anywhere between 60 and 300 gal; the smaller tanks being mounted on the tractor integrally with the mechanism. But for capacities of 100 gal or more, either the tank is trailed separately or the whole machine, mechanism and tank together, is in the form of a trailed unit.

Agitation of the tank contents may be by mechanical stirrer or by recirculation or a combination, but mechanical stirring is much more common on orchard sprayers than on ground crop machines.

**Manual knapsack sprayers.** One to three gallons of spray can be carried in a knapsack sprayer and the contents of such a machine may be pumped out either by the operation of an attached hand pump, the second hand being used to direct a nozzle, or by a charge of pressurized air above the liquid in the container. In the latter type the air capacity is usually adequate to discharge the liquid contents without repumping and, if pressure-regulating valves are fitted, the liquid may be sprayed out at constant pressure. In one type of this appliance, the spray is put into the tank and air is then pumped in above by a fitted pump or a compressed air supply may be used. In a second type the air is retained and the sprayer is recharged by pumping in the liquid to be sprayed. Pressures of 20–80 p.s.i. are normally used. These machines, with a single nozzle or with three or four small nozzles on a short boom are capable of excellent spraying where the scale of work involves the application of only a few gallons of spray and are particularly useful on small areas where the lay of the land precludes the use of vehicles.

Although these machines are manufactured with an adequate safety factor, it is well to subject old, fatigued or corroded sprayers to a periodic hydraulic pressure test. The recommended pressure should never be exceeded.

**Motorized knapsack sprayers.** In recent years lightweight two-stroke engines driving high speed fans have been fitted to knapsack sprayers. The fan creates a stream of air into which an air-shear nozzle is fitted and the spray container is pressurized to supply the nozzle at a liquid pressure of only 1 or 2 p.s.i.

## 3.2 SPECIFICATION OF MAIN COMPONENTS

**Tank.** In all spray operations it is important to have equipment which is adequate for the job. Unless the tank capacity is suitable, more time may be spent on mixing and filling than on spraying. A simple calculation will show how far a tankful will go at the normal rate of application (divide

F

tank capacity by volume/acre) and how long it will last (divide capacity by output of machine). The operation will become tedious unless the tank provides for 15 or 20 minutes' spraying. If repeated fillings are needed, it is important that the water supply should be nearby and adequate, the water flow should be enough to fill the tank in about 5 minutes.

Galvanized steel is the commonest and cheapest material used for the construction of the tank but, after a number of years, corrosion may be troublesome. More expensive alternatives, such as p.v.c., resin-bonded glass or stainless steel, will give a longer life but are not widely used except for smaller tanks. Tanks should have a strainer in the filler hole which should be of generous dimensions. They must also be readily drained and cleaned.

**Agitation.** It is necessary to maintain, in the spray tank, a uniform dispersion of suspensions and emulsions throughout the time that the machine is in operation. Mechanical agitation is generally more efficient than recirculation though, with both, it is necessary to ensure that the floor of the tank is thoroughly scoured without areas of 'dead' liquid. Hence the tank should be of regular shape with rounded corners. A danger from agitation dependent on recirculation is inadequacy when the nozzle output is near the maximum of the pump capacity. Some formulations cause frothing particularly if the stirrer or return flow is above the level of the tank contents. Also cavitation at the suction outlet must be avoided as this will reduce the throughput of the pump and may cause water hammer.

**Pumps.** The capacity of the pump must be adequate to maintain a constant working pressure and should therefore be selected for the duty it has to perform. A variety of types is available.

Plunger pumps are favoured for pressures of 150–600 p.s.i. They draw in liquid via a valve on the suction stroke; this valve closes on the pressure stroke when the outlet valve opens and the spray is forced into the delivery circuit. This type of pump must be stoutly made and has a number of component parts; it is therefore expensive. Two or three cylinders are commonly used and an air vessel is needed in the outflow to even out the pulses of pressure.

Diaphragm pumps operate on the same principle as plunger pumps; pressure is applied to the rear of the diaphragm by a reciprocating plate which moves like the plunger. Since the spray fluid is sealed from the moving parts, abrasion and corrosion of close fitting parts are eliminated. The performance of these pumps is limited by the strength and movement of the diaphragm, usually of synthetic rubber, and output per 'cylinder' is accordingly limited; nor are the pressures attainable as high as with the plunger pump.

Gear and vane pumps: most ground crop and many orchard machines are equipped with continuous delivery pumps, which may be either positive displacement or centrifugal. By positive displacement is meant any pump which, like the plunger pump, takes in a definite volume of liquid from the inlet and transfers it, without possibility of escape, to the outlet. The centrifugal pump, by contrast, takes in liquid at the axis and throws it by centrifugal force to the periphery where it is delivered. If the outlet of such a pump is restricted or blocked the outflow decreases or stops and there is no risk of damage to the pump by the build up of pressure. In like circumstances with the displacement pump, damage may result unless a relief valve comes into operation or a safety plug blows.

There are three types of continuous positive displacement pumps, namely, gear, vane and roller vane. In the gear pump, a pair of gears run together in mesh in a casing; liquid enters between the teeth as they come out of mesh and is carried round between casing and teeth to be discharged at a later point before the teeth enmesh once more. Such pumps are relatively less expensive but are of limited output and are subject to heavy wear particularly if used for dispersions of wettable powders.

Vane pumps operate similarly but with a single rotor. The space between the rotor and casing is divided into sections by vanes in the rotor which reach to the casing. Liquid enters in at one portway in the casing to fill each sector and is transferred round to another point where it is ejected. The simple vane pump has most of the drawbacks of the gear pump but the roller vane, which is a development of it, has certain advantages. Vanes are replaced by rollers which greatly reduce the wear from abrasive materials. Most medium output sprayers now have this type of pump which is suitable for direct drive and mounting on tractor power take-off shafts. Outputs are of the order of 10 gal/minute at 50 or more p.s.i.

Positive displacement pumps are fitted with pressure relief valves but performance tends to fall off as clearances become excessive. Some may be damaged if run dry for any prolonged period for liquid is relied on to reduce friction.

Centrifugal pumps are usually free from wear except where the shaft of the impeller enters the casing. For a given output they are relatively small and cheap. They need to be run at a fast speed and high pressures can be obtained only by the use of multi-stages. They are often used to deliver the spray at low pressure to the air-shear nozzles of orchard sprayers.

**Nozzles**—fan and cone. The hydraulic nozzles used on sprayers are of two types. For most ground spraying, fan nozzles are used. These depend on a shaped orifice, usually formed in a ceramic disc, through which the spray is emitted under pressure to give a fan-shaped pattern of droplets.

Cone or swirl nozzles give a cone-shaped pattern and consist of two basic elements, the disc and the swirl plate. The disc is a circular ceramic or hard metal plate with a central hole. The spray, before passing through this orifice, is given a swirling motion in a chamber which it enters by angled slots in the swirl plate. In most designs throughput and pattern can be altered by change of disc and swirl plate; cone nozzles are thus more versatile but more expensive than fan nozzles and are generally used on orchard hydraulic sprayers.

British Standard 2968 : 1958 suggests standards for both fan and cone spray nozzles. The specification relates to performance and to the means of attachment of nozzles, the main purpose being to provide for inter-changeability of nozzles of like size and type made by different manu-facturers. The following permissible variations of discharge rate are specified for fan and cone nozzles for crop spraying: $\pm 10\%$ up to and including 30 Imperial gallons per hour; $\pm 5\%$ above this rate. The included angle of spray for nozzles of both types to be used for ground crops is specified as $65°$ for general purposes and $80°$ for special wide angle spray, both within a tolerance of $3°$.

The angle of the fan or cone and the spray spectrum are important when adjusting the nozzles on a boom. Fan nozzles and small cone nozzles give a distribution across the swath greatest under the centre and tailing off toward each edge, which must be borne in mind when adjusting overlap. Swirl nozzles, as used on booms, give a hollow cone of spray and the bulk of the output is around the periphery.

The output from hydraulic nozzles varies as the square root of the applied pressure, i.e., doubling or halving pressure will result in roughly $40\%$ increase or $30\%$ decrease in flow. It is therefore necessary to change the nozzle (or the disc where provision is made) to obtain big changes of output, though pressure regulation will give a final adjustment.

A required output can be obtained from a large aperture working at low pressure or a smaller one at a higher pressure; the choice depends on the size of droplets required, higher pressures form smaller drops. Drop size affects cover and the liability to drift (see Operation, p. 96).

The simplest nozzle used in airblast sprayers is an orifice across which air is blown at high speed to shear off the emitted liquid as small droplets. Drop size is dependent on air speed but the spray becomes coarser with increased throughput.

Rotary nozzles give a more compact range of droplet size than the nozzles so far described. With them, drop size is dependent on the speed of rotation of a disc or drum from which the liquid is thrown. Small drops are obtainable but only at high rotational speeds of upwards of 10,000 r.p.m. When more liquid is supplied than can leave the periphery of the atomizer in discrete drops, it is thrown off in sheets and performance rapidly deteriorates. This type of nozzle is often used on aircraft, and for ULV treatments from ground equipment, when only small volumes are applied as a close spectrum of small droplets.

**Fan.** On airblast orchard sprayers, the fan is a critical component and, for a given horse power, it is generally better to produce air at 5 in w.g. pressure (about 100 m.p.h. in the outlet) than at 20 in. w.g. (near 200 m.p.h.). The lower air speed requires the use of pump and nozzles for droplet formation. A common compromise is the sprayer with a centrifugal fan or fans working at about 12 in w.g. discharging directly to one or both sides. This machine requires hydraulic pressure nozzles at rather less than 100 p.s.i., but the airstream is fast enough to break down the larger drops.

The choice of machine in respect of fan type is usually dependent on the suitability of the arrangement of air outlet for the crops to be sprayed rather than on considerations of fan type, air pressure and volume. But it is important that the fan be run at its rated speed and that it matches the power output of the tractor. On the matching of machine to tractor power and power take-off drive speed, the sprayer manufactuerer or his agents should be consulted.

## 3.3 OPERATION

**Maintenance.** The manufacturers' instructions for lubrication, cleaning and adjustment should be learnt and followed. Since much of the deterioration due to corrosion occurs when the machine is *not* in use, it must be washed, drained and dried before putting into store and stored where it will keep dry. Hoses should be stored in the dark and not left kinked. If the outside of the machine is wiped over with oil before use, the deposit of spray chemical can more easily be cleaned off; this oil film should be restored after cleaning before putting the machine into store. Wearing parts such as pump glands, valves, agitator shaft seals and bearings should be checked at the end of each season and replacements arranged well in time for the next.

**Planning.** It is first necessary to know what pests and diseases are likely to be met and the pesticides to be used for each, the acreage to be sprayed for each, the volume to be applied and the timing of the applications. The records of past results are so helpful in furnishing this information that details of all pesticide operations should be recorded for future reference. Regulations made under the Agriculture (Poisonous Substances) Act require that certain such records be made in respect of certain pesticides (see p. 88) and these should be extended to all pesticides and include: amount of pesticide used, volume applied, the date time and crop stage. The infestation or infection conditions and the degree of control achieved should be, if possible, noted. Interruptions by weather or breakdown should also be noted.

Simple arithmetic will often provide useful guidance in planning spray operations. As an example, consider the times taken to spray 10 acres at 200 gal/acre and at 50 gal/acre with a machine of tank capacity of 150 gal. Let the time taken to refill be 20 minutes (10 minutes for mixing and filling and 5 minutes each way to and from the filling point), and assume that the working rate is 12 minutes per acre at low volume and 16 minutes per acre at high volume (i.e. the pump capacity is $12\frac{1}{2}$ g.p.m.). The schedule in each case will be:

| 200 gal/acre | | 50 gal/acre | |
|---|---|---|---|
| Time minutes | acreage | Time minutes | acreage |
| 0–20   filling | | 0–20   filling | |
| 20–32   spraying | $\frac{3}{4}$ | 20–56   spraying | 3 |
| 32–52   refilling | | 56–76   refilling | |
| 52–64   spraying | $1\frac{1}{2}$ | 76–112 spraying | 6 |
| 64–84   refilling | | 112–132 refilling | |
| and so on for 13 loads | | | |
| 404–416 spraying | $9\frac{3}{4}$ | 132–168 spraying | 9 |
| 416–436 refilling | | 168–188 refilling | |
| 436–440 spraying $\frac{3}{4}$ acre | 10 | 188–200 spraying 1 acre | 10 |

The time taken under these conditions is therefore about $7\frac{1}{2}$ hours at high volume whereas, at low volume, it is only $3\frac{1}{2}$ hours. Similar calculations may be made to estimate the value of increased tank capacity for various circumstances.

If a large program of work is involved, it is advisable to consider the complete routine, including herbicide work if this will involve the same operator, to reach a decision on machine capacity. At the peak of the program it is usual to budget for the treatment of the required acreage in about 5 working days if a 10 day routine is aimed at, which leaves an adequate margin for breaks at weekends and for unsuitable weather.

**Warnings.** The choice of spray materials depends to some degree on surroundings and the risk of spray drift. Drift is generally less of a problem with insecticides and fungicides than with herbicides so far as damage to neighbouring crops but the hazards described in Chapter 4 must be taken into account when choosing materials and methods of application. Drift from tree spraying is unavoidable and, in the first hundred yards downwind it will be appreciable whatever application method is used. On ground crops it can be reduced by using pressures of 10 or 15 p.s.i. to give a coarse spray.

**The operation of the machine.** The subsequent notes relate to field and orchard operations. For small scale work, all that is usually needed is the application of a measured volume to a given area; both volume and area are easily measured and the machine manipulated to give the required dose. Table 3.1 may be used to convert recommendations in terms of acres (i.e. 200 gal) to operations on a few square yards.

For field work, a tractor engine speedometer is essential for accurate spray application and should always be fitted. Nozzle and machine manufacturers give instructions for the adjustment of the machine for a desired output, but this should be carefully checked the first time the machine is used. To get the correct volume per acre, it is necessary to obtain the right combination of output and speed.

**Measuring nozzle output.** Tanks should be fitted with dipsticks near the centre and measurements should be made with the machine as level as possible, particularly if the dipstick is not central. To check nozzle output: (i) fill the tank with water with machine stationary at the set tractor engine speed, (ii) spray for a fixed time long enough to reduce the tank contents by an amount which can be accurately measured, say one quarter, (iii) read amount used, (iv) reduce the tank contents by spraying or draining to between one quarter and half full, (v) repeat (ii) and (iii). The average of the two results may be taken if they are in reasonable agreement, but if there is a serious reduction in output with the reduced head in the tank, the cause must be ascertained.

**Checking rate of work.** With ground spraying the area covered is the swath width multiplied by the speed of travel. In tree spraying the row to row distance is taken as the swath when spraying double-sided but is half of this for the single-sided operation. An engine speed indicator will usually be used to set tractor speed but even if this is marked in m.p.h. the setting should be checked, preferably under actual spraying conditions. Measure the speed at set engine revolutions and in the gear or gears which will be used by timing over a measured distance. The calculation of the rate of work in acres/hour from the speed in miles/hour and the swath in feet can be done by the use of scales 1, 2 and 3 of the nomogram on p. 73.

**Setting volume per acre.** If the output of the machine has been fixed by the selection of nozzles and pump pressure, volume is adjusted by changing the speed of travel. Suppose, for example, that an application of 30 gal/acre is needed with a machine using ten nozzles on a 15 ft boom, with a nominal output of 21 gal/hour operated at 40 p.s.i. The output of the machine, measured as described above is 19 and $17\frac{1}{2}$ gallons in two 5 minute periods. The actual output is therefore 3·7 gal/minute. Now refer to the nomogram on the opposite page. Find 30 gal/acre on scale 5 and 3·7 gal/minute on scale 4. Lay a ruler across these two points and read

TABLE 3.1

*Equivalent amounts for small plots*

| Volume: | | | | | | | | | |
|---|---|---|---|---|---|---|---|---|---|
| gal/acre | 5 | 10 | 20 | 30 | 50 | 100 | 200 | | |
| square yd/gal | 968 | 484 | 242 | 161 | 97 | 48 | 24 | | |
| High volume concentrations: | | | | | | | | | |
| pints/200 gal (i.e. per acre) | 1 | 2 | 3 | 4 | 5 | 6 | 8 | 10 | |
| fl. oz/gal | 0·1 | 0·2 | 0·3 | 0·4 | 0·5 | 0·6 | 0·8 | 1·0 | |
| ml/gal | 2·8 | 5·7 | 8·5 | 11·4 | 14·2 | 17·0 | 22·7 | 28·4 | |
| lb/200 gal | 1 | 2 | 3 | 4 | 5 | 6 | 8 | 10 | 12 |
| oz/gal | 1/12 | 1/6 | 1/4 | 1/3 | 2/5 | 1/2 | 2/3 | 4/5 | 1 |
| gram/gal | 2·3 | 4·5 | 6·8 | 9·1 | 11·3 | 13·6 | 18·1 | 22·7 | 27·2 |

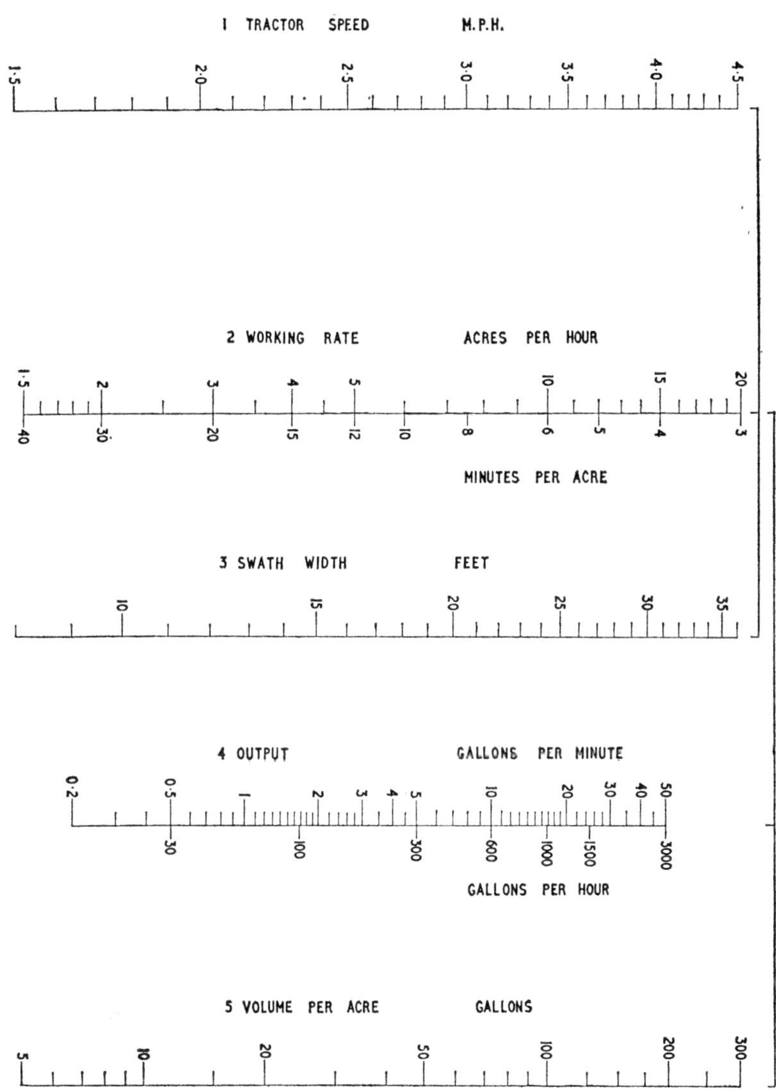


FIG. 3.1 Nomogram
Place straight edge across scales 1, 2 and 3 or 2, 4 and 5.

where this ruler crosses scale 2, namely, at 7·4 acre/hour. Now find the swath width on scale 3. Join this to the point found (7·4) on scale 2 and read where the ruler crosses scale 1 which is at 4 m.p.h.; this is the required speed.

For a second example, suppose that the tractor engine speed is fixed by fan requirements and, having selected the appropriate gear, it is necessary to adjust nozzle output by changing liquid pressure to set to a particular volume per acre. The trees to be sprayed single-sided are in rows 28 ft apart and 150 gal/acre are to be applied from an airblast machine with engine speed set at 1800 r.p.m., at which revolutions the ground speed is 2¾ m.p.h. The effective swath is therefore 14 ft, the point to find on scale 3, join this point and 2¾ on scale 1 with a ruler which will be found to cut scale 2 at 4·7 acre/hour. Now join this point on scale 2 to 150 on scale 5 and read off the required output on scale 4, namely, 11·8 g.p.m. The machine should now be set for this output and the actual volume delivered checked by the procedure given on p. 71.

**Checking application rate.** Whenever a new formulation is used, it is advisable to check the rate of application. Calculate the distance driven to spray an acre by dividing 4,840, the number of square yards in an acre, by the swath width in yards. Check the volume of spray used over this distance and repeat as the tank empties. Rough checks on the first tankful will detect gross misapplication and smaller deviations will become apparent after a larger acreage has been sprayed. The amounts used should be recorded.

**Filling and mixing.** The label of the pesticide container must be read carefully, the precautions observed and any recommended mixing procedure should be followed. The statutory requirements for handling scheduled chemicals are given in Chapter 4 and in literature available at any M.A.F.F. office.

The necessity of an adequate and clean water supply is obvious. Pre-creaming is recommended, especially when preparing the spray from wettable powders and the spray tank should be partly filled with water before the precreamed concentrate is added. Readily dispersed materials may be placed in the spray basket of the tank and washed in with the water. When the mixing of different pesticides is required, make sure that they are compatible especially when mixing the products of different manufacturers for, even though the active ingredients may be compatible, the formulating agents may be incompatible (see p. 26).

**Cleaning.** After the spraying operation has been completed, the machine should be drained, the filters taken out and cleaned, the tank flushed out and water pumped through to clean the lines and nozzles. Finally drain the tank and wash down the machine. If possible, do not use, for the application of insecticides or fungicides, a machine which has been used for herbicides, especially of the growth-regulating type. If unavoidable, the machine should be decontaminated by the procedure described in the Weed Control Handbook.

## 3.4 NOZZLE ARRANGEMENTS AND PROCEDURES FOR PARTICULAR CROPS

**Field crops.** The normal arrangement of the nozzles on boom sprayers is for them to point directly downwards though other arrangements, such as directing them rearwards at an angle to the vertical or fitting them to droplegs so that they are carried between the rows of plants, are sometimes used. The nozzle spacing on the boom is fixed by the nozzle design and the amount of overlap between the nozzle spray patterns is adjusted by the nozzle or by altering the height of the boom above the crop. Nozzles with wide-angled patterns can be operated at smaller boom heights than those with narrow-angled patterns.

Two types of patterns are recognized, namely triangular and rectangular. In the first type the heaviest deposit is under the centre of the nozzle and the amount decreases gradually towards each edge. This is the most suitable for use on booms where patterns are to be overlapped and is readily produced by fan spray nozzles. The rectangular distribution, i.e. equal amounts deposited across the entire width, can be achieved, approximately, by a special design of fan spray nozzle or by cone nozzles. Such a pattern is suitable for band spraying, where a single nozzle is used to cover a strip of ground. Small changes of nozzle height can result in strips receiving double or no dose if this distribution is used for boom nozzles. Triangular pattern nozzles should be set to overlap so that the edge of one fan of spray reaches the crop or ground directly beneath the centre of the next nozzle and the distribution across the swath will then not be very sensitive to small changes in boom height. An equally serious source of unevenness is that of variation, both in throughput and pattern, between individual nozzles which nominally are the same. Where a high degree of uniformity is required, B.S. kite-marked nozzles should be used conforming to the B.S. specification mentioned on p. 68.

When selecting boom height, it should be remembered that, under most field conditions, the boom will not always remain parallel with the crop because unevenness of the ground causes tilting, and bumps cause boom tips to flap. Hence the boom should be set high enough to ensure that its lowest end will not be lower than the height necessary to obtain overlap. Excessive height should be avoided because it increases the risk of spray drift. When it is important to reduce drift and fan nozzles are being used, the nozzles can be pointed rearwards at an angle to the vertical enabling overlap to be obtained from a lower setting of the boom height.

Drop legs are used when it is necessary to obtain a better cover of the under parts of plants. One drop leg equipped with two nozzles is required for each row to be sprayed. When using drop legs, care must be taken not to set the boom in so low a position that the nozzles dig into the ground for blockage or damage will result. The nozzles should be set at an angle to the crop rows so that the spray cone may develop before the spray reaches the main part of the target.

Airblast sprayers are rarely used for ground crops mainly because of the difficulty of estimating the effective swath width.

The main points to be borne in mind when spraying field crops are the adequate overlapping of adjacent swaths, and the avoidance of missed areas between one tank load and the next. When row crops are sprayed, the correct place to start the second and subsequent swaths is determined merely by counting the rows but, for cereal crops, the swaths should be paced and marked to enable the tractor driver to follow the correct track across the field. If possible, stops to refill should be avoided between the headlands but, if not, the stopping place should be clearly marked, for example, by immediately turning out of line into the previously sprayed area. After refilling, the return should be made along the last set of wheel marks and spraying should be resumed immediately before the point where they curve. Anti-drip devices are available and may be operated should a blocked nozzle or other cause make it necessary to stop during a bout, but their use is not quite as important in insecticide and fungicide application as in the spraying of herbicides.

**Plantation crops** present a greater diversity of target than field crops and the nozzle arrangement should be suited to the shape and composition of the target. The adjustment is usually a matter of experience but advice on difficult problems may be sought of the N.A.A.S. The addition of a fluorescent dye to the spray to give a picture of the deposition under ultra-violet light is one of the techniques used by specialists in the Advisory Service to help with setting up machines for spraying various crops.

For smaller bushes, cane fruit and for apples and pears grown on intensive systems, arrangements of hydraulic nozzles spaced over the vertical height of the crop are satisfactory but, where the leaf canopy is dense, as in blackcurrants, gaps in spray cover are unavoidable because of the screening of leaf by leaf.

For larger trees, the critical area is in the centre and top of a spreading canopy. The bulk of the spray should be directed to the top of the trees but, if the underside of leaves and the wood must be covered, extra spray should be directed, from as close as possible, in to central and lower parts of the tree. When airblast machines are used, both the airstream and nozzle position should be adjusted: the airstream should normally be centred on the diagnonal up through the tree but the spray should be concentrated in the higher and lower part of the arc according to canopy requirements. When adjustment from the tractor seat is possible and there is a cross wind, the nozzles spraying downwind should be elevated and those spraying upwind should be depressed. With low volume applications, heavy doses on the skirt of the tree near the machine outlets are difficult to avoid.

The sprayer output must be fully on when the first tree or bush is reached and the spraying should be to a set pattern so that rows are not missed. A skilled operator, making repeated application to a block, will be able to save much time at high volume rates by learning how far a full tank will go and so avoiding running-out of spray at points which involve a long journey to the filling point. It is often quicker to refill a tank not quite empty, especially in long rows affording no opportunity of turning.

One-third of the final deposit may be from spray settling on downwind rows. An additional traverse of the upwind headland at high speed to give, say, half the normal application rate, is therefore a wise precaution against failure to protect the upwind trees adequately.

**Horticultural crops,** when grown outdoors on a scale large enough to warrant tractor powered sprayers, are covered by the foregoing remarks. In nurseries and market gardens, the small ground crop sprayer can be adapted for some crops and spray pumps are available as accessories to some garden tractors. Automatic spraying from booms is sometimes possible. The alternative is the hand-held lance, more laborious and variable though capable of a good job if skilfully used.

**Glasshouse crops** may be sprayed, at high volume, with the hand-directed lance. Adequate lengths of hose are needed and the spray machine, if operated by a petrol motor, should be outside the door. Up to 400–600 gallons of spray are needed per acre of tomatoes or cucumbers to wet the foliage completely.

With certain crops, low volume applications may be made with a compressed-air paint spray-gun, a motorised mist-blower or other machine capable of dealing with concentrated aqueous dispersions of pesticides. From 4–10 gal per acre are required at low volume and more frequent applications are usually necessary than with H.V. spraying, the air compressor output and pressure must be adequate for the type of spray-gun used—normally 10–15 cu. ft/min at 50–70 p.s.i.

Aerosols are generated from atomizable formulations using paint spray-gun type sprayers, preferable suction-fed types fitted with a calibrated glass container, or with integral-motored aerosol generators from which the atomized droplets are carried in a stream of air. The droplet stream is directed into the air of the glasshouse while the operator walks backwards the length of the glasshouse. The quantity of pesticide applied is regulated by adjusting the output rate of the machine or spray-gun to the speed of walking through the glasshouse. Inhalation of aerosols or low volume spray mists must be avoided.

## 3.5 DUSTS AND DRY APPLICATIONS

Dry formulations may, for convenience, be divided into (a), light dusts for use mainly in plantation crops and in situations where the tall growth retains the dust cloud; (b), heavier dusts which fall rapidly to the target which is usually a field crop; (c), granular pesticides. But before discussing the types of machine available for dusting operations, it will be well to outline the advantages and disadvantages of dusts as opposed to sprays, for often the alternative is open to the grower.

DUSTING VERSUS SPRAYING

The main advantages of dusting rest in the fact that the problems of water supply vanish. As a consequence, dusting appliances are lighter in weight and more easily handled in difficult terrain than sprayers; the dust must be purchased ready for use whereby mixing tanks become unnecessary and handling is reduced though toxicity hazards to operators may be especially serious because of the greater inhalation risks, see Chapter 4, p. 88. It may be said that, by and large, dusting is less costly in appliances, time and labour than spraying.

On the other hand, spraying usually gives better results than dusting for the dust deposit is less adherent to foliage than the spray. For this reason and because of the greater drift hazard of dusts, it is often preferable to dust only at dawn or dusk when there is moisture on the target and the air is still.

GRANULES

Insecticides formulated in the form of granules have been introduced in recent years for certain specific purposes. Their main use has been in the application of systemic materials on broad leaved crops where the granules can be lodged in the foliage to give a gradual release of insecticide over a period. Although broadcast application can be made from the air or by ground machine, special applicators have more generally been used to apply granules to each row of plants in crops like sugar beet and brassicae. These can be fitted on a tool bar to combine the insecticide application to the crop with cultivation of the soil between the rows. The increased cost of formulating in this way can thus be offset by economies from the greater persistence and more accurate placing of the insecticide.

TYPES OF MACHINES

All machines used for applying dry formulations consist essentially of a hopper which usually contains an agitator, an adjustable orifice or other metering mechanism and delivery tubes. In the case of dusters a fan or bellows provides the conveying air and the dust may join the airstream before or after the fan.

Although dusts are commonly applied from aircraft in some overseas countries, this type of operation is rare in the British Isles. Ground machines range in size from hand-held to tractor-powered and each type of sprayer has its counterpart with dust outlets in place of spray nozzles. Also, coarse dusts and granules can be distributed broadcast by a spinning disc type of fertilizer spreader. Older machines were often driven from a ground wheel but this form of power is not often used now except sometimes to operate a metering mechanism. In this way application may be directly related to the ground covered and independent of travelling speed.

HINTS TO OPERATORS

1. Keep your powder dry! Even if the sacks are polythene-lined, store them clear of concrete floors.

2. Empty the hopper at the end of the day's work; otherwise store under cover or sheet down carefully.

3. Keep all foreign material such as string, pieces of paper, oil, grease, out of the dust; a coarse screen can sometimes be useful inside the hopper.

4. As with sprays, application rates must be accurate and constant throughout the treatment. To achieve this, the operator requires, in addition to a suitable machine, an accurate calibration chart to give him the initial setting, a means of checking the rate actually applied in the field, and an indication of his forward speed and of the speed of the mechanism so that, once set, both can be held constant. Many older types of duster, although tractor-mounted, were fitted with their own engine as a means of holding the fan speed constant. Now that most tractors are fitted with an engine speed tachometer or speedometer, the fitting of an engine is no longer necessary on these grounds. The checking of application rate of a duster is more difficult than that of a sprayer because once the dust has passed into the airstream it cannot be easily recovered for weighing. Some machines are so designed that the dust can be caught in a container before it joins the airstream and the weight delivered in a given time can then be measured. Otherwise checks can only be made in the field by noting the level of material in the hopper, dusting a known area, and weighing the dust required to refill the hopper to the previous level, see Tables 3.2 and 3.3.

5. In addition to maintaining the engine or p.t.o. speed constant during work, it is important that no slip should occur in the drive to the fan. The drives to dust fans often run at high speeds and belt tension should be carefully watched.

6. Lubricate carefully, but not excessively, all bearings.

TABLE 3.2

*Equivalent rates of application*

| Rate lb/acre | Approx. area covered by 1 lb | | | Rate per ft² | |
| --- | --- | --- | --- | --- | --- |
| | ft² | yd² | A square with .. ft sides | oz | gram |
| 1 | 43,560 | 4,840 | 209 | — | 0·01 |
| 10 | 4,356 | 484 | 66 | — | 0·10 |
| 15 | 2,904 | 323 | 54 | — | 0·16 |
| 20 | 2,178 | 242 | 47 | 0·01 | 0·21 |
| 30 | 1,452 | 161 | 38 | 0·01 | 0·31 |
| 40 | 1,089 | 121 | 33 | 0·01 | 0·42 |
| 50 | 871 | 97 | 30 | 0·02 | 0·52 |
| 60 | 726 | 81 | 27 | 0·02 | 0·63 |
| 96 | 454 | 50 | 21 | 0·04 | 1·00 |
| 100 | 436 | 48 | 21 | 0·04 | 1·04 |
| 112 | 389 | 43 | 20 | 0·04 | 1·17 |
| 2,722 | 16 | 1·8 | 4 | 1·00 | 28·36 |

TABLE 3.3

*Equivalent rates of application*

| Recommended rate lb/acre | oz/100 ft of row in | | | Feet of row dusted per lb in | | |
|---|---|---|---|---|---|---|
| | 12″ rows | 20″ rows | 28″ rows | 12″ rows | 20″ rows | 28″ rows |
| 10 | 0·37 | 0·61 | 0·86 | 4,356 | 2,614 | 1,867 |
| 15 | 0·55 | 0·92 | 1·29 | 2,904 | 1,740 | 1,245 |
| 20 | 0·73 | 1·22 | 1·71 | 2,178 | 1,307 | 933 |
| 30 | 1·10 | 1·84 | 2·57 | 1,452 | 871 | 622 |
| 40 | 1·47 | 2·45 | 3·43 | 1,089 | 653 | 467 |
| 50 | 1·84 | 3·06 | 4·29 | 871 | 523 | 373 |
| 60 | 2·20 | 3·67 | 5·14 | 726 | 436 | 311 |
| 100 | 3·67 | 6·12 | 8·57 | 436 | 261 | 187 |
| 112 | 4·11 | 6·86 | 9·60 | 389 | 233 | 167 |

C

7.   Tap all ducting at regular intervals to dislodge caked dust.

8.   At one time, dusters suffered from a tendency for the delivery rate to fluctuate violently according to the amount of material in the hopper, a fluctuation amounting, in some cases, to as much as two or three times the required rate. Although this fault is less frequent in recent models, it is still wise to refill the hopper before the agitator in the base is fully exposed, because it is at this level that the greatest variation in rate occurs.

9.   Adjust the height of the nozzles to suit crop and wind conditions.

10.   When changing from one dust to another, clean out the machine and re-check the application rate, for the same setting will not necessarily give the same rate with the new material.

11.   Take all the required precautions to protect yourself and others from toxicity hazards and, even when the dust has no poisonous ingredient, wear a mask especially when filling the hopper.

# CHAPTER 4

# THE SAFE AND EFFICIENT USE OF PESTICIDES

As might be expected, the large number of chemicals now available to the farmer and grower for dealing with their pest and disease problems includes some compounds which could be harmful if not properly used. The precautionary measures needed by users of toxic chemicals in agriculture were studied in detail by a Working Party appointed by the Minister of Agriculture and Fisheries in 1950, and as a result of recommendations, on promoting the safety of workers applying such materials contained in its first report, the Agriculture (Poisonous Substances) Act was passed in 1952. The first Regulations under this Act were introduced in the following year.

The second report of the Working Party, concerned with possible risks to consumers of treated crops, resulted in the formation of the Advisory Committee on Poisonous Substances used in Agriculture (now known as the Advisory Committee on Pesticides and other Toxic Chemicals) and its Scientific Subcommittee. The third report, in 1955, dealt with the possible effects of such chemicals on the natural fauna and flora of the countryside. Two years later, the Notification of Pesticides Scheme (now known as the Pesticides Safety Precautions Scheme) formally came into being as the result of negotiations between Government Departments and the Industrial Associations concerned. An account of how these safeguards to operators, consumers, and other parties are provided, forms the first part of this Chapter. This is followed by a description of another Scheme—The Agricultural Chemicals Approval Scheme—which is concerned with the biological efficiency of proprietary products based on particular formulations of the active chemical ingredients.

## 4.1A THE PESTICIDES SAFETY PRECAUTIONS SCHEME

The second report of the Working Party (Toxic Chemicals in Agriculture: Residues in Food, H.M.S.O. 1953) contained a recommendation that manufacturers, importers and distributors should notify Government Departments of new toxic chemicals and formulations of pesticides before putting these on the market. As a result, the voluntary Notification of Pesticides Scheme was agreed between Government Departments and the Industrial Associations concerned (the Association of British Chemical

Manufacturers: the Association of British Insecticide Manufacturers [now the Association of British Manufacturers of Agricultural Chemicals] and the Industrial Pest Control Association).

Negotiations were conducted by the Agriculture and Health Departments for England, Scotland and Wales, acting on advice from the Advisory Committee (which includes representatives of the Board of Trade, the Laboratory of the Government Chemist, the Agricultural and Medical Research Councils and the Nature Conservancy) and its Scientific Sub-committee, the latter being composed of official scientists selected for their specialized knowledge of various aspects of pesticides.

As experience with the safe use of pesticide products increased, it became clear that the Scheme needed to be brought up to date and its requirements strengthened. Accordingly, the Industrial Associations and those Government Departments concerned negotiated a revised Scheme which was finally accepted in May 1964 under its new title of 'The Pesticides Safety Precautions Scheme'.

Under this Scheme, a manufacturer, importer or distributor is required to notify a product, if it contains a new chemical or comprises a new formulation or new use of a chemical already on the market, for use as a pesticide in agriculture (including horticulture and home garden use) or food storage (including rodenticides).

Notification of products for agricultural use is made to the Director, Plant Pathology Laboratory, Ministry of Agriculture, Fisheries and Food, Hatching Green, Harpenden, Herts., or to the Director, Infestation Control Laboratory, Ministry of Agriculture, Fisheries and Food, Hook Rise South, Tolworth, Surbiton, Surrey, if it involves a rodenticide or a product for use in food storage practice. Each notification must be supported by extensive data and evidence to justify the claims concerning the safe use of the product. The information the notifier is asked to supply includes full details of the composition of the product, its proposed method of use, mode of action, toxicity, persistence and any other relevant data; and the hazards likely to result from its use, to those handling the product, consumers of treated produce, domestic animals and wild life.

A notification may be processed in one of two ways. For those which the Director considers do not present special problems a 'quick procedure' is adopted and the notification is dealt with through his Chief Chemist and an official toxicologist. If they agree that clearance should be given, the notification is sent to the appropriate Division of the Ministry of Agriculture, Fisheries and Food (which acts as the co-ordinating Department for the Scheme) for acceptance by Departments and the notifier is then informed of the decision.

The other way is for the notification to be processed through the 'committee procedure'. This generally applies to all new chemicals, new uses of the more toxic ones and those that fail to receive clearance by the quick procedure. Under the committee procedure, the information provided by the notifier and any other evidence available is sent to the Scientific Subcommittee and the notifier has the opportunity to be present when it is being considered. The Subcommittee makes recommendations to the Advisory Committee which, in turn, advises Departments. If Departments recommend precautions in the use of the chemical they will not be publicized until the notifier has accepted them and has confirmed that his proposal is no longer confidential. The recommendations are issued as supplements to a loose-leaf booklet entitled 'Chemical Compounds used in Agriculture and Food Storage in Great Britain—User and Consumer Safety—Advice of Government Departments'.

These Recommendations Sheets have a wide circulation and the substance of the recommendations must be included on the appropriate labels of pesticide products offered on the market. The advice given for a chemical cleared for agricultural use is concerned with three aspects: operator safety, consumer safety and wild life safety.

On operator safety, Departments may consider that the chemical is either too toxic to be used at all or that its use must be regulated by law under the Agriculture (Poisonous Substances) Act (see following section). On the other hand, it may be considered to offer insufficient hazard to justify regulating, but advice is given as to certain precautions that should be taken when handling it; or, again, that no special precautions are needed other than common sense general advice to treat all chemicals with care.

On consumer safety, the recommendations aim at ensuring that no unacceptable residues remain when the treated crop is harvested. The use of the chemical may be restricted to certain named edible crops and there might be limitations on the amount of active ingredient applied, and the time of such application(s). A minimum interval between the date of the last application and the time the crop is harvested is also usually stipulated. Unlike operator protection, which is provided under the Agriculture (Poisonous Substances) Act, there is, with two exceptions, no direct legislation for consumer protection, although indirect protection is afforded by the Food and Drugs Act and regulations made thereunder. The exceptions are arsenic and lead for which maximum levels in various food stuffs are prescribed under The Arsenic in Food Regulations, 1959 (Statutory Instrument 1959, No. 831) and The Lead in Food Regulations 1961 (Statutory Instrument 1961, No. 1931), respectively.

The recommendations for the protection of livestock, wild life, and the general public usually take the form of advice such as, for example, avoiding harm to bees by not applying the chemical when plants are flowering, and by ensuring that flowering weeds are kept down in orchards; avoiding harm to fish by not spraying near ponds or waterways, or contaminating them with concentrate, washings or empty containers; preventing animals, pets and children from having access to the chemical by storing full or part-full containers tightly closed in a safe place and thoroughly washing out empty containers. The recommendations may also take the form of restrictions on certain uses such as those for aldrin and dieldrin, or even complete prohibition as for alkali arsenites, selenium and cadmium compounds.

Copies of the current Recommendations Sheet for any particular chemical can be obtained free of charge from the Ministry of Agriculture, Fisheries and Food (Pesticides Branch) Great Westminster House, Horseferry Road, London, S.W.1. With a voluntary arrangement like the Pesticides Safety Precautions Scheme, emphasis is placed on the advantages to be gained by early informal consultations between prospective notifiers developing a new chemical or extending the uses of an established one, and the scientists who are concerned with the Scheme in an official capacity. Preliminary enquiries for further details relating to products for use on growing crops or as rodenticides, or intended for use in food storage should be sent to the addresses given on p. 84.

The success of these arrangements depends on the co-operation and goodwill extended by the Industrial Associations, and there is every indication that this has been achieved since the formal inception of the original Scheme in 1957. Nevertheless, an additional safeguard exists should a product that is considered to offer a user, consumer or other hazard, appear on the market. Departments can ask the marketing agent to provide evidence in support of the claims that the product does not give rise to such risks when used according to the directions. From this stage the matter is then treated as a normal notification. Although most of the work occupying the time of the officials of the Scheme and the Committees is taken up in dealing with new chemicals, those already on the market are examined from time to time. As a result, new or revised Recommendations Sheets are issued to keep the recommendations for the safe use of pesticide products in step with the latest practices in agriculture.

## 4.1B THE AGRICULTURE (POISONOUS SUBSTANCES) ACT, 1952

The purpose of Regulations made under this Act, covering users in England, Scotland and Wales, and similar Acts* applying to Northern Ireland,[1] the Channel Islands[2] and the Isle of Man,[3] is to protect employees

---

* Intending users in these countries must consult the relevant Regulations in force.

1.  1. The Agriculture (Poisonous Substances) Act (Northern Ireland), 1954.

2. The Poisonous Substances (Guernsey) Law, 1958 and The Agriculture (Poisonous Substances) (Jersey) Law, 1961.

3. *Isle of Man.* The Agriculture (Poisonous Substances) Act, 1961.

from poisoning by the more dangerous chemicals. Users of such compounds listed in these Regulations are required by law to take certain precautions, including the wearing of the prescribed protective clothing specified for particular operations.

The provisions of the Act do not extend to self-employed persons using these chemicals, or to members of the general public, but are restricted to employees of farmers, growers and contractors. Under the terms of the Act, the Minister of Agriculture, Fisheries and Food and the Secretary of State for Scotland are authorized to make Regulations, which may be amended or revoked as necessary. Before making Regulations, the two Ministers are required to consult with the representative organizations of the industries concerned, although the advice received is not legally binding in any way. A further provision of the Act is for the appointment of a sufficient number of inspectors, whose duty is to enforce the Act and any Regulations resulting from it. These inspectors have rights of entry on to land and can also enforce the production of certain documents and take statements and sample for independent analysis. The certificate issued by an approved analyst in such circumstances is a valid document in any legal proceedings which result, becoming admissible in evidence without the need for the analyst to appear in person as a witness. These inspectors will also give advice and assistance in connection with the precautions to be observed under the Regulations, and may be able to give vital help if cases of poisoning or suspected poisoning are reported to them without delay.

The present Regulations (which have replaced all earlier ones) in force under the Act, are the Agriculture (Poisonous Substances) Regulations, 1966 to 1967 (Statutory Instruments 1966 No. 1063; 1967 No. 1860) and the chemicals to which Regulations apply are those listed in the Second Schedule. Part I, II, III and IV, as follows:

PART I   demeton; dimefox; and mazidox;

PART II   amiton and its salts; dinoseb and its salts[1]; disulfoton; DNOC and its salts[1] endosulfan; endothal and its salts; endrin; fluoroacetamide[2]; medinoterb and its salts; mevinphos; potassium arsenite[3]; schradan; sodium arsenite[3]; sulfotep; TEPP (including HETP); and thionazin;

PART III   azinphos-ethyl; azinphos-methyl; chlorvenfinphos demetonmethyl; demeton-S-methyl; dichlorvos[1]; ethion; fentin acetate; fentin hydroxide; mecarbam; nicotine and its salts[1]; oxydemeton-methyl; phenkapton; phosphamidon; vamidothion;

PART IV   organomercury compounds (only when used as aerosols).

**Note** (1) The Regulations do not apply to (*a*) substances used exclusively as insecticides and which contain not more than 5% by weight of dinoseb or DNOC or the equivalent of their respective salts, and no other specified substances; (*b*) substances which contain not more than 7·5% by weight of nicotine or its salts and no other specified substance; (*c*) aerosols which contain not more than 0·4% dichlorvos and no other specified substance; (*d*) impregnated resin strips which contain not more than 20% by weight dichlorovos and no other specified substance.

(2) Under the Poison Rules 1968, fluoroacetamide may not now be purchased by farmers and growers.

(3) The use of potassium arsenite and sodium arsenite as potato haulm destroyers is banned by agreement with all the national organizations concerned and their sale for this purpose is prohibited under the Poisons Rules.

The Regulations are designed to take into account the fact that one method of using a chemical may be inherently more dangerous to the operator than another; thus, other factors being equal, soil or granular application is safer than ordinary spraying, which in turn is less hazardous than the use of aerosols under glass. It should be noted that in the Regulations, 'spraying' does not include 'soil application' when pesticides are applied to the soil in unbroken liquid form, nor does it include 'granular placement' when particles in granular form are deposited on or in the soil or on plants.

The Regulations specify 18 scheduled operations and list the type of protective clothing which must be worn, according to the chemical being used, depending on its classification as a Part I, II, III or IV substance (p. 87). Table 4.1 summarizes the present position for those chemicals in current use as insecticides and fungicides, with details of the clothing to be worn.

The Regulations impose obligations on the employer, who must provide the prescribed protective clothing, and make certain that the worker wears it, and on the employee, both of whom may be charged with infringements. Various other matters included in the Regulations are the maximum number of hours workers may carry out scheduled operations; the age at which they may be employed; precautions when working in greenhouses; the provision and maintenance of protective clothing; the provision of washing facilities for employees; the notification of sickness; the training and supervision of workers carrying out scheduled operations; the provision of drinking water and vessels; ensuring that tanks and containers for storing the substances are securely closed when not in use; and the keeping of a register containing details of all scheduled operations carried out on ground crops, bushes, climbing plants (including hops) and trees, or in greenhouses within specified acreages.

An inspector may grant a certificate of exemption from some or all of the provisions of the Regulations if he is satisfied either that the worker can be protected adequately by other precautions or that the provisions are

TABLE 4.1

*Protective Clothing Requirements*

| Part I Substances | Jobs for which protective clothing must be worn | Clothing, etc., to be worn |
|---|---|---|
| Dimefox | Opening a container, or diluting, mixing or transferring from one container to another | Rubber gloves, rubber boots, respirator and either<br>(*a*) an overall and rubber apron, or<br>(*b*) a mackintosh |
| | Washing or cleansing soil application apparatus | Rubber boots, face-shield and either<br>(*a*) an overall and rubber apron, or<br>(*b*) a mackintosh |
| | Soil application (other than in a greenhouse) by<br>(*a*) unaccompanied driver of tractor-*drawn* apparatus, or<br>(*b*) driver of tractor-*mounted* apparatus accompanied by on-foot operators*<br>(*c*) any operator on foot | Overall, rubber gloves and rubber boots<br><br><br><br><br><br>Overall, rubber apron, rubber gloves and rubber boots |

*Note:* A driver, so accompanied, of tractor-*drawn* apparatus is exempt from the regulations so long as he is driving and not performing any of the scheduled operations listed above.

| Part II Substances | Jobs for which protective clothing must be worn | Clothing, etc., to be worn |
|---|---|---|
| Dinoseb<br>Disulfoton<br>DNOC and its salts<br>Endosulfan<br>Endrin<br>Mevinphos<br>Parathion<br>Phorate | Opening a container, or diluting, mixing or transferring from one container to another except<br><br>1 Dinoseb, or DNOC or its salts when used as insecticides (see *Note* 1 on p. 87) | Rubber gloves, rubber boots, face-shield and either<br>(*a*) an overall and rubber apron, or<br>(*b*) a mackintosh<br><br>Rubber gloves and either a face-shield or eye-shield |

| Part II<br>Substances<br>(continued) | Jobs for which protective clothing must be worn | Clothing, etc., to be worn |
|---|---|---|
| Schradan<br>Sulfotep<br>TEPP (HETP)<br>Thionazin | 2. A granular formulation | Rubber gauntlet gloves and either an overall or mackintosh, with sleeves worn over the cuffs of the rubber gauntlet gloves |
| | Washing or cleansing spraying apparatus, soil application apparatus or granule placement apparatus | Rubber boots, face-shield and either<br>(a) an overall and rubber apron, or<br>(b) a mackintosh |
| | Spraying any ground crop, except from aircraft or in a greenhouse | Overall, hood, rubber gloves, rubber boots and either a face-shield or a dust-mask |
| | Spraying bushes, climbing plants (other than hops), or trees | Rubber coat, rubber gloves, rubber boots, sou'wester and face-shield |
| | Spraying hops | Rubber coat, rubber gloves, sou'wester and face-shield |
| | Handling hops previously sprayed within 4 days or within 24 hours in the case of mevinphos or TEPP | Rubber gloves |
| | Placing granules by hand or hand-operated apparatus | Rubber gauntlet gloves and either an overall or mackintosh, with the sleeves worn over the cuffs of the rubber gauntlet gloves |
| | Placing granules by mechanically-operated machinery (excluding aircraft)<br>or<br>operating other machinery mounted on or drawn by a tractor when granule placement apparatus, mounted on or drawn, either directly or indirectly, by the tractor, is being used | Either an overall or mackintosh |

| Part II Substances (continued) | Jobs for which protective clothing must be worn | Clothing, etc., to be worn |
|---|---|---|
| | Spraying in a greenhouse (except where an aerosol or smoke-generator is used) | Rubber gloves, rubber boots, hood, face-shield and either an overall or mackintosh |
| | Applying aerosols in a greenhouse | Overall, hood, rubber gloves and respirator |
| | Acting as a groundmarker for spraying ground crops from aircraft | Overall, hood, rubber gloves, rubber boots and face-shield |
| | Acting as a groundmarker for aircraft applying granules | Hood, face-shield and either an overall or mackintosh |
| | Soil application (other than in a greenhouse) by (a) unaccompanied driver of tractor-*drawn* apparatus or (b) driver of tractor-*mounted* apparatus accompanied by on-foot operators* | Overall, rubber gloves and rubber boots |
| | (c) any operator on foot | Overall, rubber apron, rubber gloves and rubber boots |
| | Soil application in a greenhouse | Overall, rubber apron, rubber gloves and rubber boots |
| | Bulb dipping or steeping with thionazin, handling wet bulbs, disposing of the solution and washing the dipping or steeping apparatus | Rubber gauntlet gloves, rubber boots, an overall and rubber apron |

* See *Note* on p. 89 in Part I above

| Part III Substances | Jobs for which protective clothing must be worn | Clothing, etc., to be worn |
|---|---|---|
| Azinphos-ethyl Azinphos-methyl Chlorfenvinphos Demeton-methyl Demeton-S-methyl | Opening a container, or diluting, mixing or transferring from one container to another except | Rubber gloves and face-shield |
| | A granular formulation | Rubber gloves |

| Part III<br>Substances<br>(continued) | Jobs for which protective<br>clothing must be worn | Clothing, etc., to be worn |
|---|---|---|
| Dichlorvos<br>Ethion<br>Fentin acetate<br>Fentin hydroxide<br>Mecarbam<br>Nicotine<br>Oxydemeton-<br>  methyl<br>Phenkapton<br>Phosphamidon<br>Vamidothion | Applying aerosols in a green-<br>house | Overall, hood, rubber gloves<br>and respirator |

| Part IV<br>Substances | Jobs for which protective<br>clothing must be worn | Clothing etc., to be worn |
|---|---|---|
| Organomercury<br>compounds<br>(when used as<br>aerosols) | Applying aerosols in a green-<br>house | Overall, hood, rubber<br>gloves and respirator |

unnecessary under the proposed conditions of use. The certificate is granted only on specified conditions, binding upon employer and/or employee.

A person guilty of an offence under the Act or the Regulations is liable to a fine not exceeding fifty pounds and, in respect of an offence continued after conviction, to an additional fine not exceeding ten pounds for each day on which the contravention is continued.

The Ministry of Agriculture, Fisheries and Food and the Department of Agriculture and Fisheries for Scotland issue a leaflet APS/1 'The Safe Use of Poisonous Chemicals on the Farm', which includes a valuable summary in non-legal terms of the main provisions of the Regulations, as well as much general advice on the safe use of pesticides in relation to persons, livestock and wild life; the cleansing and maintenance of respirators and dust-masks; notes on the symptoms of poisoning by various chemicals, and the necessity for constant medical supervision of workers. This leaflet also contains a list of addresses of regional and divisional inspectors appointed under the 1952 Act.

Another of several leaflets, aimed at advising the user of pesticides of the necessary safety measures, is entitled 'Take Care when you Spray'.

All these leaflets may be obtained from the main offices of these Departments.

Chemicals included in the Agriculture (Poisonous Substances) Regulations are usually also subject to the provisions of the Pharmacy and

Poisons Act, 1933 and Poisons Rules, which restrict their sale and impose certain labelling requirements and conditions under which listed poisons may be bought, packed, transported and stored on shop premises.

## 4.2 THE AGRICULTURAL CHEMICALS APPROVAL SCHEME
### (Insecticides, Fungicides and Herbicides)

This is a voluntary Scheme under which proprietary formulations of chemicals used in crop protection can be officially approved and its purpose is to enable users to select, and advisers to recommend, efficient and appropriate crop protection chemicals and to discourage the use of unsatisfactory products.

The chemicals covered by the A.C.A.S. are those used for the control of plant pests and diseases, for the destruction of weeds, for growth regulation and other crop protection purposes, but those used as rodenticides or for the protection of stored products or for veterinary or domestic uses, are not included.

Participation is open to manufacturers and their agents and also to the authorized agents of overseas manufacturers. The Scheme is operated on behalf of the Agricultural Departments of the United Kingdom* by the Agricultural Chemicals Approval Organization, at the Plant Pathology Laboratory, Hatching Green, Harpenden, Herts. and a member of the Organization specializing in herbicides is attached to the A.R.C. Weed Research Organization, Begbroke Hill, Kidlington, Oxford. Full support is given by the Association of British Manufacturers of Agricultural Chemicals, the National Association of Corn and Agricultural Merchants and the National Unions and Associations of farmers and growers in the United Kingdom.* Under this Scheme, it is intended that, where possible, products containing new chemicals will be given official approval at the time of marketing.

Approval is granted to products by the Organization for specific uses under United Kingdom* conditions when the recommendations made on the labels are supported by satisfactory evidence from the manufacturer's field trials, supplemented in appropriate cases by the results of work carried out by the Advisory Services, or by independent Research Stations.

A Certificate of Approval, which is subject to annual renewal, is granted to each product approved. The label of each approved product carries the identification mark shown in the Figure 4.1.

No product may receive approval under this Scheme until its safety in use has been considered under the Pesticides Safety Precautions Scheme

* For the purposes of this Scheme the United Kingdom includes England, Scotland, Wales, Northern Ireland, the Channel Islands, and the Isle of Man.

(see p. 83) and recommendations for its safe use have been issued by the appropriate authorities.

A List of Approved Products is published in February each year. Copies of the current List may be obtained free of charge from the Ministry of Agriculture, Fisheries and Food (Publications), Block C, Tolcarne Drive, Pinner, Middlesex, or from any of the Ministry's Regional and Divisional Offices. Copies can also be obtained from the main offices of the Agricultural Departments in Scotland, Northern Ireland, the Channel Islands and the Isle of Man. This booklet lists the proprietary names of approved products under the headings of the active ingredients they contain and is complementary to this Handbook which deals with pest and disease control solely in terms of active ingredients. To select a suitable approved proprietary product for a particular purpose, the List of Approved Products should be consulted.

FIG. 4.1. The official mark of the Agricultural Chemicals Approval Scheme.

Products approved too late for inclusion in the current List are announced in the journal 'Agriculture' and the trade press, before the next List appears.

Correspondence concerning the Scheme should be addressed to the Secretary, Agricultural Chemicals Approval Organization, Plant Pathology Laboratory, Hatching Green, Harpenden, Herts. Tel. 0582 75241.

## 4.3 PESTICIDES—CODE OF CONDUCT

The official provisions for the safeguarding of users and public from the use of pesticides become ineffective if the pesticides are wrongly used. The Joint Association of British Manufacturers of Agricultural Chemicals/ Wild Life Education and Communications Committee, representing manufacturers, distributors and users of pesticides, the M.A.F.F., Nature Conservancy and voluntary conservation organizations, have issued a 'Code of Conduct' to promote this proper use. The responsibilities of the agricultural and horticultural users are summarized thus:

### A  Identification of the Problem

i   to become familiar with the appearance or symptoms of common pests, diseases and weeds:

ii  to make a balanced assessment of the economic benefits that may be expected from using the pesticide.

### B  Choice of Product and Method of Use

to choose, with the best available advice,

i   a correct pesticide to combat the pest;

ii  the effective dosage, time and conditions of application.

### C  Compliance with Instructions and Recommendations

i   to become familiar with government requirements and to observe them strictly;

ii  to READ THE LABEL and all relevant literature, observing especially the dosage rates recommended and the safety precautions; to adhere strictly to the interval that must elapse between applying the pesticide and harvesting the crop.

### D  Safety to Operators

i   to make certain that the operator knows the correct method of applying a pesticide and all the safety precautions that must be observed, including the use of protective clothing if necessary;

ii  to ensure that operators wash well immediately after working with pesticides and before eating, drinking or smoking;

iii to warn operators of the danger of clearing blocked nozzles by mouth;

iv  to notify a doctor immediately if an operator shows any symptoms of illness during or after working with pesticides and, subsequently, to notify the safety inspector of one of the agricultural departments if a pesticide was shown to be the cause of the symptoms;

v  to ensure that equipment and protective clothing is frequently cleaned
   and stored safely;
vi  to check the condition of protective clothing and equipment and
   replace when necessary.

## E  Safety for Others

to apple pesticides carefully and to AVOID
 i  spray drift on surrounding areas;
ii  spraying blossom and flowers which would harm bees and other
   beneficial insects;
iii  pesticides draining into water courses, which might harm fish or other
   wild life, bees, or farm stock.

## Some important safety points:

Train operators properly.
Keep machinery in good order.
Always read the product label.
Store both full and partly used containers in a locked store.
Make certain the partly used containers are tightly closed and clearly
   labelled.
Never transfer pesticide to other containers.
Wash out containers after use and return washings to spray tank.
Never wash or cleanse equipment in water courses.
Return, burn, or flatten and bury used containers.
Keep an accurate record of quantities and places where pesticides are
   used.

# CHAPTER 5

# PEST AND DISEASE CONTROL
# IN CEREALS

Healthy seed, drilled at the right time and to the right depth in a firm and adequately manured seedbed, is most likely to give a satisfactory yield in spite of pests and diseases. The higher the potential yield, however, the more is to be gained from pest and disease control, and the greater are the chances that the use of insecticides and fungicides will be profitable. Good husbandry and chemical control should not be thought of as alternatives, but as working together to give the best results.

It should not be assumed that cultural control invariably costs nothing. The early ploughing of grass, which prevents or greatly reduces damage by the 'ley pests', may entail the forfeiture of several weeks' grazing or of a hay or silage cut. Crop rotation avoids trouble from many serious pests and diseases but, where economic conditions favour intensive corn growing, it may require the continuance of less profitable enterprises.

Because of the large annual acreage involved, the risks of 'side effects' from chemicals applied to corn crops are especially important. For this reason, as well as for the low cost, the development of control by seed treatment is of the greatest importance.

Seed dressing dusts are usually applied at about 2 oz, and liquids at between ¾ and 3 fl. oz per bushel of seed. Overdressing can injure the seed and underdressing will give poor control. It is very important that the manufacturers' instructions are followed implicitly, and that the seed is of good quality, dry (under 16% moisture content), and undamaged. Many of the compounds are extremely poisonous, and need careful handling. Treated seed should never be used for feeding stock or humans.

Some pathogens confine their attack to one cereal, others may attack two or more. In the following list multiple pathogens are listed under each crop heading with appropriate cross references. Pests and diseases are listed alphabetically by the generally accepted common name and under the appropriate host. Symptoms which are not caused by pathogens, but whose appearance is similar, are described, to aid more positive diagnosis. Fuller information about cereal diseases and pests and their control is given in M.A.F.F. Bulletins 129 'Cereal Diseases' and 186 'Cereal Pests'.

There is little or no information available on many aspects of pest and disease damage to cereals. So far as is possible, however, the following



H

recommendations take into account both the economic and biological factors.

## 5.1 WHEAT

**Pests**

****Aphids.** Pests of the greenfly group may cause serious injury to corn crops, particularly after a mild winter has been followed by a fine spring. They are important as carriers of virus diseases. The four most common species are:

Bird-cherry aphid (*Rhopalosiphum padi*), a greenish brown species which is usually restricted to the lower leaves. It often overwinters in the summer form on grasses and cereals.

Rose-grain aphid (*Metopolophium dirhodum*), a light green species with a darker green stripe down the back. It overwinters on rose and may be found on cereal leaves in summer.

Fescue aphid (*Metopolophium festucae*), a species very like the rose-grain aphid but lacking the darker stripe down the back. Overwinters on grasses and cereals and in occasional years causes severe and widespread damage.

Grain aphid (*Macrosiphum (Sitobion) avenae*), a large green or reddish species that is often noticed on the ears of corn. Usually it causes little or no damage.

Plants affected by aphids are dwarfed and of poor colour, the tips of the leaves being often reddish, purple, or brown and withered. Dis-coloration may be due to virus infection. In severe cases large areas of the crop may be completely withered, so that they can be fired with a match. Aphids and their cast skins are very numerous. Severe attacks are most likely in hot, dry weather.

Control is obtained with demeton-S-methyl 3·1 oz a.i. (6 fl. oz 58% e.c.)/acre *or* dimethoate 5 oz a.i. (12 fl. oz 40% e.c.)/acre *or* formothion 7 oz a.i. (16 fl. oz 43% e.c.)/acre *or* oxydemeton-methyl 3·4 oz a.i. (6 fl. oz 57% e.c.)/acre.

Treatment against direct injury to the crop is likely to be profitable when bird-cherry and/or fescue aphids are numerous enough to be easily noticed. When parasites and predators are also numerous, spraying may not be required.

See also Barley yellow dwarf virus, p. 105.

***Cereal cyst eelworm**—see under Oats, p. 107.

***Frit fly** (overwintering generation)—see under Oats, p. 107.

***Gout fly**—see under Barley, p. 104.

****Leatherjackets** (A.L. 179). Certain crane-flies (*Tipula* and *Nephrotoma* spp.) lay eggs in grassland during summer and autumn; these eggs hatch in the autumn. When grass is ploughed for cereals the larvae, known as leatherjackets, may feed on the seedling corn. These insects, unlike wireworms, complete the life cycle in one year, so that injury is normally restricted to the first year after ploughing temporary or permanent grass.

Cereal plants are damaged at or below ground level; injured tissues appear torn rather than cut. Identification should be confirmed by finding leatherjackets in the soil.

Cultural control depends on early ploughing of grassland (July or early August) before the main egg-laying period. In certain areas early ploughing may bring a risk of wheat bulb fly attack.

Chemical control is by bait or spray. (*a*) poison bait. To 28 lb broad bran add gamma-BHC 4 oz a.i. (*not* if potatoes due next year) *or* DDT 8 oz a.i. (may injure rye and some varieties of barley) *or* Paris green 1 lb (also controls slugs). Add just enough water (usually about $1\frac{1}{2}$ gal) to make the bait crumbly, mix very thoroughly, then broadcast by hand or suitable distributor at the above rates per acre. Paris green is effective but is a scheduled poison and should be kept in a safe place; cover any cuts or abrasions on the hands before touching this substance.

(*b*) spray. Gamma-BHC 8 oz a.i./acre (*not* if potatoes to be grown within 18 months) *or* DDT 1 lb a.i./acre (may injure rye and some varieties of barley). Where heavy leatherjacket attack is expected, spraying the seedbed during preparation may be advised.

***Slugs** (A.L. 115). Several species damage cereals, the commonest being the field slug, *Agriolimax reticulatus*. The most serious injury occurs below soil level; germinating grains are hollowed out, leaving only the outer skin. Above ground, slugs eat narrow longitudinal strips from the leaves and in a bad attack leaves may be completely shredded. Damage is most common on the heavier soils and after grass, clover, or a crop (e.g. peas) which leaves in the soil abundant partially decomposed organic matter.

Control is by the use of poison bait made as for leatherjackets above. To 28 lb broad bran add Paris green 1 lb (damp conditions, winter) *or* metaldehyde $\frac{1}{2}$ lb (drier conditions, late spring onwards) *or* Paris green 1 lb + metaldehyde $\frac{1}{4}$ lb (intermediate conditions). Broadcast over 1 acre. Alternatively use a metaldehyde prepared bait (e.g. pellets) or spray where extra cost is justified.

Little or no control of the underground attack is to be expected from treatment after sowing. This treatment is most likely to be profitable when the plant stand, having been thinned by underground attack, is still sufficient for a full crop provided that the plants are protected from further damage. With a normal full plant stand, yield is unlikely to be affected unless leaf injury is very severe.

*****Wheat bulb fly** (*Leptohylemyia coarctata*) (A.L. 177). Eggs are laid in bare soil from mid-July until the beginning of September. They hatch from late January to early March and the young larvae enter the shoots of wheat, barley and sometimes rye. Oats are not attacked. In early spring the centre leaves of damaged plants become yellow and die. Small untillered plants may be completely killed. Shoots often have a very small hole at the base (pull off the outer leaves, *cf.* wireworms), and a proportion of damaged shoots contain a white maggot.

Damage is normally confined to certain areas in the eastern part of the country and is usually restricted to cereals after fallow, bastard fallow, or after a crop (e.g. potatoes) in which there is bare soil between plants during the egg-laying period. It may also follow patches of poor growth in any crop.

Crops sown before the middle of October in a good seedbed should be able to withstand any normal attack. Top dressing in spring helps a thinned crop. Spring sowing, on land where trouble is expected, should be delayed until mid-March.

Chemical treatment depends on the time of sowing: if sown before mid-October—no treatment; if sown mid-October to end December dress seed with (*a*) aldrin 26% (liquid) 3 fl. oz/bushel *or* dieldrin 60%, 2 oz/bushel (for sowings mid-October to end November) *or* (*b*) gamma-BHC 40%, 2 oz/bushel (December sowings) *or* (*c*) chlorfenvinphos 32% (liquid) 3 fl. oz/bushel *or* ethion 67%, $2\frac{1}{2}$ oz/bushel. For spring sowings before March 1st and spring patching of partial failures from wheat bulb fly, dress the seed with gamma-BHC 40% 2 oz/bushel. If required, spray during early March with dimethoate 10 oz a.i. (24 fl. oz 40% e.c.)/acre *or* formothion 15 oz a.i. (32 fl. oz 43% e.c.)/acre.

Seed dressings containing aldrin or dieldrin should only be used on autumn-sown grain and then only where the crop is in danger of attack from this pest. Such dressings should not be used at all for spring-sown grain. Seed dressings of aldrin, dieldrin and gamma-BHC will also give protection against wireworm attack. Lower strength gamma-BHC seed dressings, as used against wireworms, will *not* give adequate protection against wheat bulb fly.

Forecasts of the likelihood of attack are obtainable from the N.A.A.S. and are usually available by the beginning of October.

***Wheat shoot beetle** (*Helophorus nubilus*). Small larvae, often difficult to find, which cause damage somewhat similar to that of wireworms or wheat bulb fly larvae (small lateral hole at base of shoot). Attacks occur only after grass, usually ley, and commence in very early spring. Some control is obtained by early ploughing of grass for autumn-sown cereals.

*****Wireworms** (A.L. 199). These tough, yellow, long-lived larvae of certain click beetles (mainly *Agriotes* spp.) are normal inhabitants of old grassland. Damage may be expected during the first 4 years after ploughing up grass that has been down for 4 years or longer. On some farms, usually on moderately heavy soils, wireworms may be troublesome on old arable land, particularly if it has been allowed to get weedy. Attacked plants become yellow and die. Shoots are damaged below soil level, there being often a distinct hole in the side of the plant at the base or the shoots are chewed and frayed just above the old seed. Sometimes only the centre leaf becomes yellow. Damage is often less severe on headlands. Attacks are usually seen in autumn and spring but, unlike wheat bulb fly and wheat shoot beetle, not as a rule in very early spring.

Attack is reduced by sowing in a firm seedbed, which may be rolled if an attack develops. For chemical control use either (1) gamma-BHC 10 oz a.i./acre before sowing (also controls leatherjackets), (2) seed dressing, gamma-BHC 20% 2 oz/bushel. Follow manufacturer's instructions for peat soils on which higher rates may be required.

Where wireworm populations are expected to be high, use (1) but only if potatoes or canning carrots are *not* to be grown within 18 months of treatment. For lower populations use (2).

Seed dressings of aldrin, dieldrin and gamma-BHC used against wheat bulb fly will also protect against wireworms (but gamma-BHC at wireworm strength will not protect against wheat bulb fly). *Do not use* wheat bulb fly dressings for protection against wireworms alone. After treatment (1) there should be no need for further control measures (even for potatoes) until the land has again been under a long ley.

## Diseases

*Barley yellow dwarf virus.** Symptoms similar to those described under barley (see p. 105) but the leaves of affected plants of some varieties may develop a purplish tint. For control, see Barley, p. 105.

**Black mould** (*Cladosporium herbarum*). This fungus, and other sooty moulds often cause blackening of the ears, stems and leaf sheaths, especially in wet seasons. Poor plants with poorly filled heads are most likely to show this effect, but the fungi are secondary. They cause respiratory troubles in man and are sometimes mistaken for the smut diseases.

**Black point** (*Alternaria* spp.). In this condition (unlike black mould) the plumpest grains are usually affected. The embryo end of the grain shows a brown or black discoloration, and there may be some shrivelling of the grain. Discoloured grains lower the market value of the sample, but germination is usually unimpaired and no control measures are used.

****Brown foot root** and **Ear blight** (*Fusarium* spp.). Young plants may be killed outright (seedling blight) resulting in a poor braird. Mature plants show rotten roots, brown rotten stem bases, often with pink or white fungus pustules on the lesion. Leaves may show a light coloured spot with a dark red margin, often become yellowed and sometimes wither.

Ears are poorly filled or empty (whiteheads). The grain may bear red or pink fungal masses on the surface, the ear blight stage, and some penetration of the grain may occur. The disease is common in some seasons and organomercury seed dressings give partial control, though the pathogens are also found in the soil.

***Bunt** or **Stinking smut** (*Tilletia caries*). Affected plants produce bluish ears, on shorter stalks, and with shorter and plumper grains than healthy plants. The grains contain a mass of black greasy spores, smelling of rotting fish. The spores adhere to the seed and on germination penetrate into the young seedling. This disease is now rare and is readily controlled by organomercury seed dressings containing 0·6–2% mercury and applied at ¾–2 oz/bushel. Another effective material is hexachlorobenzene (40%), which, unlike organomercury dressings, controls no other seedborne diseases of cereals.

***Ergot** (*Claviceps purpurea*)—see under Rye, p. 109.

****Eyespot** (*Cercosporella herpotrichoides*) (A.L. 321). Infection is seen in young plants as light necrotic brown bordered oval lesions on the leaf sheaths. If the attack is severe the young plants are killed outright, but otherwise the fungus works its way in to attack the lower part of the stalk, resulting in an eye-shaped brown bordered spot with a black spot in the middle, from which symptom the disease is named. Later in the season the stalk, weakened at the eyespot lesion, may topple over, giving the condition known as 'straggling'.

Yields are reduced and a lot of tail corn is produced. Whiteheads may occur and become secondarily invaded by black mould (*q.v.*). Whiteheads may also be due to an attack of take-all (*q.v.*) or other causes. Occurs widely in the British Isles especially on heavy fertile soils.

Good crop rotations, eradication of couch *Agropyron repens*, moderate but not excessive manuring and adequate drainage effect some control. Stiff-strawed varieties are less affected. Cappelle Desprez, Joss Cambier, Maris Ranger, Maris Widgeon and West Desprez have some resistance to the disease.

****Glume blotch** (*Leptosphaeria* (*Septoria*) *nodorum*). Dark brown spots on the ears, leaves with pale yellow spots with brown margins, the spots often bearing small black fruit bodies (pycnidia). Browning of the stem nodes may occur in wet seasons. The disease is common and seedling losses may be serious, and it is not adequately controlled by organomercury seed dressings.

***Grey leaf** (Manganese deficiency). Not common, but sometimes locally troublesome—see under Oats, p. 108.

***Leaf spot** (*Septoria tritici*). Indefinite light coloured spots on the leaves bearing small black fruiting bodies (pycnidia). Common some seasons, and of consequence. Control measures are usually ineffective.

**Loose smut** (*Ustilago nuda*). Black spore masses on the diseased heads are very conspicuous in a crop as the ears emerge. The spores are readily dispersed by the wind and eventually just a bare stalk remains.

Seed should be taken from smut free crops. Hot water treatment of the seed, at 32°C (90°F) for 4 hours and then 10 minutes at 52–54°C (126–129°F) or $2\frac{1}{2}$ hours at 46°C (115°F) is a cure. Sow 'certified' or 'field approved' seed.

**Mildew** (*Erysiphe graminis*). Seen as off-white or brown patches of mycelium on the lower leaves. As the season and the disease progresses, the lower leaves become yellow and shrivel, and the upper leaves become affected. The mycelial patches become dark brown with age and small black spore cases (perithecia) become apparent on the patches. In a severe attack mildew is found on the ears and glumes and may discolour and shrivel the grain.

Although mildew attacks wheat, barley, oats, rye and pasture grasses, the races of mildew are specific to each crop and the disease does not spread from one type of corn to another. Excessive use of nitrogen is to be avoided, nitrogen should be applied early and spring crops sown early. Some of the newer varieties of spring wheats show some resistance to mildew at present.

**Scab** (*Gibberella zeae*). Similar to brown foot rot and kills young plants; older plants show a foot rot and pinkish fungal masses on the ears. Small blue-black fruit bodies (perithecia) often occur on ears and stubbles. Grain is shrunken and affected grain should not be fed to pigs, dogs or humans. The disease is not common but isolated outbreaks are occasionally reported. Organomercury seed dressings and crop rotations usually prevent the disease becoming of economic importance.

**Sharp eyespot** (*Corticium* (*Rhizoctonia*) *solani*) (A.L. 321). This fungus often attacks cereals at the seedling stage causing the roots to turn brown, whilst distinct lesions may occur on the older thicker roots. The leaves are erect, narrow and rolled and the foliage may be purple in colour (purple patch). Badly affected plants succumb, others may survive but are stunted and yield poorly.

The fungus also attacks the lower part of the stalks producing a lesion somewhat like that of eyespot (*q.v.*). Sharp eyespot can be distinguished from eyespot in that the lesions may be much elongated, have a much sharper border and outline and do not show the black spot in the middle. The disease is often troublesome after a ley or a non-cereal arable crop especially in cold soils, so sow winter crops early and spring crops late.

**Take-all** (*Ophiobolus graminis*) (A.L. 304). The fungus attacks the roots of the plant resulting in either the complete death of young plants or yellow stunted plants. The disease is however usually noticed after heading when patches of affected plants are seen, with bleached empty heads—the whitehead condition. These whiteheads often become covered with a dark fungal growth, the black mould condition (*q.v.*) but this colonization is secondary. The plants with whiteheads often show rotten roots and the base of the stalk may be covered superficially with black mycelium of the take-all fungus. Barley is also attacked as is couch grass (*Agropyron repens*). Oats are unaffected and ryegrass is resistant. However there is a strain of the take-all fungus (*Ophiobolus graminis* var. *avenae*) common in oat-growing areas, which can attack wheat, barley, oats and couch grass. The disease is favoured by alkaline and light soils.

No chemical control method is known but these precautions should be taken: avoid over-liming; control perennial weed grasses which can harbour the disease; do not include wheat and barley too often in the rotation; follow autumn sown crops of wheat and barley with spring sown ones, and not vice versa; cultivations which make for quick breakdown of stubbles should be practised.

**Yellow Rust** (*Puccinia striiformis*) (A.L. 527). This disease can be very harmful, especially in the south and east of England. Recently, new strains of the fungus which are capable of attacking most varieties of winter and spring wheats have become widespread. At the time of writing, the winter wheats, Maris Ranger, Cama, West Desprez and Elite Lepeuple, and the spring wheat Maris Ensign, show resistance. The fungus also occurs on barley, rye and some grasses, but as distinct strains, so that these do not pass from one kind of cereal or grass to another.

Symptoms of yellow rust are seen initially as parallel lines of lemon yellow pustules on the leaves, but the stems and ears may also become affected. Later in the season the pustules turn black. Mild winters and cool moist weather in the spring and early summer favour the disease, but a hot dry spell will often arrest the disease spectacularly. Control is by growing resistant varieties.

## 5.2 BARLEY

**Pests**

**Aphids**—see under Wheat, p. 98.

*Cereal cyst eelworm**—see under Oats, p. 107.

*Frit fly** (overwintering generation)—see under Oats, p. 107.

**Gout fly** (*Chlorops pumilionis*) (A.L. 174). Larvae live in shoots of barley, wheat and rye (*not* oats). There are two generations each year, causing damage in the autumn and during early summer. Attacked shoots are usually swollen ('gouty') and at first of a rich green colour—later they rot away. Shoots attacked at a late stage of growth are not swollen but the ear may be damaged on one side and there is a groove down one side of the stem below the ear.

To reduce attack, avoid sowing autumn corn too early and spring corn too late.

***Leatherjackets**—see under Wheat, p. 98. Note that many varieties of barley may be injured by DDT (e.g. Deba Abed, Impala, Inis, Julia, Mosane, Vada, Zephyr, and the winter barley Senta) whereas others are not affected (e.g. Maris Badger, Maris Otter, Proctor and Sultan).

Aldrin may be used to control leatherjackets only on those varieties of spring barley sensitive to DDT; use aldrin $1\frac{1}{2}$ lb a.i./acre as a spray.

**Slugs**—see under Wheat, p. 99.

*Wheat bulb fly**—see under Wheat, p. 100.

***Wireworms**—see under Wheat, p. 100.


## Diseases

**Barley yellow dwarf virus** symptoms vary in severity, depending upon the strain of the virus, the age of the plant on infection and the variety of the cereal. The virus is transmitted by some cereal aphids. Young plants on infection show a golden-yellowing of the leaf tips and gradually this colouring extends down the leaf. Plants are stunted and may be killed. Surviving plants, or those infected later in life, show golden-yellow colouring of the leaves, heads may not emerge, ears may be blasted, yield is reduced. Common some seasons when conditions are favourable for aphids, i.e. a mild autumn and winter, especially if followed by a warm and dry spring.

When a crop, particularly an autumn sown one, becomes infested soon after brairding with the bird-cherry aphid, *Rhopalosiphum padi*, spraying with an insecticide (see page 98), to minimize further spread of the virus, is probably worthwhile.

*Black point** (*Alternaria* spp.)—see under Wheat, p. 101.

**Brown foot rot** and **Ear blight** (*Fusarium* spp.)—Common some seasons—see under Wheat, p. 101.

*Brown Rust** (*Puccinia hordei*). This disease, seen as small brown spore pustules on the leaf and leaf sheaths, has gained in prominence in recent years. Attacks usually occur late in the season and the effect on yield is probably slight.

*Covered smut** (*Ustilago hordei*). At ear emergence, affected plants produce black ears containing masses of black spores enclosed by the coat of the grain. The coat generally remains intact, but at threshing it ruptures releasing the spores. No longer common except in the remote areas. The disease is controlled by organomercury seed dressings.

*Ergot** (*Claviceps purpurea*)—Infrequent—see under Rye, p. 109.

**Eyespot** (*Cercosporella herpotrichoides*)—Common—see under Wheat, p. 102.

*Grey leaf** (Manganese deficiency)—Locally common some seasons, leaves pale green, brown streaks and spots interveinally—see under Oats, p. 108.

****Halo spot** (*Selenophoma donacis*). Brown spots with a purple border, a grey centre and bearing black fruiting bodies (pycnidia) occur on the leaves. Severe attacks result in the death of leaves and scorched patches in the crop. Disease is possibly seedborne, but is not controlled by seed dressings. More common in Wales and south-west England than elsewhere. Special measures are unnecessary.

****Leaf blotch** (*Rhynchosporium secalis*). Dark grey lesions with dark brown margins occur on the leaves. Frequent on both winter and spring barleys, often damaging, especially in south-west England. Also occurs on rye. Destruction of volunteer plants and wide separation of autumn sown and spring sown crops helps control.

***Leaf spot** (*Septoria passerinii*). Affected leaves have brownish-yellow striped areas bearing the black fruit bodies (pycnidia). Recorded from Scotland and England. Rare and not of economic importance.

***Leaf stripe** (*Pyrenophora (Drechslera) graminea*). Emerging leaves show pale striping and seedlings are often killed. Plants which survive often have their leaves striped, first pale green, changing later to yellow and finally to brown and may split. Poorly filled and sometimes discoloured heads are produced, or the plants may not head at all. The disease is controlled by grading out the thin severely affected grains, organomercury seed dressings and crop rotation.

****Loose smut** (*Ustilago nuda*). Symptoms similar to loose smut of wheat (*q.v.*). This disease is not controlled by organomercury seed dressing and only seed from healthy crops should be sown. Infected seed should be hot water treated. Winter varieties—4 hours in cold water followed by 10 minutes in water at 51–52°C (124–126°F). Spring varieties—4 hours in water at 32°C (90°F) followed by 10 minutes in water at 51–52°C (124–126°F). Alternatively all varieties may be treated for $2\frac{1}{2}$ hours at 42·5°C (109°F). Sow 'certified' or approved seed. A test to establish the level of loose smut infection in seed is carried out at the Seed Testing Stations at Cambridge and Edinburgh on request, and payment of a fee of thirty shillings.

*****Mildew** (*Erysiphe graminis*)—see under Wheat, p. 103. Often severe on late sown crops. Of the new varieties of spring barley, Sultan and Vada show marked resistance to mildew at present.

***Net blotch** (*Pyrenophora teres*). Seen on leaves as irregular brown blotches, rectangular or elongated. These blotches show a network of dark brown lines, on top of a paler background. Of little consequence, and is controlled by organomercury seed dressings.

****Scab** (*Gibberella zeae*)—see under Wheat, p. 103.

****Sharp eyespot** (*Corticium (Rhizoctonia) solani*)—see under Wheat, p. 103.

****Take-all** (*Ophiobolus graminis*)—see under Wheat, p. 103.

***Yellow Rust** (*Puccinia striiformis*). This disease has become more prevalent on barley in recent years although generally the attacks develop late and thus the effects on yield are minimized. The strains of this fungus on barley are distinct from those on wheat, rye and certain grasses; and spread does not occur from one kind of cereal or grass to another. For symptoms and more detail see under Wheat, p. 104.

## 5.3 OATS

**Pests**

****Aphids**—see under Wheat, p. 98.

*****Cereal cyst eelworm** (*Heterodera avenae*) (A.L. 421). Microscopic nematodes enter the roots, interrupt the flow of plant food and water and cause excessive root branching; infected plants are pale and stunted. The crop most seriously affected is oats, but wheat and barley may also be damaged; rye is highly resistant. Mature female eelworms form pinhead-sized cysts, at first white but becoming dark brown. They are full of eggs and may remain in the soil for several years. Heavily infested fields should be grassed down for at least 3 years.

*****Frit fly** (*Oscinella frit*) (A.L. 110). There are three generations each year; they arise from eggs laid in spring, summer, and autumn. Larvae of the first generation damage the shoots of spring sown oats, causing the centre leaf to become yellow and die. Small plants may be killed but larger ones usually survive and may produce an excessive number of short tillers. Each damaged shoot contains a very small white maggot. Except in Scotland the second generation larvae damage oat grains, causing the kernels to be shrivelled and black, but having little effect on the husks. Larvae of the third (overwintering) generation live in grasses but may transfer from ploughed grass to a following cereal crop. Symptoms resemble those of the first (spring oats) generation.

The first and second generation attacks are largely avoided by sowing oats early (before 1 April); damage from the overwintering generation is avoided by ploughing grass early and leaving an interval of at least 4–6 weeks between ploughing and sowing.

For the chemical control of the first generation of frit fly on spring oats use DDT 1 lb a.i./acre twice (usually 1st and 3rd weeks of May). Treatment is likely to be profitable only in certain years and then only on oats sown after 1 April.

*****Leatherjackets**—see under Wheat, p. 98.

****Slugs**—see under Wheat, p. 99.

****Stem eelworm, Tulip root** (*Ditylenchus dipsaci*) (A.L. 178). Microscopic nematodes enter shoots of oats and rye, causing swelling, distortion, and often death by rotting. Host plants of the oat race include mangolds, sugar beet, beans, onions and rhubarb, and also several weeds. There is no cyst stage (*cf* cereal cyst eelworm). Some oat varieties are highly resistant.

Control is possible by crop rotation, weed control and the use of resistant varieties (e.g. Manod, Peniarth, Maris Quest).

***Wireworms**—see under Wheat, p. 100.

### Diseases

**Barley yellow dwarf virus.** Leaves of affected plants become purplish-red. Other symptoms similar to those seen with infected wheat or barley (*q.v.*).

**Brown foot rot** and **Ear blight** (*Fusarium* spp.) is common some seasons—see under Wheat, p. 101.

*Covered smut** (*Ustilago hordei*). At ear emergence, affected plants show black ears containing masses of spores, which sometimes remain within the coat of the grain. Now and again the spore masses are exposed, making differentiation between covered smut and loose smut of oats (see below) difficult except by microscopic examination. Sometimes black stripes occur on the topmost leaves. Rarely seen in recent years; organo-mercury seed dressings effect control.

**Dark leaf spot** (*Leptosphaeria* (*Septoria*) *avenaria*). Rounded or lozenge shaped orange bordered dark brown spots, with fruiting bodies (pycnidia) appearing translucent or faintly pink when held up to the light, occur on the leaves and leaf sheaths. Under wet conditions spotting of the panicle occurs and the stalk may be attacked, resulting in rotting and breaking of the straw. Under these conditions the glume can also be attacked and the underlying grain is infected and sometimes discoloured. The disease can then be carried over on the seed. Widely occurring in the oat growing areas of north England, Scotland, Ireland and Wales, and some-times damaging, particularly in wet seasons. No practical means of control are known.

*Ergot** (*Claviceps purpurea*). Very rare—see under Rye, p. 109.

*Eyespot** (*Cercosporella herpotrichoides*). Frequent, but not usually serious in Great Britain. Common and serious, however, in the Republic of Ireland, see under Wheat, p. 102.

**Grey leaf** (Manganese deficiency). Light green spots occur in the leaves, often about a month after emergence. These spots and sometimes streaks develop, becoming greyish or brown in colour and may have purple margins. When these spots meet across a leaf, the blade collapses at this point and the distal end hangs down. Plants are stunted and badly affected ones succumb. The early stages are often mistaken for halo blight (*q.v.*). To control the disease, avoid overliming and spray plants with 8–10 lb manganese sulphate/acre plus a wetter, in 20–100 gal/acre.

*Halo blight** (*Pseudomonas coronafaciens*). Seen on leaves as a spot with a dead brown centre surrounded by a pale green area, circumscribed by a halo. The spot may fan out in different directions but the distinctive halo generally remains at the edges. This bacterial disease is widely distributed in northern and western areas. Sprinkling the grain with formalin (40% formaldehyde), 1 pint/40 gal of water, gives partial control.

***Leaf (stripe) spot** (*Pyrenophora* (*Drechslera*) *avenae*). Affected seedlings show narrow brown stripes on the leaves and the first leaf may be distorted and twisted. Plants can die at this stage, or even before they emerge from the soil (pre-emergence blight). Less severely attacked plants survive, showing brown spots on the leaves. Subsequently the younger leaves and the spikelets become infected and other plants in the crop can be secondarily infected. Organomercury seed dressings are not always effective as mercury tolerant strains of the fungus exist.

*Loose smut** (*Ustilago avenae*). When the ears emerge, black spore masses are seen replacing the grain in the ears. The spore masses may sometimes be partly or completely enclosed by the coat of the grain, resembling covered smut of oats (see above). Sometimes black stripes occur on the topmost leaves. The disease is rare for, unlike the loose smuts of wheat and barley, it is controlled by organomercury seed dressings.

***Mildew** (*Erysiphe graminis*) is a common and often severe disease, though resistant varieties of oats are available. One or two varieties of spring oats which showed resistance now seem to be more susceptible to mildew. See under Wheat, p. 103.

**Scab** (*Gibberella zeae*). Not common—see under Wheat, p. 103.

*Sharp eyespot** (*Corticium* (*Rhizoctonia*) *solani*). Occasional, rarely severe—see under Wheat, p. 103.

**Take-all** (*Ophiobolus graminis* var. *avenae*). This fungus is distinct from *Ophiobolus graminis* (see under Take-all of Wheat) in that it attacks wheat, barley and oats, and is common in oat-growing areas but occasional elsewhere. To reduce the disease, control perennial grass weeds and avoid cereals too close in the rotation—see also under Wheat, p. 103.


## 5.4 RYE

The major pests are aphids, leatherjackets, slugs, wheat bulb fly and wireworms (see under Wheat); gout fly (see under Barley); stem eelworm (see under Oats). All varieties of rye may be injured by DDT. Of diseases, black point, brown foot rot and ear blight, which is common some seasons, bunt, which is rare, eyespot, and sharp eyespot are described under Wheat, p. 101. For leaf blotch, which is frequent but not important, see under Barley, p. 106. Rye is fairly resistant to take-all (see under Wheat).

**Ergot** (*Claviceps purpurea*) (A.L. 548). The ergot is seen as a horn-shaped body protruding from a diseased spikelet, where it completely replaces the grain. It can be as much as ¾-inch long. The same fungus is often to be found in the spikelets of grasses. The strains which attack grasses are not always distinct from those attacking cereals. Ergots contain toxic alkaloids derived from ergotine, many of which, though used in medicine, are capable of causing acute illness in animals and humans. The dangers of corn samples containing ergots must be emphasized. Some control is effected by rotation of crops and deep ploughing.

## 5.5 STORED CEREAL PESTS

***Saw-toothed grain beetle** (*Oryzaephilus surinamensis*) (A.L. 492). The most important pest of farm stored grain. It destroys the germ and causes grain to heat and to become caked and mouldy (A.L. 404). Only a few insects are required to start an infestation; they are not brought in from the field but from premises previously infested. Cooling of grain below 63°F (17°C) prevents breeding. Provided that the granary or bins are reasonably gas tight, infested grain can be fumigated. Use a 3 : 1 mixture of ethylene dichloride and carbon tetrachloride for bulks of grain under 6 ft deep, or a 1 : 1 mixture if over 6 ft in depth, at a dosage of 1 gal/5 ton. Farmers should not attempt to treat quantities in excess of about 20 tons without professional assistance.

Infested buildings should be thoroughly cleaned and sprayed with 1·5% malathion w.p. Any spaces which cannot be reached by sprays should be treated with DDT/gamma-BHC smoke generators. If there is a risk of clean grain being infested it may be mixed with malathion at a rate not exceeding 10 ppm.

*Grain weevils** (*Sitophilus* spp.) (A.L. 219). *S. granarius* is the commonest species. Larvae eat endosperm, hollow out grains and cause heating (A.L. 404). They cannot breed below 55°F (13°C). For fumigation of infested grain see under saw-toothed grain beetle. If only *Sitophilus* is present, spraying of infested buildings is best done using 1·0% gamma-BHC w.p., use gamma-BHC smoke generators for inaccessible places. If a mixed infestation of weevils and saw-toothed grain beetles is present, spray with malathion or malathion/gamma-BHC mixture.

**Mites** (*Acarus siro*). *A. siro* eats the germ and may cause tainting. Other species may be present but are not important. Mites are liable to develop in grain stored for any length of time at over 15% moisture content. Drying is the best treatment for infested grain, but turning of bulk grain is helpful. Infested granaries should be sprayed with 1·0% gamma-BHC w.p. or dusted with 1·0% gamma-BHC powder. For treatment of infested bags see A.L. 469.

# CHAPTER 6

# PESTS AND DISEASES
# OF POTATOES

## 6.1 PESTS

***Angleshades moth** (*Phlogophora meticulosa*). The caterpillars of this moth feed on many herbaceous plants. On potatoes they bite irregular holes in the leaflets and occasionally destroy the flowers. They may be controlled, where necessary, by a DDT spray at 1 lb a.i. (3 pints 25% e.c.)/ 20 gal/acre.

***Ants** (*Lasius flavus, L. niger*). Ants often shelter within tubers in holes made by other pests. They may cause superficial damage to tubers, their excretory materials producing small sunken pits in the skin. Chemical control is rarely necessary even in heavily infested fields or gardens.

**Aphids**—Plant lice, greenfly (A.L. 139, 278). Aphids cause damage (*a*) by direct feeding injury as they suck plant juices (*b*) by increasing plant susceptibility to damage from anti-blight copper fungicides (see p. 121) and (*c*) by acting as vectors of certain potato virus diseases.

****Bulb and Potato Aphid** (*Rhopalosiphoninus latysiphon*). Although of no importance as a vector of potato viruses, this species can kill shoots of sprouting seed tubers in the chitting house. In medium and heavy soils, particularly in dry summers, it colonizes roots and underground stems of the growing plants, which may wilt and die or have a reduced tuber yield.

Chemical control in the chitting house is by BHC/DDT smoke giving *ca.* 3·5 g DDT and 1 g BHC per 1,000 ft³ *or* dimethoate spray at 5 oz a.i. (12 fl. oz 40% c.c.)/100 gal *or* formothion spray at 7 oz a.i. (16 fl. oz 43% conc.)/100 gal using hand lances with long extensions *or* malathion atomizing conc. at ⅓ fl. oz of 25% solution per 1,000 ft³ *or* sulfotep smoke giving *ca.* 11·4 g a.i. per 5,000 ft³. No measure has yet been developed for chemical control in the growing crop.

*****Peach-Potato Aphid** (*Myzus persicae*), *****Potato Aphid** (*Macrosiphum euphorbiae*), *****Glasshouse—Potato Aphid** (*Aulacorthum solani*) and *****Buckthorn-Potato Aphid** (*Aphis nasturtii*). In the chitting house, all except *A. nasturtii* may be present and cause damage to the young sprouts; in addition *M. persicae* may transmit potato viruses, the plants being very susceptible to infection at the chitting stage. For chemical control—see *R. latysiphon* above. In the growing crop, numbers of each species vary considerably between seasons. *M. euphorbiae, A. solani* and *A. nasturtii* are of little importance as field vectors of potato viruses, but they may cause physical damage to the plant foliage with consequent yield loss; *M. euphorbiae* may be associated with 'top roll' symptoms occasionally seen in Majestic and other varieties. *M. persicae* is rarely present in sufficient numbers to cause direct feeding damage, but is important as the vector of leaf roll and rugose mosaic diseases.

Other aphid species, such as the leaf-curling plum aphid (*Brachycaudus helichrysi*) and the black bean aphid (*Aphis fabae*), may colonize potato foliage in mid-summer. At least one (*B. helichrysi*) plays some part in the transmission of potato viruses.

For chemical control in crops grown in areas where *M. persicae* is usually present and from which seed is to be saved, insecticidal cover is necessary from about 80% plant emergence until senescence or haulm destruction. Unless disulfoton, phorate or menazon is used in the seedbed or as a menazon tuber treatment, the first spray application should be a systemic organophosphorus aphicide, e.g. demeton-S-methyl 3·4 oz a.i. (6 fl. oz 58% e.c.)/40 gal/acre *or* demephion at 3·6 oz a.i. (12 fl. oz 30% e.c.)/40 gal/acre *or* dimethoate at 5 fl. oz a.i. (12 fl. oz 40% e.c.)/40 gal/acre *or* ethoate-methyl at 4·8 oz a.i. (24 fl. oz 20% e.c.)/20–100 gal/acre *or* formothion at $7\frac{1}{2}$ fl. oz a.i. (16 fl. oz 43% conc.)/20–100 gal/acre *or* menazon at 4 oz a.i. (10 fl. oz 40% e.c.)/40 gal/acre *or* oxydemeton-methyl at 3·4 oz a.i. (6 fl. oz 57% e.c.)/40 gal/acre *or* phosphamidon at 3·2 oz a.i. (16 fl. oz 20% e.c.)/40 gal/acre. The second spray application should be made 10 days later and subsequent ones at 14 day intervals. A maximum of four to five sprays is usually necessary, of which the first three are generally most important; late spread of potato viruses occurs in some seasons, and later spray applications are advocated to counter this spread.

Menazon pre-planting tuber treatment at 1 lb a.i. ($1\frac{1}{4}$ lb 70% w.p.)/$2\frac{1}{2}$ gal/ton seed should be followed 6 weeks after emergence by one or more of the foliar sprays described above.

Disulfoton at 16·8 oz a.i. (14 lb 7·5% granules)/acre *or* phorate at $1\frac{1}{2}$ lb a.i. (15 lb 10% granules)/acre may be applied to the open furrow at or before planting maincrop potatoes, followed in July by one of the foliar spray treatments described above. Similar rates of granules are used for early potatoes when a similar period of persistence to that obtained on maincrops is required; where shorter persistence is acceptable, e.g. for early lifting after planting in March, disulfoton rate may be reduced to 12 oz a.i. (10 lb 7·5% granules)/acre and phorate rate to 1 lb a.i. (10 lb 10% granules)/acre. The above granules rates apply to mineral soils; on highly organic (fen) soils the rates for maincrop potatoes are disulfoton 24 oz a.i. (20 lb 7·5% granules)/acre *or* phorate 2 lb a.i. (20 lb 10% granules)/acre. Wireworm control is also achieved using this high rate of phorate. Note that phorate is not recommended on soils containing over 30% organic matter.

The insecticidal treatments listed above may effectively reduce the spread of leaf roll within the crop. They cannot prevent entry of this disease into the crop, nor the entry and spread of virus Y (rugose mosaic).

For ware crops, insecticidal cover may be unnecessary in years when aphid numbers are low. If required, use one of the above listed chemicals *or* DDT spray at 1 lb a.i. (3 pints 25% e.c.)/100 gal/acre, using drop nozzles to wet the undersurface of leaves *or* malathion spray at 18 oz a.i. (1½ pints 60% e.c.)/100 gal/acre. Application should be made (in anti-blight sprays if compatible) before aphid numbers build up and in July to reduce the numbers of migrating *M. persicae*.

Demeton-S-methyl, demephion, disulfoton, oxydemeton-methyl, phorate, phosphamidon and sulfotep are included in the Agriculture (Poisonous Substances) Regulations.

Cultural control is effected by growing crops intended for seed in as much isolation as possible, and certainly not in areas where leaf roll and virus Y are prevalent. Only healthy seed should be used at planting followed by careful roguing of virus-infected plants during the early part of the growing season. Encourage the build-up of aphid parasites and predators, including ladybirds and hover fly larvae, when these are present by using chemicals having selective toxicity towards aphids. Burn off or mechanically destroy the haulm of seed crops as early as possible.

****Capsid (Mirid) Bugs.** The three main potato-infesting species are the common green capsids, *Lygocoris pabulinus*, the potato capsid, *Calocoris norvegicus* and the tarnished plant bug, *Lygus rugulipennis*. These, together with occasional feeders like *Dicyphus errans*, feed on many herbaceous and woody plants. Extensive brown necrotic spots may be caused on potato foliage, the brown tissue later collapsing to leave holes. Young shoots and foliage may die or become distorted under heavy attacks. Damage is usually confined to plants on the headlands.

Chemical control is seldom necessary and may be confined to treatment of headlands with a DDT spray at 1–2 lb a.i. (3–6 pints 25% e.c.)/100 gal/acre. Another less effective material is nicotine spray at 8 fl. oz 98% conc./100-gal/acre. Phorate granules applied at or before planting to control aphids will also control capsid damage.

***Chafer grubs** (A.L. 235). Grubs of chafer beetles may attack potatoes planted after old pasture. The chief offenders are the larvae of the cockchafer, *Melolontha melolontha* and the garden chafer, *Phyllopertha horticola*. In some areas grubs of the summer chafer, *Amphimallon solstitialis*, the rose chafer, *Cetonia aurata* and the brown chafer, *Serica brunnea* may be important.

Chemical control can be obtained with the insecticides listed below for cutworms. For cultural control, thoroughly cultivate and disc soil before

I

ridging. Do not grow potatoes immediately after old pasture in areas where chafer damage is frequently seen.

****Cutworms** (A.L. 225). Cutworms are the caterpillars of a number of Noctuid moths and may bite roots and stems of potatoes near ground level or more importantly, later tunnel into tubers. Attacks are more prevalent on lighter soils and in warm, dry summers.

For chemical control apply DDT dust at $2\frac{1}{2}$ lb a.i. ($\frac{1}{2}$ cwt of 5% dust)/acre *or* DDT spray at 1 lb a.i. (3 pints 25% e.c.)/20–50 gal/acre *or* insecticide/bran (see leatherjackets, page 115). Dusts and sprays should be well worked into the soil before or soon after planting. The materials listed above suffice for leatherjacket control if these are also present.

For cultural control, keep the land free from weeds as these encourage egg-laying and provide food for caterpillars.

**Death's head hawk moth** (*Acherontia atropos*). Larvae of this moth are occasionally found eating potato foliage, especially of garden and allotment crops.

**Earwigs** cause damage by biting neat holes in potato foliage; in severe attacks leaves may be stripped and left ragged. Chemical control is unnecessary.

*****Eelworms** (Plant nematodes). Potato cyst eelworm (*Heterodera rostochiensis*) (A.L. 284) is an important pest in the major ware-growing areas and is present in most gardens and allotments. Advice on the frequency of planting potatoes on infested land is obtainable through the M.A.F.F.'s free soil sampling service; in Northern Ireland cropping is controlled by legislation. Resistant potato varieties are now available but should be grown only after consultation with the advisory service, as the resistance factors may not cover the particular 'races' of the eelworm present in the soil.

For chemical control, dazomet prill granules at 340 lb/acre incorporated into the soil 4–5 weeks before planting have given promising results. D-D mixture injected into the soil at 400 lb/acre has been used successfully for P.C.E. control before planting early potato crops in the Channel Isles.

Potato tuber eelworm (*Ditylenchus destructor*) (A.L. 372) may be introduced within infested seed tubers. It persists in fields where field mint and creeping sowthistle are prevalent, these weeds being the major hosts. No effective chemical control measures are known.

Stem eelworm (*Ditylenchus dipsaci*) (A.L. 178) is of little importance to the potato crop.

Migratory free-living nematodes include species shown to damage potato roots in south-west England and the Scilly Isles, and one species is the vector of tobacco rattle virus, which causes 'spraing' disease in potatoes. A proprietary mixture of dichloropropane and dichloropropene

('D-D mixture') has given useful control of migratory nematodes when injected into soils in south-west England at 400 lb/acre in three-year rotations, and at 200 lb/acre in one-year rotations.

*Flea Beetle* (A.L. 109). The potato flea beetle (*Psylliodes affinis*) feeds on potato leaves and its larvae mine the roots, but serious damage rarely occurs. For chemical control if required, use a DDT spray at 1 lb a.i. (3 pints 25% e.c.)/20 + gal/acre.

*Leafhoppers.* Green leafhoppers (*Empoasca decipiens* and *E. flavescens*) and potato leafhoppers (*Eupteryx aurata and Typhlocyba jucunda*) frequently suck juices from potato foliage, sometimes causing speckling, browning or wilting of leaves. The damage is never serious and the insects do not transmit potato viruses in the U.K. Chemical control if required is given by sprays of DDT 1 lb a.i. (3 pints 25% e.c.)/20 gal/acre *or* malathion 18 oz a.i. (1½ pints 60% e.c.)/100 gal/acre.

*Leatherjackets.* These are the grubs of craneflies (daddy-longlegs) including *Tipula* spp. and *Nephrotoma maculata*. They are usually found in grassland but will feed on roots and stems of many crops including potatoes, on which they are of little importance. Populations of up to one million larvae per acre are of no economic consequence on potato crops. If necessary use DDT spray at 1 lb a.i. (3 pints 25% e.c.)/100 gal/acre *or* DDT dust at 1 lb a.i./acre (1 oz 5% dust/15 yd$^2$) *or* surface baits consisting of 28 lb bran plus either 1 lb DDT *or* 1 lb Paris green/acre.

Paris green/bran bait gives some control of slugs and cutworms if these are also present. The other materials control cutworms but not slugs. Note that Paris green is a poison included in Part II of the 2nd Schedule of the Poisons List (1962).

*Millepedes* (A.L. 150). The spotted millepede (*Blaniulus guttulatus*), the black millepede (*Cylindroiulus londinensis*) and the flat millepede (*Polydesmus angustus*), commonly shelter in holes in potato tubes, being attracted to tissues decaying as a result of attacks by other soil organisms. They can, however, act as primary feeders, scabbing the tuber surface and even tunnelling into the flesh.

A tentative recommendation for chemical control is to dip seed tubers in DDT at 5 pints 15% emulsion/100 gal *or* DDT spray at 1 lb a.i./acre (5 pints 15% emulsion/100 gal/acre) *or* nicotine spray at ½ pint 98% conc./ 100 gal/acre.

**Slugs** (A.L. 115). Slug damage may be serious in some years in heavier soils, including clays and heavy silts. The garden slug (*Arion hortensis*) is the chief species affecting potatoes on a field scale, although the grey field slug (*Agriolimax reticulatus*) may also be responsible. The keeled slugs, *Milax gagates* and *M. budapestensis*, are occasionally important while the black slug (*Arion ater*) is often a serious pest in gardens and allotments. Although slugs are often secondary feeders, enlarging holes in tubers already bored by wireworms, etc., they can also penetrate the tuber skin as primary feeders. Damage, which appears before the tubers are mature but greatly increases in severity before lifting, is worst in late autumn following mild, wet summers. Varietal differences in susceptibility to damage are marked; Ulster Glade, Maris Piper and King Edward tubers are much more susceptible than Majestic, Pentland Dell and Pentland Crown.

Chemical control after planting is of little value. Pre-planting application of high rates of metaldehyde or methiocarb have given promising results in recent experimental work. For cultural control, lift maincrop potatoes, especially of susceptible varieties, as soon as possible after tubers become mature.

**Springtails.** Both the garden springtail (*Bourletiella hortensis*) and *Onychiurus* spp. may cause slight holing of the young leaves lying close to the soil surface; the latter species are very common in fen oils. Chemical control is unnecessary.

****Swift Moths** (A.L. 160). Caterpillars of the ghost swift moth (*Hepialus humuli*) and the garden swift moth (*H. lupulina*) may feed readily on potato roots when the crop follows an infested pasture. For chemical control use DDT dust at 2 lb a.i. (40 lb of 5% dust)/acre *or* DDT soil drench at 7–10 oz a.i. ($1\frac{1}{2}$–2 pints 25% e.c.)/100 gal/acre. For cultural control, cultivate the soil frequently after ploughing old pasture and before planting potatoes.

***Symphylids** (A.L. 484). Damage by the glasshouse symphylid (*Scutigerella immaculata*) is occasionally seen in field crops in southern England and the Channel Isles. Feeding marks on roots and root hairs may provide entry for secondary disease organisms.

Chemical control, if necessary, is given by diazinon spray on seed tubers in the furrows before splitting back ridges, at 3·2 oz a.i. (16 fl. oz 20% e.c.)/100 gal/acre *or* parathion spray, as for diazinon at 2 oz a.i. (10 fl. oz 20% e.c.)/100 gal/acre. Note that parathion is included in the Agriculture (Poisonous Substances) Regulations.

*****Wireworms** (A.L. 199). Wireworms, the larvae of certain click beetles, are especially numerous in the first years after ploughing old turf, but populations as low as 30,000 per acre, often found in arable fields, can damage the potato crop. Some soils habitually carry higher numbers of wireworms than others. Early attacks on seed tubers and sprouts are not usually of much consequence. Later, holing of newly-formed tubers affects quality and provides access for slugs, millepedes and other soil organisms; damage increases in late autumn in the few weeks prior to lifting maincrops. Earlies are not generally affected.

For chemical control use aldrin dust at 3 lb a.i. (2 cwt of $1\frac{1}{4}$% dust)/ acre, broadcast and worked in before ridging *or* aldrin spray at $2\frac{1}{2}$–3 lb a.i.

(6–8 pints 30% e.c.)/30 gal/acre worked thoroughly into the soil before ridging, after ridging and before splitting back the ridges or after the attack on sets is noticed when the ridges are sprayed, harrowed and set up again. Materials which have given equally good results as aldrin are dyfonate 2 lb a.i./acre *or* parathion 2 lb a.i. (20 lb 10% granules)/acre *or* phorate 3 lb a.i. (30 lb 10% granules)/acre, applied into the furrow along with the seed tubers at planting.

Higher rates of aldrin as shown above should be used on peat soils. When potatoes follow old grass, leys or stubble, aldrin can be sprayed immediately before ploughing in the autumn preceding planting potatoes, as this affords aldrin longer time to exert its maximum effect. Phorate granules applied in the furrow also control potato aphids. Do not use BHC or gamma-BHC for wireworm control in potatoes, and do not plant potatoes on land treated with BHC or within 2 years of treating with gamma-BHC as off-flavour may be imparted to tubers.

Parathion and phorate are included in the Agricultural (Poisonous Substances) regulations, and dyfonate is a recently introduced chemical which is likely to be similarly regulated.

Cultural control rests in growing resistant crops after infested turf before taking the first potato crop. Grow early varieties, where damage is suspected, or lift maincrop varieties as soon as the tubers are mature and their skins are set. Cultivate the soil well before ridging.

## 6.2 DISEASES

***Blight** (*Phytophthora infestans*). The haulm and tuber symptoms are described in A.L. 271. As it is particularly important to recognize the earliest signs of blight it should be noted that in the very susceptible variety King Edward the first lesions are often on the upper part of the stem, just below the growing point, and not on the expanded leaves.

Blight can be reasonably well controlled by spraying, but only if the fungicide applications are suitably timed; to achieve this a knowledge of the course of development of an epidemic is needed.

The most important sources of blight each season are:

(1) Imperfectly cleared clamp-sites and dumps of discarded potatoes. The latter are particularly dangerous, because they tend to occur in damp and sheltered sites such as ditches, where sporulation of the fungus is likely to take place earlier than in the open field and where the usually dense growth of potato shoots favours rapid build-up from the first affected plants to form an 'initial focus' (see below).

(2)  Slightly infected seed tubers of early varieties planted with the crop. Early varieties are in general more susceptible than maincrops, and the earlier development of the foliage encourages earlier spread of the disease. Furthermore, since the yield of earlies is seldom affected by blight it often happens that no measures are taken to check its spread, even though the earlies are growing alongside a susceptible maincrop.

(3)  Slightly infected seed tubers of susceptible maincrop varieties, planted with the crop.

The relative risk from these sources undoubtedly varies from year to year and from place to place, but recent evidence suggests that infected seed tubers planted with the crop are not so consistently of first importance as was previously thought, except in areas where early varieties predominate.

Slightly infected seed tubers are almost impossible to detect, but there is no excuse for permitting the sources of type (1) to exist. Clamp-sites should be kept clean throughout the period of riddling; if there is any growth of potato plants on the site in the spring it should be destroyed by rotovating, which should be done before the new crop emerges. If necessary the rotovating should be repeated to make sure the clamp-site is kept clear of growth. It should not be over-planted with any crop—least of all with potatoes. Dumps, once in existence, are less easy to deal with, but the worst place to make them is where they will not be noticed. The best plan is to make any quite unavoidable dump in some obvious place where any growth in the spring can be seen and destroyed by the use of desiccant herbicides.

There are usually three stages in the development of an epidemic, though it should be realized that these so-called stages are simplifications of what is in reality a complicated process in which there are often no clear-cut distinctions between the various phases.

The stages are:

(*a*)  Focus-building—the spread by rain-splashed spores from an 'initial infector' plant (i.e. one which has grown from an infected tuber) to a group of neighbouring plants, the whole group forming an 'initial focus' of blight. The special case of the development of an initial focus on a dump has already been mentioned.

(*b*)  Short-range spread—the spread by wind-borne spores from an initial focus on a dump into a nearby potato field, or from a focus in a crop to other parts of the field or into adjacent fields. In either case most of the spread will probably be confined to within a short distance of the initial focus, though a few infections may occur some hundreds of yards down-wind. The new infections develop into 'daughter foci', each of which will, in favourable conditions, grow and act as a centre for the further dispersal of the disease. This short-range spread seldom occurs before June except in south-west England and west

Wales. If it is delayed until July or later this stage may not be easily recognizable: in the more favourable infection conditions of late summer, blight may spread very quickly from the daughter foci so that they do not remain distinct for long. In such cases stage (*b*) merges quickly into stage (*c*).

(*c*)  Long-range spread—in favourable weather the improved infection conditions and the vast number of spores now available in initial and daughter foci lead to widespread and often almost simultaneous infection over large areas. This may occur in June in south-west England and west Wales, but is unusual before July in the chief maincrop districts.

General spread of this kind occurs during a Beaumont period (48 hours during which the temperature does not fall below 50°F nor the relative humidity of the air below 75%) and the first signs of this spread—a scattering of infected plants in many crops—are usually seen in 7–14 days, though they may not appear until 3 weeks after the infection period. This general appearance of blight is what is generally termed the 'outbreak' of the epidemic, though in fact it is a fairly late stage in its history: it is forecasts of 'outbreak' in this sense, based on the recording of Beaumont periods at a network of meteorological stations, which are issued by the Ministry of Agriculture *via* the National Agricultural Advisory Service, the Press, radio and television.

Protective spraying is only really effective when the amount of disease present is small. Once blight gets a firm hold on a field and begins to increase rapidly, further protective spraying is almost useless. Spraying should therefore be concentrated against the earliest stages of infection in a crop.

In the north of England and some parts of the Midlands blight usually develops so late that protective spraying is unnecessary. The following suggestions may be useful to growers in other areas, though they should be modified according to local experience. Not all the more southerly areas are equally prone to the early development of blight, while certain fields or districts in the less susceptible zones may have a bad reputation in this respect.

In intensive potato growing areas, especially those where blight susceptible varieties predominate, the risk of being near an initial focus is great enough to warrant routine early spraying (i.e. before the broadcast warning) as an insurance against infection during stage (*b*) of the epidemic (above). Routine precautionary spraying of this kind should be started just before the crop meets across the rows. Only susceptible varieties such as King Edward and Bintje (and the second early, Craigs Royal) need be sprayed at this stage.

In less intensive areas it is usually safe to wait for the forecast warning; when this is received, susceptible varieties should be sprayed within the next few days. Less susceptible varieties such as Majestic need not be sprayed until blight has been seen in neighbouring susceptible varieties or, of course, in the crop itself.

If in doubt about the interpretation of the M.A.F.F. blight warning (which does not include advice on spraying), growers should consult their local N.A.A.S. adviser. Growers who have already applied one or more precautionary sprays should nevertheless spray again as soon as possible after the forecast warning.

The usual practice for later applications is to continue spraying at 10–14 day (or dusting at 7–10 day) intervals unless the weather turns really hot and dry.

An alternative method is to apply sprays (after the first) only when about $\frac{1}{2}$ in of rain has fallen in a running period of 5 days. This system has the advantage that spray deposits washed off by heavy rain will be replaced at once, and it prevents the wasteful application of sprays in very dry weather. In humid weather, however, blight (once in a crop) can develop fast even in the absence of appreciable actual rainfall; in such conditions a routine spray might have delayed its progress considerably.

High volume (60–100 gal/acre) and low volume (20–40 gal/acre) applications are about equally effective, and most modern materials can be used equally well in either way. Ultra-low volume spraying (2–5 gal/acre) from aircraft is also satisfactory, provided that the special formulations necessary for this work are used, that the field is suitably shaped and situated, and that the work is done by a reliable firm. In open country there is little difference in efficiency between fixed-wing aircraft and helicopters, but in difficult country and with irregular and badly obstructed fields helicopters are to be preferred.

Dusting is also satisfactory, providing that allowance is made for the lower tenacity of dusts. In wet or windy weather particularly, more applications, in the proportions of about five dusts to three sprays, have to be made. Moreover it is safest to assume that anything like a heavy rain will have removed most of a dust deposit; it should then be replaced as soon as possible. Modern dusters enable this to be done quickly and easily and in some ways the more frequent applications necessary with dusts are an advantage in that there is less unprotected new growth than with the more widely-spaced sprays.

In view of the importance of correct timing, growers who normally have their spraying done by contract are well advised to have machines and materials available so that in emergency they can do the work themselves if the contractor is held up by bad weather.

The fungicides recommended for blight control are as follows:

**Bordeaux Mixture.** This is made by mixing a solution of copper sulphate with a suspension of slaked lime. It is a good fungicide and has the valuable characteristic that the weathered remains of earlier deposits assist the adhesion of the later applications.

Directions for making up the standard mixtures for H.V. use—e.g. 10 : 12½ : 100 (10 lb $CuSO_4$, 12½ lb $Ca(OH)_2$, 100 gal water)—are given in A.L. 271. Because of the great excess of lime, this mixture is only suitable for use in H.V. machines, but new formulations are now available in which the lime has been reduced (e.g. to 8 : 4 : 20) so that the same amount of copper per acre can be applied through an ordinary L.V. farm sprayer.

One of the farmer's objections to Bordeaux mixture is the comparative difficulty of preparing the suspension, which must be freshly made. The dissolving of the copper sulphate, which is the time-consuming part of the operation, should therefore be done in advance, using large wooden barrels or plastic containers.

**Burgundy Mixture.** This differs from Bordeaux in containing washing soda or soda ash instead of lime. Directions for making up the standard H.V. mixtures—e.g. 8 : 10 : 80—are given in A.L. 271.

As with Bordeaux mixture, more concentrated mixes such as 8 : 5 : 20 (where the second figure refers to soda ash—anhydrous sodium carbonate) can now be made.

**Other Copper Compounds.** Various inorganic compounds of copper —e.g. basic copper chloride (copper oxychloride), basic copper carbonate and cuprous oxide—are the essential ingredients of many formulated w.ps for blight control. They are usually compared on the basis of the quantity of metallic copper present. In most cases the recommended rate of application is 1½–2½ lb metallic copper/acre, though the so-called colloidal brands, which are more finely divided, are used at ¾–1 lb metallic copper/acre.

Copper oxychloride is probably the most widely used of these materials, but they are all efficient fungicides which are easy to handle and can be used at high, low and ultra-low volume.

Copper dusts for blight control contain copper oxychloride equivalent either to 9% or to 35% metallic copper. The former is applied at 20 lb/acre, the latter at 5 lb/acre.

Copper compounds can cause phytotoxic effects ranging from a barely perceptible 'hardening' of the plant to marked yellowing of the haulm and/or a brown scorching of the leaves. In some cases there is a noticeable loss of yield. Damage is particularly likely to occur in industrial districts where air pollution is prevalent, but it may also occur in rural areas, especially in dry years or when much damage by aphids or wind is present.

The risk of loss of yield from copper injury diminishes as the rate of growth of the crop declines towards the end of the growing season. Except in high rainfall areas copper fungicides should therefore be kept for the later applications, when full benefit can be obtained from their persistence, which is rather greater than that of the dithiocarbamates.

All copper compounds are harmful to fish and livestock.

Copper compounds are sometimes combined with a small quantity of an organomercury compound. Used at the recommended rate of $1\frac{1}{2}$ lb metallic copper/acre, these mixtures generally give much the same control as an equal quantity of the copper compound alone, though they may have a slight advantage in some seasons if the timing of the spraying is exactly right. Inhalation of the powder or its contact with the skin must be avoided.

**Dithiocarbamates and allied fungicides.** Zineb and maneb give much the same degree of control of blight as do the copper compounds, but have the advantage of being practically non-phytotoxic. When they are first applied they are rather more active fungicidally than the copper compounds, but they are slightly less stable and persistent. They are available as w.ps: the proprietary zineb formulations normally contain 70% of the active ingredient and are used at about $1\frac{1}{2}$ lb a.i. (2 lb 70% w.p.)/acre; maneb preparations usually contain 80% a.i. and are used at about $1-1\frac{1}{2}$ lb a.i. ($1\frac{1}{2}-2$ lb w.p.)/acre.

A widely used dithiocarbamate fungicide is mancozeb, which is available both as a w.p. and as a dust. The formulated w.p. contains 80% a.i. and is used at the rate of about $1\frac{1}{4}$ lb a.i. ($1\frac{1}{2}$ lb w.p.)/acre. The dust contains not less than 32% w/w manozeb and is used at the rate of $5\frac{1}{2}$ lb/acre. Metiram is formulated as a 70% w.p. and used at the rate of about $1\frac{1}{2}$ lb a.i. (2 lb w.p.)/acre, while propineb (mezineb), a 70% w.p., is applied at about $1\frac{1}{2}$ lb a.i. (2 lb w.p.)/acre. Other dithiocarbamate-complex fungicides, available as w.ps contain respectively manganese, zinc and iron, and manganese, zinc, copper and iron. The former is used at $1\frac{1}{2}-2$ lb w.p./acre, the latter at $1\frac{1}{2}$ lb w.p./acre.

Inorganic copper compounds mixed with maneb are also available, used at 2 or 3 lb w.p./acre. They are less phytotoxic than the straight copper compounds while the benefit of the extra persistence of copper is to some extent retained. All the dithiocarbamates can be irritating to the skin, eyes and nose, while the copper-dithiocarbamate mixtures are harmful to fish and livestock.

**Organotin fungicides.** These are now used on a considerable scale, and in some circumstances give a measure of control of blight in the tubers as well as on the haulm. They are rather more phytotoxic than the dithiocarbamates, but less so than the copper compounds. They should not be mixed with oil-based insecticides. The Agriculture (Poisonous Substances) Regulations apply to the use of these compounds. Fentin hydroxide can be obtained as a w.p.; some formulations contain 20% a.i. and are used at 4 oz a.i. (1¼ lb w.p.)/acre while one contains 50% a.i. This too is used at 4 oz a.i. (8 oz w.p.)/acre. A w.p. mixture containing fentin hydroxide (7%) and maneb is also available and is used at 1½ lb w.p./acre. Fentin acetate, the other tin fungicide in use, can only be obtained in the form of a mixture with maneb (60% fentin acetate, 20% maneb). It is applied at the rate of 7 oz w.p./acre. All the organotin compounds are dangerous to fish and livestock.

**Haulm destruction.** Haulm destruction reduces the risk of tuber blight by removing the source of infection (the infected but still living stems and leaves); it also facilitates lifting by destroying weed growth.

To achieve anything like complete control of tuber blight, haulm destruction must take place soon after blight is first seen in the field, but in practice this would often involve an unacceptably heavy loss of crop weight. On the other hand it is rarely worth destroying haulm already half dead with blight.

Factors to be considered before deciding to 'burn off' the haulm are: (1) the amount of blight on the leaves and stems, (2) the crop already formed; it is not worth risking an already good crop for the sake of a little extra weight, (3) the rate of bulking; it is useful to do weekly sample liftings to get some idea of this, (4) the nature of the soil and the state of the ridges; some soils seem to encourage tuber blight, either because they crack in the ridges or because they tend to retain water—most growers know the record of their fields in these respects, (5) the tuber susceptibility of the variety; it is rarely necessary to 'burn off' a variety as tuber resistant as Majestic for the control of tuber blight, though it may be desirable to do so to prevent 'second growth' or for the control of weeds.

Whenever the haulm is destroyed, the crop should not be lifted for at least 10 days.

The materials in common use for haulm destruction are:

**Sulphuric acid.** The quickest kill, especially of stems, is obtained by spraying with concentrated (70%) sulphuric acid (B.O.V.) at the rate of about 20 gal/acre. Acid-proof machines and protective clothing are necessary, so that haulm destruction by this material is essentially a contractor's method.

**Diquat.** The proprietary formulation of this material is used at a rate of 4 pints/acre in 20–50 gal water. Only one application at this rate must be made to any one crop, and it must not be used unless the soil around the potato roots is thoroughly wet. It is particularly important to check this after a dry spell, even if a good deal of rain has fallen.

**Dinoseb.** Formulations containing about 10% W/V dinoseb as an emulsifiable oil solution are used the rate of about 2 gal in 40–100 gal water/acre. The higher volumes should be used if there is a heavy growth of haulm or weeds, but run-off should be avoided.

***Dry Rot** (*Fusarium* spp.). Dry rot in stored seed or ware potatoes can be controlled by the use of tecnazene (TCNB) dust applied at the rate of 10 lb/ton at lifting. For clamped potatoes it is best to put on a layer of straw with a light covering of soil, adding the full winter cover as soon as practicable. Potatoes stored indoors should for preference be covered with straw, but sacks or tarpaulins may be used.

Tecnazene is a sprout depressant, and seed for first crops in early districts should be chitted in boxes in November. All other seed tubers should be aired for 6 weeks before planting.

Dry rot and some other tuber diseases, e.g. skin spot (*Oospora pustulans*) and black scurf (*Rhizoctonia solani*) can be controlled to a considerable extent by dipping in a dilute solution of certain organomercurial compounds immediately after lifting. This treatment is less reliable as a control for gangrene (*Phoma exigua* var. *foveata*) but good results are often obtained. The tubers are usually washed free of soil before being treated and the process unfortunately seems to involve some risk of increasing the bacterial disease blackleg (*Erwinia carotovora* var. *atroseptica*). Because of the risks to operators, dipping in organomercurial solutions should be carried out only in establishments registered under the Factories Act. Attempts to find a satisfactory and less poisonous substitute for organomercurial compounds have not yet been successful.

Tubers treated with organomercurials should on no account be used for human food or fed to stock.

At present there is no effective fungicidal control for any of the other fungal diseases of the potato.

*****Virus diseases.** The two main virus diseases of potatoes, Leaf Roll and Severe Mosaic (Virus Y), are transmitted by aphids and in certain circumstances the spread of Leaf Roll can be controlled to some extent by the use of insecticides (see p. 112).

# CHAPTER 7

# PESTS AND DISEASES OF SUGAR BEET, FODDER BEET AND MANGOLDS

### (see M.A.F.F. Bull. 153, 'Sugar Beet Cultivation')

About 440,000 acres of sugar beet are sown annually under contract with the British Sugar Corporation. Three quarters of the crop follows cereals and a clause in the contract enforces a minimum rotation of 1 sugar beet crop in 3 years except after a ley of at least 3 years. Normally sugar beet is sown in late March and early April at 2–5 lb seed/acre in 18–24 in rows. The plants are singled to about 25,000 per acre on average, between late April and early July, and approximately half the acreage is singled by the end of May. A rapidly increasing proportion of the acreage is sown with pelleted monogerm seed at about 5 in spacing (8½ lb/acre of pellets, equivalent to slightly less than 2 lb actual seed/acre), obviating the necessity for singling. The crop is harvested from the end of September to about the end of December. Most roots are delivered to the sugar factories within a week or two on a permit system, but some may be stored in clamps for 8 weeks or more before delivery.

Mangolds (40,000 acres) and fodder beet (a few thousand acres) are grown in a similar way to sugar beet but are stored in clamps or barn, for use as stock feed mainly in spring and early summer. Stecklings for about 3,000 acres of sugar beet and mangold seed crop are sown either under cereal cover-crops in April, or in July–August in the open in areas where root crops are few. They are grown-on *in situ* or transplanted in October–March for harvesting the following summer.

## 7.1 PESTS

### (see M.A.F.F. Bull. 162, 'Sugar Beet Pests')

***Beet carrion beetle.** The beet carrion beetle (*Aclypea opaca*) has damaged beet rarely in recent years, but used to be an occasional, serious pest. Both the adults and larvae feed on the foliage and leave very characteristic blackened, ragged edges. The materials recommended for flea beetle will readily control this pest.

****Birds.** Game birds, sparrows and other small birds defoliate seedlings, sometimes so severely as to kill them; the seedlings are usually very small at the time that significant damage is caused, and spraying a repellent such as anthraquinone is unlikely to be of value.

***Black bean aphid** (A.L. 54). The black bean aphid or blackfly (*Aphis fabae*) is one of the most serious pests of sugar beet. Severe and widespread epidemics occur in some years, causing considerable damage to the seed and root crop. The aphid transmits beet yellows virus but the direct physical damage it causes when feeding on the plant is probably more important. The aphid overwinters in the egg stage, mainly on spindle (*Euonymus europaeus*), and migrates to beet in May and June. The primary infestation develops rapidly in hot, dry weather, especially when predators of the aphid are few, and the plants lose yield as soon as there are more than 2 aphids per leaf. When left unchecked, infestations can blacken the crop and cause large losses of yield, especially on late-sown crops in drought.

Early treatment is essential and only systemic insecticides should be used, in not less than 20 gal of water per acre. Suitable materials are demephion at 3·6 oz a.i. (12 fl. oz 30% e.c.)/acre, *or* demeton-S-methyl at 3·4 oz a.i. (6 fl. oz 58% e.c.)/acre, *or* dimethoate at 4·8 oz a.i. (12 fl. oz 40% e.c.)/acre, *or* formothion at 7·5 oz a.i. (16 fl. oz 43% e.c.)/acre, *or* menazon at 6 oz a.i. (15 fl. oz 40% e.c.)/acre, *or* oxydemeton-methyl at 3·4 oz a.i. (6 fl. oz 57% e.c.)/acre *or* phosphamidon at 3·2 oz a.i. (16 fl. oz 20% s.c.)/in 35 gal/acre. For larger plants the dosage rate recommended should be increased by at least 25%, especially of dimethoate and menazon. The application of 16·8 oz a.i./acre of disulfoton (14 lb 7·5% granules) or 16 oz of phorate (10 lb 10% granules), with the granules concentrated in bands over the rows, controls black aphids for longer periods and is preferable to spraying when the plants are heavily infested or late in the season; alternatively, these amounts may be split into two applications, and are then approximately equivalent to two sprays. Treatment with insecticide after the middle of July is unlikely to be worthwhile. For control of black aphids in seed crops see p. 136.

***Capsids.** Sugar beet can be damaged by several species of Miridae, especially the potato capsid (*Calocoris norvegicus*) and the tarnished plant bug (*Lygus rugulipennis*).

Damage by the potato capsid is confined to crop edges close to hedgerows, woods and orchards from where the nymphs migrate from woody winter hosts to feed on the beet. The active bugs are seen among the beet and injury shows as necrotic spots, puckering of the lamina, and general distortion and yellowing of the leaf, especially at the tip. Spraying with DDT at 16 oz a.i. (e.g. 4 pints 20% e.c.) in at least 20 gal of water per acre controls the pest; damage is rarely severe and, at most, only the field margins need treatment. Phorate granules, at the rate recommended for aphids, also control the pest. The tarnished plant bug migrates as an adult into the beet field very early in the season and feeds on the growing point of the young seedlings, producing blind distorted seedlings. These symptoms do not show for some time, too late for control measures. Preventive control with DDT might be worth while where attacks are expected, but attacks cannot be forecast.

***Chafer Grubs** (A.L. 235). Larvae of the cockchafer (*Melolontha melolontha*) and occasionally of the summer chafer (*Amphimallon solstitialis*) sometimes damage beet, especially in well wooded areas where the soil is light. They feed entirely below the soil surface, at depths of as much as 1 ft. Control is impossible in the growing crop, and soil treatment before drilling impractical because of the difficulty of incorporating the insecticide deep enough and of anticipating damage.

***Cutworms** (A.L. 225). The two main species are caterpillars of the turnip moth (*Agrotis segetum*) and the garden dart moth (*Euxoa nigricans*). Turnip moth caterpillars hatch in mid-summer, and feed just below soil level for the rest of the season. Damage is sometimes extensive (e.g. autumn 1959) but rarely severe enough to require control measures. The garden dart moth is local in the fenland regions. The caterpillars hatch in early spring and then feed until mid-June, sometimes killing many plants and leading to thin stands. Attacks cannot be forecast, so preventive measures are not usually possible. When damage is occurring apply DDT in the late afternoon or evening, preferably under moist conditions which favour surface feeding by the cutworms. Use a dust at not less than 2 lb a.i. (40 lb 5% dust)/acre, or a spray at not less than 1 lb a.i. (3 pints 25% e.c.)/acre in 20 gal water. The insecticide should be concentrated along the rows and worked into the soil by steerage hoeing. Alternatively, 4 oz DDT (1 pint 25% e.c.), mixed thoroughly with 28 lb of bran and sufficient water to moisten, should be applied as a poison bait at 28–42 lb/acre during the late afternoon or evening; this is a convenient method for small areas but may not be as effective as dusting or spraying. Gamma-BHC, either as spray or bait, may well be an effective alternative to DDT, provided that 8 oz a.i./acre is not exceeded and potatoes or carrots are not to be grown within 18 months.

*****Eelworms.** Beet cyst eelworm (*Heterodera schachtii*) (A.L. 233) is a potential menace to the beet industry but causes little actual yield loss. 'Beet sickness' (crop failure or severe damage due to root eelworm) occurs rarely in Great Britain, largely because crop rotation is controlled in two ways: (1) The contract made between the grower and the British Sugar Corporation specifies that beet may be grown only after a minimum of 2 years of non-host crops (i.e. crops other than *Cruciferae* and *Chenopodiaceae*) except following a ley of at least 3 years; (2) The Beet Eelworm Order enforces a 3-year rotation of all susceptible crops within a scheduled area of the fens where beet eelworm is prevalent and a 4-year rotation at least in every field where beet eelworm is known to exist. No practical method of chemical control is available.

Free-living eelworms (principally *Trichodorus* and *Longidorus* spp.) (S.T.L. 66) damage seedling root systems and stunt plant growth on light sandy soils, causing a trouble known as Docking disorder. Damage is usually patchily distributed in affected fields, being most severe on the sandiest areas, and is also characterized by nitrogen and magnesium deficiency symptoms in foliage, poor root development, frequently a fangy (forked) tap root, and occasional plants showing symptoms of tobacco rattle virus ('yellow blotch') or tomato blackring virus infection ('ringspot'). Rainfall, cultivations, herbicide and fertilizer usage, and previous cropping can affect the incidence of the trouble which, in the worst cases, halves or even quarters yield. A liberal application of nitrogen in the seedbed obviates very severe yield depression, but control of nematode damage is necessary for the best yield and root shape. D-D mixture is tentatively recommended at 1 ml/ft of row applied 6 to 8 in deep in the rows where the beet are to be drilled not less than two weeks later; beet must not be drilled if the smell of D-D mixture is still detectable in the soil. Recent trials with granular nematicides applied in the furrow with the seed suggest that very small quantities may give adequate seedling protection; approximately $\frac{1}{2}$ lb a.i./acre of one new compound was needed in 1967 trials for outstanding yield increases on sites heavily infested with *Trichodorus* spp., whilst in 1968 smaller yield increases were obtained on sites with few eelworms. The use of this compound, 2-methyl-2-(methylthio)propionaldehyde *O*(methylcarbamoyl)oxime, is still in the experimental stage. It has not been given a common name.

The Northern Root-Knot eelworm (*Meloidogyne hapla*) occasionally attacks beet and other crops such as potatoes and carrots on light sandy soil in East Anglia. Another species (*M. naasi*) occurs in Wales and the western counties of England, where it is most often found on the roots of cereals and grasses but also attacks beet; this species severely stunts beet on light sandy soil in Belgium. No control measures can be recommended, or are known to be necessary.

Stem eelworm (*Ditylenchus dipsaci*) (A.L. 178) is endemic, but rarely causes severe damage. Seedlings are invaded by the eelworms, which cause galling, bloating and distortion of the stem or petioles and mid ribs, and sometimes death of the growing point. Obvious symptoms of infection are absent during the summer but reappear in the autumn as crown canker—a dry, corky canker in the region of the lower leaf scars—which develops rapidly and eventually invades the whole crown. There are many hosts for the stem eelworm, especially oats and onions, and beet should not immediately follow these crops if they were infested.

***Leatherjackets** (A.L. 179). Leatherjackets are usually a minor pest, occurring mainly in wet soils and immediately after grass or ley but damage was more extensive in Spring, 1968. The larvae feed on the plants just below, at, or just above soil level. The commonest of several species is the larva of the marsh crane fly, *Tipula paludosa*. Control is by spraying, dusting or baiting with DDT at the rates recommended for cutworms; or by spraying gamma-BHC at 8 oz a.i. (10 fl. oz 80% suspension)/acre, provided the land will not be cropped with potatoes or carrots within 18 months; or by using gamma-BHC in a bran bait at 4 oz a.i. (e.g. 5 fl. oz 80% suspension)/28 lb bran/acre.

***Mammals.** Hares and rabbits defoliate seedling beet and, later, feed on the petioles and crowns: repellents have not been tested and damage is best minimized by local control of these pests.

****Mangold flea beetle** (A.L. 109). The mangold flea beetle (*Chaetocnema concinna*) occurs in root crops, but the amount of damage caused differs considerably from year to year. Outbreaks are commonest in the drier parts of the country and are particularly prevalent in sheltered fields during a cold, dry spring. Diagnosis is sometimes confused by damage caused by birds, but whereas birds nip off part or all of the cotyledons, flea beetles chew irregular pits and holes in the leaves. The most serious damage is caused to seedlings in the cotyledon stage and insecticide must be applied promptly; use DDT at 8–16 oz a.i. (e.g. 2–4 pints 20% e.c. or 10–20 lb 5% dust)/acre, depending on the severity of the attack. Even 5 gal spray/acre is adequate.

****Mangold fly** (A.L. 91). Attacks on sugar beet by the larvae of the mangold fly (*Pegomyia betae*) vary considerably from district to district and year to year; there is a tendency for damage to be more on lighter soils, especially near the coast, and the pest is favoured by cool, moist conditions. White eggs are laid singly or in groups on the underside of the cotyledons and true leaves; the larvae hatch within a few days and burrow straight into the leaf from the underside of the eggs, producing at first linear and later blotch mines between the upper and lower leaf surface. There are two or three generations each year but only the first generation attack, during May and early June, is worth controlling, and only when the attack is severe. When the number of fresh eggs plus living larvae exceeds the square of the number of rough leaves, yield is likely to be diminished unless the pest is controlled. This is achieved by spraying, preferably at 20 or more gal/acre with one of the following materials: dimethoate at 1·2 oz a.i. (3 fl. oz 40%) e.c./acre, *or* 1·8 oz a.i. of formothion (4 fl. oz 43% e.c.)/acre, *or* 3 oz a.i. of parathion (e.g. 15 fl. oz 20% e.c.)/acre, *or* 6·4 oz a.i. of trichlorphon (8 oz 80% s.c.)/acre are all suitable for any level of attack or stage of larvae, and all kill rapidly. Treatment should be when egg-laying seems complete and when the first mines are appearing.

The circumstances that lead to severe losses—poor, backward growth with large infestations—are not common in England. Probably no crop past the eight-leaf stage is worth spraying against mangold fly alone, but in years when aphids invade early, the two pests can be simultaneously checked by using dimethoate at 4·8 oz a.i. (12 fl. oz 40% e.c.)/acre, which controls mangold fly excellently and green aphids well, although an increased dosage is necessary to control *A. fabae*; alternatively, formothion at 7·5 oz a.i. (16 fl. oz of 43% e.c.)/acre *or* phosphamidon at 3·2 oz a.i. (16 fl. oz 20% s.c.)/acre may be used. A mixture of 3·2 oz trichlorphon and 3·4 oz demeton-S-methyl (4 oz 80% s.c. plus 6 fl. oz 58% e.c.)/acre controls both pests excellently and virus yellows well. Trichlorphon should not be used without the addition of an aphicide when any aphids are present in areas where virus yellows becomes prevalent.

'Band-spraying' is suitable for sugar beet in May and early June if done carefully, and it avoids wasteful use of insecticide.

****Millepedes** (A.L. 150). Serious attacks by millepedes on seedling beet are sporadic and the conditions leading to them are unknown; the acreage affected appears to be increasing and the pest has recurred in fields where damage was noted in the previous beet crop in the rotation. The spotted millepede, *Blaniulus guttulatus*, is the species most commonly responsible for damage, and the flat millepede, *Brachydesmus superus*, occurs occasionally.

The standard seed treatment possibly gives some protection and soil treatment with gamma-BHC before drilling may increase protection. There is no treatment that can be recommended to control the pest when it has started to attack the seedlings. Some growers have obtained partial control by applying gamma-BHC at 16 oz a.i. (20 fl. oz 80% suspension) in as large a quantity of water as possible and preferably concentrated in a band along the rows; inter-row cultivation should follow treatments to help mix the insecticide into the surface soil. Potatoes or carrots cannot be grown during the succeeding 3 years because of the risk of gamma-BHC taint.

*****Peach-potato aphid** (*Myzus persicae*). This green aphid is the most important vector of sugar beet viruses; its biology and control are described under **Virus Yellows** (p. 134).

****Pygmy mangold beetle.** Small, blackened pits in the hypocotyl of young beet plants are the characteristic form of damage by this pest, *Atomaria linearis*. Seedlings attacked when small may be killed, either directly or because fungi invade through the wounds, but once the stem starts to thicken this pest does little harm. In wet weather the beetles also feed on the young leaves. The pest migrates in fine weather in April, and later, from the previous year's beet fields to the new ones, and is thus most common in intensive beet-growing areas. In the 1930's damage was common due to the frequency of beet in the crop rotation; it was grown in successive years. Sugar beet is no longer grown so frequently (see p. 125), but mangolds may be and this may well lead to damage from pygmy mangold beetle and eelworm (see p. 127). Seed treatment with gamma-BHC (see below) probably gives very little protection; soil treatment with gamma-BHC might be more effective, but too little is known for any definite recommendation to be made. Damage to the leaves is rarely severe enough to warrant treatment but may be prevented by spraying with DDT at 16 oz a.i. (e.g. 4 pints 20% e.c.)/acre in sufficient water to give slight run-off.

***Rosy rustic moth.** The caterpillars of this moth (*Hydraecia micacea*), burrow inside the swelling root of beet plants from late May onwards and may kill them. Damage is usually negligible since so few plants are attacked.

***Sand weevil.** Damage caused by sand weevil (*Philopedon plagiatus*) is confined to the Breckland regions of Norfolk and Suffolk. The adults feed on the foliage during May and early June and can be controlled by the insecticides recommended against flea beetle.

****Slugs** (A.L. 115). Injury by slugs, usually the grey field slug (*Agriolimax reticulatus*), and sometimes species of *Arion*, is recognized only rarely, but is probably fairly common on the heavier soils. Seedlings can be seriously damaged, either above or below ground, and prompt action with metaldehyde bait is necessary; $\frac{1}{2}$ lb of metaldehyde in 28 lb of bran is a standard bait for up to 1 acre and should be moistened before distributing it uniformly, or a proprietary slug bait of metaldehyde and pelleted carrier may be used. The compound methiocarb controls slugs even more effectively than metaldehyde and is less affected by rainfall; the proprietary pellets should be used at 3·2 oz a.i. (5 lb of 4% pellets)/acre.

***Thrips.** Injury by thrips (*Thrips angusticeps*) is not common. The thrips overwinter in the soil as brachyapterous (virtually wingless) adults and are found on the seedlings in April and May, mainly feeding on the still-curled heart leaves; when these expand they are elongated and even straplike, roughened, with irregular and partially reddened or blackened margins and tips. Small silvery lesions on the leaf surfaces are also usually evident. Control by spraying with DDT at 16 oz a.i. (e.g. 4 pints 20% e.c.)/acre in at least 30 gal of water.

****Wireworms** (*Agriotes* spp.) (A.L. 199) are not at present a serious pest of beet, but the steadily diminishing seed rate per acre, and the increasing use of genetical monogerm seed, is likely to magnify the importance of small populations of this pest, and indeed of all other seedling pests.

Seed treatment with gamma-BHC in combination with fungicide was introduced in 1948 and dieldrin seed treatment was later introduced also, as an alternative. From 1954 to 1960 growers demanded 75% of their seed so treated. In 1961 the British Sugar Corporation decided to supply only treated seed to growers and dieldrin became the preferred material because of gamma-BHC's occasional phytotoxicity. Dieldrin (9 oz 40% dressing/

cwt of seed) is used for rubbed and graded seed and for monogerm seed, whether pelleted or not. Seed is treated by the merchant before dispatch to the sugar factories or to the pelleting plant, and thence to the growers.

Current seed rates apply less than 1/5 oz dieldrin, as seed dressing, per acre and, where the wireworm population exceeds 500,000 per acre, additional treatment of the soil with gamma-BHC is needed unless a high seed rate and/or late drilling is preferable; apply 8 oz a.i. (10 fl. oz 80% suspension) in at least 20 gal/acre and work thoroughly into the seedbed. Increase the application rate to 12 oz a.i. on fen soil. Some growers now use extremely low seed rates so as to achieve a final population of 25,000–40,000 plants per acre with no thinning or singling. Under such circumstances soil treatment will be needed where there are only 100,000 wireworms per acre, especially for early drillings. Control of wireworms damaging the seedlings in April or May is very difficult; a tentative recommendation is to apply gamma-BHC as suggested for millepedes.

## 7.2 DISEASES

(see M.A.F.F. Bull. 142, 'Sugar Beet Diseases')

***Black leg.** Of the several fungi that can cause black leg, *Pleospora bjoerlingii* is the most prevalent and important. These fungi attack at different stages of the growth of the seedlings: germinating seeds may be killed below ground, the young seedlings may 'damp-off' soon after they come above ground, or they may survive the early infection and then die from stem girdling later on. These fungi rarely cause complete crop failure, but lead to thin and gappy stands of plants. The disease is controlled by seed disinfection with organomercury fungicides, and since 1945 all seed has been treated by the seed merchants. At first only dusts were used, usually methoxyethylmercury silicate or a mixture of phenylmercury acetate with ethylmercury chloride. They contained 1–2% metallic mercury and were usually applied at the rate of 12 oz/cwt of seed. Precision drilling at smaller seed rates demands better control. In 1961 ethylmercury phosphate (EMP) steep-treatment was used for some and, since 1962, for all rubbed seed whether monogern or multigerm, and whether to be pelleted or not. The seed is soaked in a 40 ppm water solution of EMP (72·3% Hg) for 20 minutes, drained, dried and regraded. (See Byford, W. J. *Ann. appl. Biol.* (1963), 51, 41–49.)

In some countries, seed is treated with organic fungicides, such as thiram, as a protection against soil organisms, but this appears not to be necessary in Great Britain; experiments are in progress testing row treatments with soil fungicides applied as liquids or granules.

*Cercospora leaf spot** (*Cercospora beticola*) is a major disease in warmer climates. It occurs most years in Great Britain but never attains economic importance so control measures are not needed.

***Docking disorder** (see under Eelworms, p. 128).

**Downy mildew** (*Peronospora farinosa*). This disease is prevalent in some years in areas where a cycle of infection is maintained between seed crops and root crops. Contracts for growing seed crops specify minimum distance between stecklings, seed crops and root crops, and this separation helps to check the spread of the disease. Stecklings sown in summer and early autumn, especially those under mustard cover crops, are particularly susceptible. Spraying with fungicide decreases but seldom more than halves the incidence of the disease. Materials used are copper oxychloride (equivalent to 2 lb Cu/acre), maneb ($1\frac{1}{2}$ lb 80% w.p./acre) *or* zineb ($2\frac{1}{2}$ lb 65% w.p./acre). Fentin acetate (8 oz a.i./acre) has proved effective in experiments. Spraying should start shortly after seedling emergence and be continued at fortnightly intervals; the fungicide can be incorporated in the sprays of organophosphorus insecticides used against aphids. Although the disease occurs in root crops regularly, the acreage severely affected is usually small and routine spraying of crops is therefore unnecessary. Varieties differ in susceptibility (see N.I.A.B. Farmer's Leaflet No. 5). Plants infected in June may have their root yield more than halved but infection in September has small effect on yield. Infection in July depresses sugar content from 16% to 11%, for instance, but earlier or later infection depresses it less. The purity of the root juice is greatly decreased by the disease and this adversely affects sugar extraction in the factory.

**Powdery Mildew** (*Erysiphe* spp.) is common on foliage in the late summer of hot, dry years but it seems to do little damage and control in the field has not been attempted. The disease is sometimes troublesome on experimental plants in the glasshouse and is controlled by dusting the foliage with flowers of sulphur or spraying with dinocap.

**Ramularia leaf spot** (*Ramularia beticola*) occurs most years in the south-west where it defoliates the crop in a wet summer. In trials, fentin acetate, zineb or copper oxychloride sprays (rates as for downy mildew) in late summer gave some control and increased yield considerably; the problem is a local one in an area where few beet crops are grown and no recent experiments have been made.

Some sugar beet seed crops grown recently in dense stands in narrow rows in Oxfordshire and Kent have been defoliated in July by *Ramularia*. Sprays of fentin hydroxide at 0·6 lb a.i. (3 lb 20% w.p.)/acre have delayed defoliation in trials but have not affected yield or quality of seed.

**Rust** (*Uromyces betae*) shows as reddish brown pustules on the leaves of beet which may become numerous in late summer and kill some of the older leaves, but usually too late to cause appreciable loss of root yield, so no control measures are taken.

***Silvering** (*Corynebacterium betae*)—see under Beet (p. 280).

***Violet root rot** (*Helicobasidium purpureum*). This is common in suga beet but rarely causes serious loss. The roots of affected plants have a purplish fungus growth on the surface below which the root tissues decay. The fungus survives in the soil as resting sclerotia and grows on the roots of many crops and weeds. Soil sterilization with chemicals has not proved practical and control is by crop rotation, generous nitrogen fertilizer dressings, keeping the land free from weeds and deep and thorough cultivation.

*****Virus yellows** is the disease caused by beet yellows virus (BYV) and beet mild yellowing virus (BMYV), infection with either of which turns the leaves yellow and makes them more susceptible to fungal infection. The disease depresses yields and quality of sugar beet, mangolds and red beet, to an extent depending on how early in its development the plant gets infected. The viruses are spread to healthy plants by aphids that have previously fed on an infected plant. BYV persists in aphids for a few hours but, once infected with BMYV, aphids remain infective for most of their lives. The viruses are carried into young root crops in spring or early summer by winged aphids which infect a few, usually widely scattered, plants. Aphids then spread the disease from these centres of infection to give well defined circular patches of yellow plants. These patches may increase gradually in size or, where spread is not checked, the whole crop may rapidly become yellow.

The main field vector is *Myzus persicae*, but other aphids may spread the viruses, e.g. *Aphis fabae* can also spread BYV but not BMYV. Various species of green aphids are common on beet in the spring, and some are vectors of yellows viruses, but their main significance is to indicate that heavy infestations of *M. persicae* are likely to follow soon.

Control measures are aimed against clamped mangolds, seed crops and other host plants from which the viruses spread in spring and also against aphids spreading the disease within the root crop. Insecticides are used extensively to achieve both these objectives. Early sowing, a regularly spaced, dense population of plants and quick growth tend to decrease the incidence of yellows. Varieties better able to tolerate yellows viruses are now available commercially (see N.I.A.B. Farmers' Leaflet No. 5).

On average, about one-third of the remaining mangold clamps in late April are infested with aphids, and nearly all of these contain the mangold clamp aphid, *Rhopalosiphoninus staphyleae*, and about half contain *M. persicae*. Most of the mangolds are infected with yellows viruses. Winged aphids leave the clamps from mid-April onwards and carry the viruses acquired from the mangold shoots to young root crops. Their effects are often obvious later in the season by the greater incidence of yellows around clamp sites than elsewhere, but some winged aphids doubtless fly considerable distances and infect occasional plants over a wide area. Fumigation of infested clamps with methyl bromide controls aphids excellently, but only specialist contractors can do this. The clamp is covered with a 10/1,000 in PVC or similar gas proof sheet sealed to the ground with soil and the methyl bromide is vaporized into the clamp and retained there for a minimum of 3 hours. The dose of methyl bromide is adjusted to give a CT product of approximately 100 (i.e. concentration in $oz/1,000 \, ft^3 \times$ time in hours). The operators must take great care in handling methyl bromide and the clamp must be allowed to air for 24 hours before mangolds are fed to stock. In some European countries a proprietary dust containing a sprout depressant (tecnazene) and an insecticide (gamma-BHC) is dusted on to fodder beets as they are placed in the clamp. In the U.K. the dust only partially controlled sprouting of mangolds and aphids in clamps.

Aphid infestation in clamped mangolds is best avoided firstly by cutting away all leaves and leaf stalks when the mangolds are harvested, without cutting into the flesh of the root; secondly, by delaying covering the clamps until the roots have cooled, and removing the covering in March. Where practical, all roots should be fed to stock before the end of April.

Contracts between the British Sugar Corporation and the seed merchants, and between merchants and growers of sugar beet and mangold seed, specify measures to be taken to control pests and diseases of seed crops. The stecklings are inspected in the autumn and only those certified as having less than 1% of plants with yellows and as reasonably free from downy mildew are grown in the second year to produce seed.

Stecklings grown under cover crops, especially those sown in April under cereal, are unlikely to need protection with an insecticide until the cover crop is harvested; if there is any risk of aphids infecting the seedling beets, treating the seed with menazon at 5% by wt. of 80% dust is recommended. Stecklings grown in open beds for transplanting and direct sown stecklings without cover or grown on *in situ* must be protected with systemic insecticides from the time seedlings emerge until late October, and stecklings grown under cover need protection once the cover crop is harvested. The standard recommendation is frequent spraying, at least three times, with liquid formulations of such chemicals as demephion at 3·6 oz a.i. (12 fl. oz 30% e.c.)/acre *or* demeton-S-methyl at 3·4 oz a.i. (6 fl. oz 58% e.c.)/acre, *or* dimethoate at 4·8 oz a.i. (12 fl. oz 40% e.c.)/acre, *or* formothion at 7·5 oz a.i. (16 fl. oz 43% e.c.)/acre *or* menazon at 6 oz a.i. (15 fl. oz 40% e.c.)/acre *or* oxydemeton-methyl at 3·4 oz a.i. (6 fl. oz 57% e.c.) *or* phosphamidon at 3·2 oz a.i. (16 fl. oz 20% s.c.)/acre in 35 gal/ acre. Alternatively, any two spray applications except the first may be replaced by one application of disulfoton *or* phorate at approximately 16 oz a.i. (respectively 14 lb 7·5% grannules or 10 lb 10% granules)/acre or a split application; they should be concentrated in a band along the row so that the plants retain as many granules as possible. Adequate and timely protection of the very young seedlings is difficult by foliar application, since sprays will not persist and granules are not retained, but can be achieved by treating the seed with menazon as above. This treatment should control aphid infestation until at least the first pair of true leaves have expanded; protection afterwards can be by applying liquid or granule formulations, as detailed above. Such treatment to protect the stecklings from aphids and yellows is obligatory being specified in the contract between the British Sugar Corporation and the seed merchants.

Aphid infestations of the seed crop need checking to prevent winged ones leaving the crops and spreading the viruses to root crops and, if the infestation is heavy, to prevent damage to the seed crop by the aphids' feeding. This is achieved by spraying with demephion, or demeton-S-methyl or dimethoate or formothion or menazon or oxydemeton-methyl or phosphamidon, at the rates of a.i. recommended for the stecklings but in at least 50 gal/acre, just before the crop grows too tall for tractor mounted sprayers. Band-application to the foliage of disulfoton or phorate granules at a minumum of 24 oz a.i./acre is likely to give good and lasting control of aphids when applied just before the plants exceed 2 ft in height. Infestations that develop only after the crop is too tall to be sprayed by tractor-drawn machines can be sprayed from aircraft.

Removing the sources from which yellows viruses spread in spring will diminish the initial infection in root crops but inevitably some infective aphids will eventually introduce and spread the disease. This spread can be checked by well timed treatment with systemic insecticides. The time at which aphids invade root crops differs greatly in different years and different districts so seed treatment with menazon is not recommended as a routine. However, in the south-east of England and in other areas where aphids and yellows usually arrive early, seed treatment with menazon (5% by weight 80% dust) is justified, especially following a mild winter; partial aphid control persists into June, yellows incidence is decreased and sugar yield increased.

Timing of foliage treatments is important and the British Sugar Corporation warn growers by post card when the time seems opportune for treatment to start. The issue of warning cards is decided on the basis of local and general information about the development of aphid populations. In the areas where yellows is commonly prevalent an average population of about 0·25 wingless green aphid/plant justifies the warning cards, but the prevalence of yellows sources, aphid activity, stage of the crops, weather, etc., are also taken into consideration. Growers are advised to examine their own crops regularly and treat if green aphids occur on more than 1 in 4 plants on average. In early seasons aphids may infest the crop before singling. Spraying with any of the materials listed on p. 135 rapidly checks the aphid infestation and the insecticide remains effective for a week or more. Band spraying is effective with a suitable machine provided it is done carefully enough. A second spray 2–3 weeks later may pay, but a third one rarely does.

Disulfoton or phorate granules applied to the foliage kill aphids for longer, control yellow more effectively and increase yield more than do any spray materials. The granule applicator can be mounted on the steerage hoe, thus combining insecticide treatment with inter row cultivation; the granules should be directed, in a band the width of the plants, so that as many as possible are retained by the foliage. Use disulfoton or phorate at approximately 16 oz a.i./acre (respectively, 14 lb of 7·5% granules or 10 lb of 10% granules). The granules should be applied on receiving the British Sugar Corporation's spray warning or when green aphids are found on the crop and the dosage can be split between two applications. Granules should not be used on plants before singling, especially during dry weather.

# CHAPTER 8

# PEST AND DISEASE CONTROL IN GRASS AND FODDER CROPS

## PESTS

Grass and fodder crops grown for conservation or grazing are subject to attacks by a wide range of pests; the damage so caused is often of little or no economic significance because of the high plant population density. Even where pest damage is noticeable, expensive chemical control measures cannot usually be justified for crops of such low cash value. Under exceptional circumstances, e.g. drought, a fodder crop may be vitally important to the livestock farmer and chemical control measures must then be adopted to reduce pest damage. Chemical methods of control may also be used when the pest is the vector of an important virus disease.

A suitable period of time, depending on the insecticide used (e.g. minimum period of 2 weeks for DDT), should elapse between application and allowing animals to graze treated crops.

Herbage and fodder seed crops are additionally affected by pests attacking the inflorescence and developing seed. As the cash value of seed crops is often high, chemical control measures are more justifiable on economic grounds.

Pollinating insects, especially bees, are of prime importance to many brassica and some clover seed crops and also to field beans. Chemicals which are toxic to bees should not be used on crops in open flower, and their use (if unavoidable) should be restricted to dull, cloudy days and application made in early morning or late evening. Local beekeepers should be given adequate warning before such chemicals are used, so that hives can be closed temporarily or removed from the vicinity of the area to be treated.

## 8.1 CEREALS

RYE

*****Frit fly** (*Oscinella frit*)—see under cereal pests (p. 107).

   ***Stem eelworm** (*Ditylenchus dipsaci*)—see under cereal pests (p. 107).

MAIZE

*****Frit fly** (*Oscinella frit*)—see under cereal pests (p. 107).

   ***Cereal leaf aphid** (*Rhopalosiphum maidis*)—causes little direct damage but may transmit virus disease. The chemical control methods are as for grass aphids below.

## 8.2 GRASSES

Pests common to several grasses include:

**Antler moth** (*Cerapteryx graminis*), the larvae of which occasionally reach epidemic proportions in upland areas.

The control given by DDT spray at 1 lb a.i. (3 pints of 25% e.c.)/100 gal/acre may be helpful but is a tentative recommendation.

***Aphids.** To the bird-cherry aphid, the grain aphid, the grass aphid and rose-grain aphid, described on page 98, should be added the apple-grass aphid (*Rhopalosiphum insertum*). One or more of these species may cause serious damage, particularly to grass seed crops in warm summer months following mild winters.

Control is obtained by demeton-S-methyl at 3·4 oz a.i. (6 fl. oz 58% e.c.)/40 gal/acre *or* dimethoate at 5 oz a.i. (12 fl. oz 40% e.c.)/40 gal/acre. Other materials include demephion at 3·6 oz a.i. (12 fl. oz 30% e.c.)/40 gal/ acre *or* ethoate-methyl at 4·8 oz a.i. (24 fl. oz 20% e.c./20–100 gal)/acre *or* formothion at 6½ oz a.i. (16 fl. oz 40% conc.)/20–100 gal/acre *or* menazon at 4 oz a.i. (10 fl. oz 40% e.c.)/40 gal/acre *or* oxydemeton-methyl at 3·4 oz a.i. (6 fl. oz 57% e.c.)/40 gal/acre.

Demeton-s-methyl, demephion and oxydemeton-methyl are included in the Agriculture (Poisonous Substances) Regulations.

Burning the seed crop stubbles after harvest effects a degree of cultural control.

*Cereal leaf beetle** (*Lema melanopa*). The adults and larvae of this beetle eat longitudinal strips from the leaves of grasses and cereals, and transmit a virus disease of cocksfoot (see page 153). Control measures are unnecessary.

**Chafer beetles** (A.L. 235). Grubs of garden chafer and other species destroy grass roots; the turf is then easily rolled back. Fine leafed grasses are often preferentially attacked. For chemical and cultural control, see under potato pests (p. 113).

**Common leaf weevil** (*Phyllobius pyri*). The larvae feed on grass roots chiefly on light, sandy soils. Perennial ryegrass and fescues are chiefly affected, cocksfoot less so. Adult feeding on grass foliage is of little consequence.

Chemical control is given by DDT at 1 lb a.i. (3 pints 25% e.c.)/100 gal/acre applied in early April to sward and the hedge bottoms of fields damaged the previous summer.

Rolling of the sward repeatedly with heavy rollers when damage is first seen can give some control.

*Common rustic moth** (*Mesapamea secalis*). The caterpillars live and feed within the central tillers of grasses. Control measures are not advocated.

****Frit fly** (*Oscinella frit*) (A.L. 110). The larvae destroy the central shoots of ryegrasses, fescues and bents. They may affect both establishment and development of a young ley after an arable crop or direct reseeding, causing bare patches or allowing coarse grasses to thrive at the expense of sown grass species.

***Grass and cereal flies.** A number of flies lay their eggs on various grasses and the larvae destroy the central shoots. They include *Opomyza florum*, *O. germinationis*, *Cetema* spp., *Geomyza* spp. and *Meromyza saltatrix*. Control measures in grassland have not been advocated.

***Grass and cereal mite** (*Siteroptes* (*Pediculopsis*) *graminum*). This mite is associated with a fungus in causing 'silver top' of herbage seed crops, especially of timothy and cocksfoot. The burning of infested stubbles immediately after harvesting the seed crop serves as a control measure.

***Grass thrips.** *Aptinothrips rufus*, *A. stylifer* and other species, notably *Limothrips cerealium*, feed in both adult and nymphal stages on seed heads of many grasses. The effect on yield is insufficient to warrant control measures.

****Leatherjackets** (A.L. 179). Old grassland and newly established swards may be severely affected, with bare patches appearing in spring. Less severe attacks may delay the first flush of vegetative growth in spring, a matter of some consequence where 'early bite' is required for livestock.

Spray with DDT 1 lb a.i. (3 pints 25% e.c.)/100 gal/acre *or* gamma-BHC 4 oz a.i. ($\frac{1}{4}$ pint 80% conc)/40–100 gal/acre, harrowing the soil after treatment and before direct re-seeding or sowing the nurse crop. For attacks in young leys, surface baits of bran plus insecticide (see p. 99) may give rather better results.

***Marbled minor moth** (*Procus strigilis*). Though the caterpillars bore up the central shoots of several grasses, they are of no practical importance.

***Oat spiral mite** (*Steneotarsonemus spirifex*). The feeding of this mite on grasses may result in a twisted rachis, failure of the ear to emerge, blind spikelets and reduced grain. Control measures are not warranted.

***Rustic shoulder knot moth** (*Apamea sordens*). The caterpillars feed within central shoots of many grasses. Control measures are not necessary.

****Slugs** (A.L. 115). On heavy soils the grey field slug, *Agriolimax reticulatus*, may completely destroy newly established swards or herbage seed crops. For control measures, see under cereal pests (p. 99).

***Swift moths** (A.L. 160). The caterpillars occasionally cause serious damage to meadows and newly established swards; for control see under potato pests (p. 116).

***Wireworms** (A.L. 199). These pests cause little trouble to established swards but newly sown seedlings may be destroyed. Chemical control is based on insecticidal seed dressings or sprays—see under cereal pests (p. 101).

## COCKSFOOT

**Cocksfoot aphid** (*Hyalopteroides humilis*). Although often present in large numbers, this aphid causes little damage by its direct feeding; it may transmit cocksfoot streak virus. Control is rarely necessary but see under grass aphids above.

**Cocksfoot gall midges.** White larvae or dark brown puparia of *Mayetiola dactylidis* live within swollen stem bases, resulting in a rotting of leaf and stem tissue. Chemical control is unwarranted, for attacks are usually controlled by a high degree of parasitism. Larvae of *Contarinia dactylidis* feed on developing seed within the glumes, control measures are not required in this instance nor in the case of *Dasyneura dactylidis*, *Sitodiplosis dactylidis* or *Stenodiplosis geniculati*, the larvae of which also feed within cocksfoot seed heads.

**Cocksfoot moth** (*Glyphipteryx cramerella*). The larvae feed on the seed heads and pupate in the stem. Chemical control measures are unnecessary. Burning the stubble after harvesting seed crops gives some control.

**Red-legged earth mite** (*Penthaleus major*). This causes silvering and die-back of foliage, with patches of thin tufts of grass appearing in late autumn; the pest is very rare.

## FESCUES

**Gall midges.** The yellowish larvae of *Contarinia festucae* live in large numbers within flower heads while reddish larvae of *Dasyneura festucae* live singly within florets. Neither species is of economic importance.

## MEADOW FOXTAIL

**Gall midges.** Larvae of the three species *Contarinia merceri*, *Dasyneura alopecuri* and *Stenodiplosis geniculati* feed within the developing seed heads and seriously affect yield.

Control for *D. alopecuri* and *S. geniculati* whose larvae remain within infested seed consists in dry-heating the seed to 59–60°C for 35 minutes.

## MEADOW GRASSES

**Gall midges.** Larvae of *Mayetiola schoberi* live within rotting tillers of *Poa pratensis*. Larvae of *M. joannisi* similarly infest shoots of *Poa annua*. Chemical control has not been advocated against these or larvae of *M. poae* and *Caulomyia radifica*, also found within the stems. Larvae of *Contarinia poae* and *Sitodiplosis cambriensis* feed in the flowers of *Poa* *pratensis* and *P. trivialis*, thus preventing seed formation. Attacks are of no economic significance.

RYEGRASSES

***Gall midges.** Yellow larvae of *Contarinia lolii* feed on ovules but are of no practical importance to the seed crop.

TIMOTHY

***Timothy fly** (*Amaurosoma* spp.). The yellowish larvae feed on the developing flower head before it emerges from the leaf sheath. Although damaged heads are conspicuous within the seed crop, loss of seed in most years is slight. Sprays of DDT or parathion have given variable results because of the need for precise timing when applying sprays.

***Timothy tortrix moth** (*Amelia paleana*). This moth is sometimes prevalent in low-lying, marshy areas. The caterpillars web together the leaves and feed on leaf tissue and seed heads. Control measures are unnecessary.

## 8.3 LEGUMES (EXCLUDING FIELD BEANS)

Pests common to several leguminous crops include:

****Aphids.** Large numbers of the black bean aphid (*Aphis fabae*) (A.L. 54), pea aphid (*Acyrthosiphon pisum*), vetch aphid (*Megoura viciae*) or of *Aphis craccivora* may be present on several legume plant species and affect plant growth and vigour; in addition they may transmit certain virus diseases of leguminous crops. Their chemical control depends on methods given for aphids on field beans (see below). In addition, for pea aphid control only, use DDT at 1 lb a.i. (3 pints 25% e.c.)/100 gal/acre.

***Cabbage thrips** (*Thrips angusticeps*) feed on a wide variety of plants including lucerne and clovers, the foliage becoming speckled and growth retarded. Chemical control is rarely necessary; if so, use DDT spray at 1 lb a.i. (3 pints 25% e.c.)/100 gal/acre.

***Capsid bugs.** A number of species feed on the foliage and may destroy the flowers and developing seeds, but attacks are of no economic significance.

***Clover leaf weevils** (*Hypera variabilis, H. nigrirostris*). The adult weevils and larvae feed openly on foliage, buds and seed of clover, lucerne, sainfoin, trefoil and vetches; in some years they affect seed production of these crops. Badly attacked crops should be cut as late as possible to remove pupating larvae with the hay.

For chemical control, apply DDT spray at 1 lb a.i. (3 pints 25% e.c.)/ 20 gal/acre, 10–14 days after the first cut and before flower buds form.

****Clover seed weevils** (*Apion* spp.). Larvae feed on developing ovules and seeds, particularly of red clover. Adult feeding on the foliage is less important. For chemical control, spray with DDT at 1 lb a.i. (3 pints 25% e.c.)/20 gal/acre. Apply the sprays to the headlands in early May and 10–14 days after the first cut when flower buds begin to form.

***Leafhoppers** (*Empoasca* spp.). Speckling of the foliage may result from adult and nymphal feeding. A number of other species transmit clover phyllody virus and other closely related viruses. Special chemical controls are unnecessary, the insects being killed by insecticides applied against other pests.

****Leatherjackets** (*Tipula* spp.) (A.L. 179). Legume seedlings and established plants, especially of clovers and lucerne, may be seriously affected. For control measures, see under cereal pests (p. 98).

****Pea and bean weevils** (*Sitona* spp.) (A.L. 61). Adults feed on the foliage of a wide range of leguminous crops, the larvae feeding within the root nodules. Attacks are rarely important, but for control measures see under vegetable pests (p. 295).

****Slugs** (A.L. 115). The rasping of seedling legumes may occasionally be serious, but that of foliage of mature plants is of less consequence. The chief species involved is the grey field slug, *Agriolimax reticulatus*. Control measures are given under cereal pests (p. 99).

****Stem eelworm** (*Ditylenchus dipsaci*) (A.L. 409). A number of distinct races, often with overlapping host ranges, infest legume crops and may cause complete failures. Resistant varieties should be grown and over-cropping with susceptible legumes should be avoided.

Chemical control for clovers and lucerne is possible by the methyl bromide fumigation of seed, carried out by the seed merchant in special chambers. As this operation requires careful control and methyl bromide is extremely poisonous, it should not be used by the farmer.

Pests of individual legume crops:

**Gall midges.** Larvae of the following gall midges infest the leaflets, leaf axils, florets or pods of legumes. Only four are of economic importance, namely, *Dasyneura leguminicola*, *D. gentneri*, *D. viciae* and *Contarinia medicaginis*. Chemical control may be effected by DDT sprays at 1 lb a.i. (3 pints 25% e.c.)/100 gal/acre), applied before the flower heads are formed; optimum timing of the spray is often difficult. For cultural control, the crop should be cut early for hay or silage.

The clover flower midge (*Dasyneura gentneri*) is sometimes serious on red and white clover; the clover seed midge (*D. leguminicola*) may be serious on red clover, and the clover leaf midge (*D. trifolii*) may be found on both red and white clover. Less common are the lucerne flower midge (*Contarinia medicaginis*), the lucerne leaf midge (*Jaapiella medicaginis*) and the red clover gall gnat (*Campylomyza ormerodi*), the latter being found on red clover. Other midge pests are *Dasyneura ignorata*, on lucerne, and the sainfoin flower midge (*Contarinia onobrychidis*) and the sainfoin leaf midge

(*Bremiola onobrychidis*), both on sainfoin. The trefoil flower midge (*Contarinia loti*) is rare but the vetch leaf midge (*Dasyneura viciae*) may be serious on vetches.

## 8.4 FIELD BEANS

***Black bean aphid** (*Aphis fabae*) (A.L. 54). Although both winter-sown and spring-sown field beans are attacked by this aphid, the damage caused to winter-sown beans is only rarely important. In southern counties of England, spring-sown beans are colonized almost every year and whole field preventive treatments, applied as a routine measure at early flowering before aphid colonies build up and when wheel damage is minimal, are recommended as follows. Use disulfoton at 16·8 oz a.i. (14 lb of 7½% granules)/acre *or* phorate at 1 lb a.i. (10 lb of 10% granules)/acre *or* demeton-S-methyl at 3·4 oz a.i. (6 fl. oz 58% e.c.)/40 gal/acre *or* menazon at 4 oz a.i. (10 fl. oz 40% e.c.)/40 gal/acre. A suitable applicator is required for granule treatments, but these are preferred because they are virtually non-toxic to bees. A number of other aphicides including dimethoate, formothion, malathion, mevinphos, nicotine, oxydemeton-methyl, phosphamidon and schradan give useful control of black bean aphid, but are more toxic to bees than the materials recommended above.

In midland, eastern and northern counties spring-sown field beans are less frequently attacked and routine measures are not justified. In such areas, and in the south on fields where preventive treatments have not been used, eradicant treatments may be required in years when large aphid colonies develop. Use disulfoton at 12 oz a.i. (10 lb 7½% granules)/acre *or* phorate at 1 lb a.i. (10 lb 10% granules)/acre *or* demeton-S-methyl at 3·4 oz a.i. (6 fl. oz 58% e.c.)/40 gal/acre *or* oxydemeton-methyl at 3·4 oz a.i. (6 fl. oz 57% e.c.)/40 gal/acre. As preventive treatments are likely to be applied when the crop is in flower, the risk of chemical treatments killing bees is greater, as is the amount of wheel damage likely to be incurred. Narrow wheeled high-clearance granular applicators are thus preferable for preventive treatments.

In most midland and eastern counties, routine preventive treatments of disulfoton, phorate, demeton-S-methyl, oxydemeton-methyl *or* menazon at the rates given above have been successfully applied to the headlands alone. Such headland treatment may have to be followed with whole field preventive treatment in years when *A. fabae* attacks are heavy.

***Green aphids.** The vetch aphid (*Megoura viciae*) and pea aphid (*Acyrthosiphon pisum*) colonize field beans, but are important only in their role as vectors of bean viruses. The preventive chemical control measures applied against black bean aphid (*q.v.*) serve to control green aphid infestations as well.

****Pea and bean weevils** (*Sitona* spp.) (A.L. 61). Although leaf margins are often notched by adult weevil feeding, the damage is of little consequence. Control measures if necessary can be obtained under vegetable pests (p. 295). Materials applied for aphid control (e.g. phorate) may also control weevil attacks.

## 8.5 CRUCIFEROUS CROPS

**APHIDS

**Mealy cabbage aphid** (*Brevicoryne brassicae*) (A.L. 269). This aphid colonizes most cruciferous crops and may cause serious leaf-curl especially in swedes. Its effect on brassica seed crop yield is unknown; in these crops chemical control (see under vegetable pests, p. 280) during July and August can be effected only by aerial means.

**Peach-potato aphid** (*Myzus persicae*) overwinters on many brassica crops and may damage the foliage. It transmits certain virus diseases to healthy plants. For chemical control, see under vegetable pests (p. 111).

***BEETLES

**Blossom** (= **Pollen**) **beetles** (*Meligethes* spp.). A serious pest of brassica seed crops, especially of those with a long flowering period, for both adults and larvae damage flower buds and thus reduce the number of pods set.

Chemical control for winter rape—use DDT spray at 1 lb a.i. (3 pints 25% e.c.)/20–100 gal/acre *or* DDT dust at 2 lb a.i./acre (40 lb/acre 5% dust) *or* gamma-BHC spray at 6 oz a.i. (3 pints 10% e.c.)/20–100 gal/acre, the treatment to be applied before flowering when more than 20 beetles per plant are present.

For spring rape, trowse mustard and white mustard use DDT spray *or* dust *or* gamma-BHC spray as strengths quoted above *or* malathion spray at 18 oz a.i. ($1\frac{1}{2}$ pints 60% e.c.)/30–100 gal/acre *or* azinphos-methyl spray at 2·2 oz a.i. (10 fl. oz 22% conc.)/100 gal/acre. Two treatments may be necessary, one at early green bud stage if more than 5 beetles per plant are present, and the second at yellow bud stage if more than 10 beetles per plant are then present.

To reduce the risks of off-flavour to root crops for human consumption, avoid growing these within 2 years of applying gamma-BHC sprays.

Damage to tall crops may be reduced by using high-clearance machines or those fitted with large-diameter land wheels with narrow treads. For very tall seed crops, ground spraying is impracticable. The spray boom should be adjustable to clear the crop by 12–18 inches and low tank pressures with fan-type nozzles (30–50 lb/in$^2$) should be used.

Both malathion and azinphos-methyl are very toxic to bees and should not be used on flowering crops, and in any case it should not be necessary

L

to treat against blossom beetles once the crop has begun flowering. Treatments should be applied in early morning or late evening to reduce hazards to bees.

Early drilling for both overwintered and spring-sown seed crops ensures a vigorous plant able to compensate partly for loss of buds.

****Cabbage stem flea beetle** (*Psylliodes chrysocephala*). The larvae tunnel within leaf stalks and stems of overwintering brassica plants but the adult beetles are of no importance. Control may be obtained with a gamma-BHC spray at 6 oz a.i. (3 pints 10% e.c.)/20–100 gal/acre, applied in October–November. If more than one larval mine occurs per 3 in of plant height, spraying on coleseed may be worthwhile. If gamma-BHC is used do not plant potatoes or carrots within 2 years, so reducing the risk of off-flavour.

Overwintering brassica crops should be grown as far as possible from the site of the previous year's crops.

***Chafer beetles** (A.L. 235). The grubs occasionally attack cruciferous crops planted in newly ploughed grassland; for control, see under potato pests (p. 113).

*****Flea beetles** (*Phyllotreta* spp.) (A.L. 109). The cotyledons and stems of fodder and seed crops are holed and often destroyed by adult beetles during April and May. Damage is accentuated in dry seasons on spring-sown crops. Larval feeding at the roots is unimportant. For control use a gamma-BHC/fungicide seed dressing at ½–⅔ oz of 50% material/lb of seed. If the attack is prolonged, these measures should be supplemented by DDT dust at 2 lb a.i. (40 lb 5% dust)/acre *or* DDT spray at 1 lb a.i. (3 pints 25% e.c.)/20 gal/acre *or* gamma-BHC dust at 6 oz a.i. (80 lb 0·5% dust)/acre. Because of risk of imparting off-flavour, potatoes or carrots should not be planted within 2 years of treating fields with gamma-BHC.

Sow spring crops as early as possible on a fine tilth, and provide adequate fertilizer in the seedbed.

***Mustard beetle** (*Phaedon cochleariae*). Both adults and larvae occasionally affect seed crops of mustard, turnips, rape and swedes. The second generation attack in August–September is of chief importance. For chemical control, use DDT dust at 2 lb a.i. (40 lb 5% dust)/acre *or* DDT spray at 1 lb a.i. (3 pints 25% e.c.)/20 gal/acre *or* gamma-BHC dust at 0·1 lb a.i. (40 lb 0·26% dust)/acre. Apply in May and repeat later if necessary.

As gamma-BHC may impart off-flavour to potatoes and carrots, these should not be grown within 2 years of treatment.

For cultural control, burn stubbles of seed crops and remove any cruciferous weeds; clean hedge bottoms and dykes bordering infested fields.

*****Seed weevil** (*Ceutorhynchus assimilis*) is an important pest of brassica seed crops except white mustard. The adults lay eggs in the young pods and each grub eats several seeds before boring exit holes in the pod wall during June and July.

For chemical control on winter rape, two or more weevils per plant justifies spraying from first open flower onwards with gamma-BHC spray at 6 oz a.i. (3 pints 10% e.c.)/20–100 gal/acre *or* malathion spray at 18 oz a.i. (1½ pints 60% e.c.)/30–100 gal/acre *or* azinphos-methyl spray at 2·2 oz a.i. (10 fl. oz 22% conc.)/100 gal/acre. As one or two sprays over whole fields need careful timing, and as adult weevils invade fields at different times depending on weather conditions, it may be preferable to apply a series of headland sprays rather than whole field treatments.

For spring rape, trowse mustard and other brassica seed crops (excluding white mustard), chemical control is justified when one or more weevils per plant are present just before the first flowers open. Sprays of gamma-BHC, malathion *or* azinphos-methyl as outlined above are applied at the late yellow bud stage.

Best control of seed weevil damage is obtained by spraying crops in flower, but the hazard to bees from using materials such as malathion *or* azinphos-methyl is then greatly increased. Phosalone has given promising results when used experimentally on crops in full flower, with little harm to bees. Endosulfan has also been used with success, but does not kill seed weevils at temperatures below about 18°C (64·4°F).

The precautions on spraying machinery and on tainting by gamma-BHC given under blossom beetles should be observed, as should those designed to reduce toxic hazards to bees. While self-pollinated types of rape (chiefly winter rape) and trowse mustard do not require bees for pollination, seed yields of other brassicas, especially spring rape, Brussels sprouts and most *B. oleracea*, are improved when bees are present in large numbers.

***Stem weevil** (*Ceutorhynchus quadridens*) is an important pest of brassica seed crops, especially of spring-sown ones; older hollow-stemmed varieties of white mustard are less affected than modern short-stemmed ones. The larvae tunnel through leaf-stalks and into stems during May and June, affecting plant vigour and seed yield. Adult weevils feed on the leaves without serious effect.

A partial control is effected (*a*) by the gamma-BHC seed dressing used against flea beetles (see above) using a high rate of 1 oz of 50% material/lb of seed with a kerosene sticker *or* (*b*) by the spray programme directed primarily against blossom beetles and seed weevil.

**Turnip gall weevil** (*Ceutorhynchus pleurostigma*) (A.L. 196). The larvae live within rounded root galls, and root decay following attack may be important in the seedling stage of mustard and other brassica seed crops. Control is not normally required for fodder or seed crops.

**Wireworms** (A.L. 199). Many brassica crops are highly resistant to wireworm attack and can be safely grown after ploughing infested grassland without using chemical control measures.

## **BUTTERFLIES AND MOTHS

Caterpillars of the following insects feed on a wide range of cruciferous fodder and seed crops, though attacks are rarely serious enough to warrant chemical control (see under vegetable pests, Chapter 11).

Beet webworm (*Loxostege sticticalis*), **Cabbage moth (*Mamestra brassicae*) (A.L. 69), **Cutworms (A.L. 225), **Diamond-back moth (*Plutella xylostella*) (A.L. 195), Garden pebble moth (*Mesographe forficalis*), **Green-veined white butterfly (*Pieris napi*), Large white butterfly (*Pieris brassicae*), Small white butterfly (*Pieris rapae*), all described in A.L. 69, and Swith moths (*Hepialis* spp.) (A.L. 160).

## **FLIES

**Cabbage root fly** (*Erioischia brassicae*) (A.L. 18). This fly is locally troublesome on fodder and seed crops, but special control measures are not normally adopted (see also under vegetable pests, p. 281).

## GALL MIDGES

Of the species whose larvae infest cruciferous plants, the following are of some importance:

**Brassica pod midge** (*Dasyneura brassicae*) lays its eggs in pods damaged by feeding or oviposition of seed weevil (*q.v.*) or other agencies. The midge larvae cause attacked pods to ripen prematurely and shed their seed during July and August. Overwintered seed crops are worst affected. The measures recommended for the control of the seed weevil (p. 147) effectively control midge attacks.

**Swede midge** (*Contarinia nasturtii*) attacks cruciferous fodder crops especially in the south-western and northern counties of the U.K. Its yellow-white larvae are also occasionally found in seed crops.

## LEAF MINERS

*Cabbage leaf miner** (*Phytomyza rufipes*). The larvae tunnel within leaf midribs and stems of rape and other brassica plants. Chemical control measures are rarely if ever necessary. The larvae of *Scaptomyza apicalis* form bloth mines in cruciferous plants, but control measures are unnecessary.

*Leatherjackets** (*Tipula* spp.) (A.L. 179). The larvae may cut off the stems of plants but are only occasionally troublesome in fodder or seed crops, and then at the seedling stage. If control measures are required, see under potato pests (page 115).

***Turnip root fly** (*Erioischia floralis*). Larval damage is occasionally important on turnip and swede fodder crops in Scotland and northern England. The control measures given for cabbage root fly (p. 280) are effective.

## DISEASES

The problems of disease in grass and fodder crops are of a different nature to those of most other crops. These crops may be grown in mixed swards or in pure stands, and with the grasses and legumes, compensation by unaffected plants will often obscure the effects of the disease. Only occasionally will conditions be especially favourable to the disease and severe loss be recognized; in fact the extent of damage caused by these diseases is largely unknown. Control is primarily by management, and very rarely are direct control techniques involving fungicides economically justified, and then usually as seed dressings.

## 8.6 CEREALS

**Rye** is subject to most of the cereal diseases (see Chapter 5) but, when grown for early bit or ensilage, is little affected. No control measures are normally adopted apart from the use of an organomercury seed dressing (containing equivalent of 1–1·5% mercury), on rye to control ***Stripe smut** (*Urocystis occulta*) at the rate of 0·02 oz a.i./bushel of seed (2 oz of material).

### Maize

***Damping off** (*Pythium* spp.). The use of a seed dressing of captan at the rate of 2 oz a.i. (2½ oz 75% material), *or* thiram at 2 oz a.i. (4 oz material)/cwt of seed is officially approved.

***Smut** (*Ustilago maydis*) can be troublesome on land which has been cropped with maize for a number of years. It is considered useful to dress maize seed with an organomercury seed dressing (1–1·5% mercury) at the rate of 4 oz/cwt to prevent the introduction of this disease. This dressing will also control pre-emergence damping off.

***Take all** (*Ophiobolus graminis*). Maize is susceptible to root infection by the take-all fungus, and infected maize root fragments remaining in the soil will infect wheat after 6 months.

## 8.7 GRASSES

The grower may be confronted by any of the following diseases:

**Ryegrass.** **Mildew (*Erysiphe graminis*), *foot rot and leaf spot (*Drechslera siccans*), *net blotch (*Drechslera dictyoides*), *leaf blotch

(*Rhynchosporium orthosporum*), **crown rust (*Puccinia coronata*), *brown rust (*P. dispersa*), **ergot (*Claviceps purpurea*) and **blind seed disease (*Gloeotinia temulenta*).

**Cocksfoot.** **Mildew (*Erysiphe graminis*), **leaf fleck (*Mastigosporium rubricosum*), *black leaf spot (*Phyllachora graminis*), *leaf blotch (*Rhynchosporium orthosporum*), **halo spot (*Selenophoma donacis*), **yellow rust (*Puccinia glumarum*), *stem rust (*P. graminis*), *rust (*Uromyces dactylidis*), **ergot (*Claviceps purpurea*), ***choke (*Epichloe typhina*), take-all (*Ophiobolus graminis*) and *yellow slime (*Corynebacterium rathayi*).

**Fescues.** **Mildew (*Erysiphe graminis*), *leaf spot (*Drechslera siccans*), *black leaf spot (*Phyllachora sylvatica*), **crown rust (*Puccinia coronata festucae*), *stem rust (*P. graminis*), *rust (*Uromyces dactylidis*), **ergot (*Claviceps purpurea*), **choke (*Epichloe typhina*) and *halo spot (*Pseudomonas coronafaciens*).

**Timothy.** **Leaf spot (*Mastigosporium cylindricum*), *leaf spot (*Heterosporium phlei*), *leaf blight (*Drechslera dictyoides phlei*), *leaf blotch (*Rhynchosporium secalis*), *halo spot (*Selenophoma donacis*), *blotch and char spot (*Septogloeum oxysporum*), **stem rust (*Puccinia graminis phlei-pratensis*), **ergot (*Claviceps purpurea*) and **choke (*Epichloe typhina*).

The multiplicity of the foliage diseases is evident from the lists given under the different hosts, but they can be considered under the headings of Mildew, Rust and various Leaf Spotting diseases.

**Mildew** (*Erysiphe graminis*) is destructive under certain conditions, particularly after a dry period, and gives a white powdery appearance to the upper sides of leaves. The fungus exists in many strains and attacks most of the grasses, though there appears to be no record of it attacking timothy. Potash manures will increase resistance, and the disease is worst under conditions of high nitrogen.

**Rusts.** Ryegrass can be severely attacked by crown rust (*Puccinia coronata*) in warm summers, and the rust can be particularly damaging to the aftermath, rendering it unpalatable to stock. It has been noted that leys under intensive production receiving high applications (400 + units) of nitrogen are less severely affected. Ryegrass can also be attacked by brown rust (*P. dispersa*). Timothy is frequently attacked by a stem rust, *Puccinia graminis phlei-pratensis*, which can be destructive to the aftermath in hay meadows, affects seed production and can also be responsible for the failure of newly-sown leys.

Cocksfoot is attacked by four rusts. The most destructive is yellow rust (*Puccinia glumarum*) which forms yellow stripes on leaves and sheaths and develops inside the glumes where it can severely affect the yield and quality of the seed. This rust is common on wheat and barley, but the cocksfoot strain does not attack cereals. The timothy stem rust can also cause damage to cocksfoot and another variety of stem rust, *P. graminis avenae* from oats, has been recorded in Scotland on cocksfoot. A fourth rust, *Uromyces dactylidis*, is common on cocksfoot from midsummer onwards.

Fescues are attacked by a specialized form of crown rust, *P. coronata festucae*, but in Wales the commonest rust on these grasses is a form of the stem rust of wheat, *P. graminis*.

## LEAF SPOTTING DISEASES

Leaf fleck (*Mastigosporium rubricosum*) can cause quite severe damage to cocksfoot in the autumn and early spring months. Another species, *M. cylindricum*, is recorded on timothy but does not cause much damage.

The leaf blotch disease of barley, *Rhynchosporium secalis*, has been found attacking timothy and other grasses but these have been shown to be distinct strains from those found on cereals. Another species, *R. orthosporum*, causes trouble in cocksfoot and Italian ryegrass.

*Selenophoma donacis* causes a halo spot on timothy, but is more damaging to cocksfoot, particularly seed crops, which are attacked at heading time in wet seasons, and yields are reduced.

*Drechslera siccans* causes a leaf spot of ryegrass and meadow fescue, whilst a related species *D. dictyoides* attacks ryegrass. A variety of the latter species has been described causing damage to timothy in Scotland. Another leaf attacking fungus is *Phyllachora graminis* which produces shiny black stromata on the leaves of cocksfoot in winter, and a related species *P. sylvatica* is quite commonly recorded on fescues. Timothy is also attacked by two other fungi, *Heterosporium phlei* causing a leaf spot, and *Septogloeum oxysporum*, responsible for a blotch and charspot.

## INFLORESCENCE DISEASES

**Ergot** (*Claviceps purpurea*) (A.L. 548), in which the grass seed is replaced by a fungal sclerotium, attacks most grasses. The ergots can cause gangrene and lameness in cattle or sheep, and are sometimes blamed for causing abortion, but it is very probable that abortion is seldom, if ever, a true symptom of ergot poisoning. It is most common in old pastures, and where attacks have been serious, these areas should be mown before coming into flower. When such areas are reseeded, the ground should be deeply ploughed, so that the ergots are buried more than 10 inches deep and will be unable to produce their fruiting bodies above the ground. Seed should not be taken from affected crops to avoid disseminating ergots with the seed.

**Choke** (*Epichloe typhina*). The inflorescence is converted into a fungal stroma, first white then turning yellow, and results in complete blindness and no flowers or seed are produced. Most of the grasses are attacked, though there appears to be no record for ryegrass. It is most important on cocksfoot seed stands which may be rendered unproductive after the third harvest year. It has been recorded on timothy seed stands but not to the same extent.

**Blind seed disease of ryegrass** (*Gloeotinia temulenta*) is usually recognized by poor germination figures for the seed. The latter is rotted internally though this may not be evident from outward appearance. The disease has been found throughout the country, but particularly in the wetter parts, and can be serious in Northern Ireland. Control consists mainly in not taking seed from affected crops, as infected seed sown normally will produce the fruiting bodies of the fungus, which by means of aerial spores infect the flowering heads of the new crop. Valuable nuclear stocks can be cleared of infection by immersing the seed for 20 minutes in a 0·25% solution of an organomercury compound (containing the equivalent of 1% Hg) held at 50°C.

*Yellow slime disease of cocksfoot* (*Corynebacterium rathayi*). This disease has been recorded a few times in this country, and is sometimes found in imported seed. If infected seed is sown, bare and stunted patches are caused, with failure to tiller; finally the leaves and panicles are coated with a yellow slime, and the seed becomes infected. Disinfection of the seed is not normally practised for, under the certification scheme, seed is only taken from healthy crops.


ROOT DISEASES

Take-all disease (*Ophiobolus graminis*) of cereals is known to affect grasses, but not generally to any damaging extent. However, grasses carry over take-all infection to a subsequent cereal crop; in this connection, it is the stoloniferous weedy grasses which are most important and the main herbage grasses are not considered to play a serious role.


SEEDLING DISEASES

Damping-off fungi (*Pythium* spp., *Fusarium* spp. and species of *Drechslera*) can cause pre-emergence and seedling losses. For this reason manufacturers recommend for the direct seeding of ley grasses, the use of a combined mercury/gamma-BHC seed dressing at the rate of 3 oz per acre, for seeding rates of over 20 lb/acre, to combat damping-off and wireworm damage. Field trials have sometimes shown benefit from this treatment, particularly with meadow fescue, but the final stands were indistinguishable from those from untreated seeds. The fungicidal dressing of ley grass seed is not normally practised, but the use of a captan seed dressing (75%), used at the rate of 2 oz/28 lb seed, is recommended for lawn grasses.

***Virus Diseases.** Barley yellow dwarf virus exists in several strains and can be found attacking most of the grasses which are tolerant to this virus. A recent random survey of S.24 Perennial Ryegrass fields showed 90% to be infected, mostly without symptoms, and this must serve as a reservoir of infection available to infect cereals. The virus is transmitted by several species of aphids, and is of the semi-persistent type and may therefore be controlled by insecticides though these are not normally used.

Another aphid-borne virus is cocksfoot streak, which is common in leys. Cocksfoot is also susceptible to Cocksfoot mottle virus, which can cause serious losses in seed crops. This has recently been shown to be transmitted by both adults and larvae of the cereal leaf beetle, *Lema melanopa*, and can also be transmitted by mechanical contact.

Ryegrass can be extensively infected with Ryegrass mosaic virus, transmitted by mites, and this can materially affect production. A recently discovered virus complex including Ryegrass mosaic virus has been found severely affecting S.22 Italian seed stands.

## 8.8 LEGUMINOUS CROPS

Of fungi causing disease in these crops, the following merit mention:

**Clover.** *Leaf spot (*Ascochyta trifolii*), *ring spot (*Pleospora herbarum*), **leaf spot (*Pseudopeziza trifolii*), *leaf spot (*Stagonospora meliloti*), *leaf spot (*Cercospora zebrina*), *mid-vein spot (*Mycosphaerella carinthiaca*), *black blotch (*Cymadothea trifolii*), **mildew (*Erysiphe polygoni*), *downy mildew (*Peronospora trifoliorum*), *rust (*Uromyces trifolii*), *white clover rust (*U. flectens*), **scorch (*Kabatiella caulivora*), *white clover burn (*Sphaerulina trifolii*), *black stem (*Ascochyta imperfecta*), ***clover rot (*Sclerotinia trifoliorum*), *violet root rot (*Helicobasidium purpureum*), *black root rot (*Thielaviopsis basicola*), *root rot (*Corticium solani*), *grey mould (*Botrytis cinerea*), *Anther mould (*Botrytis anthophila*), *wilt (*Verticillium dahliae*) and *bacterial leaf spot (*Pseudomonas syringae*).

**Trefoil.** *Leaf spot (*Pseudopeziza medicaginis*), *ring spot (*Pleospora herbarum*), *downy mildew (*Peronospora trifoliorum*), *rust (*Uromyces striatus*), **clover rot (*Sclerotinia trifoliorum*), *Violet root rot (*Helicobasidium purpureum*), *black root rot (*Thielaviopsis basicola*), **black stem (*Ascochyta imperfecta*).

**Lucerne.** **Leaf spot (*Pseudopeziza medicaginis*), *ring spot (*Pleospora herbarum*), *burn (*Sphaerulina trifolii*), *downy mildew (*Peronospora trifoliorum*), *rust (*Uromyces striatus*), **anthracnose (*Colletotrichum trifolii*), **black stem (*Ascochyta imperfecta*), **clover rot (*Sclerotinia trifoliorum*), **crown wart (*Urophlyctis alfalfae*), *violet root rot (*Helicobasidium purpureum*), *root rot (*Corticium solani*) and ***wilt (*Verticillium albo-atrum*).

**Sainfoin.** *Leaf spot (*Ascochyta orobi*), *leaf spot (*Ramularia onobrchidis*), *leaf spot (*Septoria orobina*), *ring spot (*Pleospora herbarum*), **mildew (*Erysiphe polygoni*), **clover rot (*Sclerotinia trifoliorum*), *black root rot (*Thielaviopsis basicola*), *grey mould (*Botrytis cinerea*), *rust (*Uromyces onobrychidis*) and *wilt (*Verticillium dahliae*).

**Tares or Vetches.** *Chocolate spot (*Botrytis fabae*), **downy mildew (*Peronospora viciae*), *rust (*Uromyces fabae*) and *clover rot (*Sclerotinia trifoliorum*).

The extent of loss caused by the many foliage diseases is largely unknown and in many instances obscured by compensation. No fungicidal spraying is normally practised. The commonest leaf spot disease is caused by two species of *Pseudopeziza* which are present in most crops of red clover and lucerne, and can attack White Clover, trefoil and other clovers. They can be quite destructive in wet autumns. In lucerne, trials have shown that an earlier second cutting can reduce leaf and protein losses and retard disease outbreak by 1 week, whilst still allowing sufficient growth after the first cut. **Black blotch** (*Cymadothea trifolii*) may be found causing extensive infection in both red and white clover in late summer, but seldom causes serious damage. **Mildew** (*Erysiphe polygoni*) may cause extensive infection in red clover, and sainfoin is sometimes severely attacked. Other leaf diseases including downy mildew, rusts and various leaf spotting fungi occur, but are seldom important.

STEM DISEASES

**Black stem** (*Ascochyta imperfecta*) is very widely distributed in Britain in these crops. It is seed-borne and attacks clovers, trefoil and lucerne, causing cankering and death of shoots. It is most severe in the early months and may affect the first cut of lucerne. Organomercury seed dressings are known to be effective against the disease, but are not normally used in this country. The disease has been shown to be completely controlled on trefoil seed by soaking the seed in 0·2% thiram solution held at 30°C for 24 hours.

**Red clover scorch** (*Kabatiella caulivora*) may cause the death of leaves and a shrivelling of flowers by forming lesions which girdle leaf and flower stalks. Temporary severe loss of foliage, and in a seed crop, a total failure of seed production may occur.

**Anthracnose** (*Colletrotrichum trifolii*) causes wilting and death of stems by forming a girdling lesion at the base of the shoots of lucerne, and can cause temporary loss of foliage.

**Clover rot** (*Sclerotinia trifoliorum*) (A.L. 266) can be serious, particularly in red clover when a short rotation is followed. Infection occurs in October from spores produced by the germinating resting bodies (sclerotia) in the ground. In mild seasons of high humidity the disease spreads killing large patches of plants, and severe loss of stand may be evident at the end of the winter. Red clover is most susceptible, but all these crops may be affected. Trefoil was once thought to be resistant, but is sometimes more severely attacked than red clover in the same field, and it is thought there may be a special strain which attacks trefoil. Lucerne is generally resistant, but the disease can cause temporary loss of crop in the early summer months. Susceptible crops should not be grown more than once in 8 years, and care should be taken to secure seed from uncontaminated sources, as the sclerotia can be disseminated with the seed. Grazing the crops off in the autumn is beneficial, as it reduces the amount of foliage available for infection, and reduces conditions of high humidity within the crop.

****Crown wart of lucerne** (*Urophlyctis alfalfae*) produces characteristic warty galls at the crown of the plant, and may lead to wilting in hot weather and loss of yield. This disease is invariably associated with poor drainage, and is often found in fields which have been 'poached' by winter grazing of cattle. Control is best effected by attending to drainage and lengthening the rotation.

*****Verticillium wilt** (*Verticillium albo-atrum*) has caused most trouble in lucerne in recent years and can infect the ground for a considerable, and as yet undetermined, number of years. The symptoms are a yellowing and wilting of the leaves and stems, followed by a shrivelling of the whole plant from the base up. Symptoms appear after the first cut and increase throughout the season. Regrowth may be stunted with shortened internodes, and eventually the whole plant may be killed and lucerne stands rendered quite worthless in their third year. A recent survey showed that in one-third of infected fields, the disease had most probably been introduced with the seed. Experiments in the U.K. have shown that the use of a (50%) thiram seed dressing at the rate of 8 oz a.i./cwt of seed appears promising. The dressing may be applied by the merchant prior to bacterial inoculation by the farmer before sowing. Although field information is lacking, it is recommended that all lucerne seed should be so dressed before sowing in unaffected land, to prevent the possible introduction of this troublesome disease, and this is now officially approved.


VIRUS DISEASES

****Phyllody** of clover, in which the inflorescence is converted to a green vegetative and sterile condition, can cause considerable losses in white clover seed crops. It is transmitted by leafhoppers (see p. 143) and can also be carried to strawberries in which it produces the troublesome green petal virus disease.

A recent survey of white clover in permanent leys revealed the presence of seven different viruses.

## 8.9 FIELD BEANS (*Vicia faba*)

The following are the more important diseases:

***Stem Rot (*Ascochyta fabae*), ***Chocolate spot (*Botrytis fabae*) *Leaf Spot (*Cercospora zonata*), *Downy Mildew (*Peronospora viciae*), *Net Blotch (*Pleospora herbarum*), *Rust (*Uromyces fabae*), **Sclerotinia Rot (*Sclerotinia trifoliorum* v *fabae*), *Root Rot (*Thielaviopsis basicola* and *Corticium solani*), **Damping off (*Pythium* spp.).

***Stem Rot** (*Ascochyta fabae*). The increasing popularity of beans as a break crop has brought this disease into prominence, and it can reduce yields to less than a ton per acre. The disease is seed borne, and no seed treatment is effective. Only seed free from disease should be sown, and farmers can have their seed tested for the presence of the disease by the Official Seed Testing Centre, Cambridge. Volunteer plants from previous infected crops in adjoining fields also carry over the disease, and these should be eliminated.

***Chocolate Spot** (*Botrytis fabae*). In wet seasons favourable to this disease, winter sown beans can be very seriously affected. Spring sown beans are not so severely attacked. Infection is carried over on debris of previous crops in adjoining fields, on self-sown seeds and occasionally on seed. An inadequate supply of potash may aggravate the incidence of the disease.

***Damping off** (*Pythium* spp.). The use of captan and thiram seed dressings to combat this trouble is officially approved for bean seed—2 oz a.i. captan ($2\frac{1}{2}$ oz 75% dust) *or* 2 oz a.i. thiram (4 oz 50% dust)/cwt of seed.

**Sclerotinia rot** (*S. trifoliorum* v *fabae*). This is a specialized form of the clover rot fungus, and follows too close cropping with beans, and the failure to eradicate ground keepers. The common form of clover rot has been found also on beans which should not be included in the break to control clover rot in red clover.

## 8.10 BRASSICA CROPS

The following are the more important diseases:

**Kale.** **Dark leaf spot (*Alternaria brassicicola* and *A. brassicae*), *leaf spot (*Ascochyta brassicae*), *grey mould (*Botrytis cinerea*), *wire stem (*Corticium solani*), *white blister (*Cystopus candidus*), *Mildew (*Erysiphe polygoni*), *light leaf spot (*Gloeosporium concentricum*), *downy mildew (*Peronospora parasitica*), *ring spot (*Mycosphaerella brassicicola*), *canker (*Phoma lingam*), **root rot (*Phytophthora megasperma*), **club root (*Plasmodiophora brassicae*), *sclerotinia rot (*Sclerotinia sclerotiorum*) *black rot (*Xanthomonas campestris*) and **soft rot (*Pectobacterium carotovorum*).

**Rape.** *Dark leaf spot (*Alternaria brassicae*), *damping off (*Phytophthora cryptogea*), *damping off (*Corticium solani*), *damping off (*Pythium* spp.) *mildew (*Erysiphe polygoni*), *downy mildew (*Peronospora parasitica*) and **club root (*Plasmodiophora brassicae*).

**Turnip and Swede.** *Leaf spot (*Alternaria brassicae*), *leaf spot (*Ascochyta brassicae*), *root rot (*Botrytis cinerea*), *white spot (*Cercosporella brassicae*), *white blister (*Cystopus candidus*), *mildew (*Erysiphe polygoni*), **downy mildew (*Peronospora parasitica*), *light leaf spot (*Gloeosporium concentricum*), ***club root (*Plasmodiophora brassicae*), **dry rot and canker (*Phoma lingam*), *sclerotinia rot (*Sclerotinia sclerotiorum*), *violet root rot (*Helicobasidium purpureum*), *damping off (*Pythium* spp. and *Corticium solani*), *scab (*Streptomyces scabies*), *soft rot (*Pectobacterium carotovorum*) and *black rot (*Xanthomonas campestris*).

The many leaf-attacking diseases are generally considered to be of little importance, and no protective spraying is normally practised. The use of zineb at $1\frac{1}{4}$ lb a.i. (2 lb 65% w.p.)/100 gal has given some control of **Downy mildew** (*Peronospora parasitica*) in field swedes, where this disease has been found troublesome in seedling crops.

**Mildew** (*Erysiphe polygoni*) gives a white powdery appearance to the leaves, attacks all these crops, but is more prevalent on turnips and swedes, the latter suffering more severely. Spraying is not economic, and there is some evidence that late sowing may reduce the trouble, though yields may be considerably affected.

***Clubroot** (*Plasmodiophora brassicae*) (A.L. 276) will attack all these crops, but is rarely seen on kale. Swedes and turnips may be seriously affected. Fungicidal alleviative measures are practised on vegetable crops, but not normally on fodder crops. The use of lime to produce an alkaline reaction in the soil, improvement in drainage, and in some instances, e.g. swedes, the use of resistant varieties are the cultural methods practised with these crops. An interval of 8 years between susceptible crops is normally sufficient for the disease to have died down sufficiently, unless there have been susceptible weeds to carry over the trouble.

A root and foot rot caused by *Phytophthora megasperma* is one of the most destructive diseases of kale, usually under conditions of poor drainage, which is the predisposing factor. Correction of the drainage is the best means of control.

Swedes and turnips may be attacked by a dry rot of the roots caused by *Phoma lingam*, which splits the roots and may rot them completely. The trouble may be picked up from the soil or introduced with the seed. It is best to adopt a rotation of 6 years between crops to allow the soil infection to die down. This fungus may also cause a serious stem canker in kale and other brassicae crops. It is seed borne, and seed can be effectively decontaminated by hot water treatment, consisting of immersing the seed in hot water at 122°F for 25 minutes, a practice usual for vegetable crops, but not normally for the fodder crops. The recently introduced warm water thiram soak treatment (see p. 154) has been shown to be completely effective for *Phoma lingam*, and also for other seed borne diseases of Brassicae, e.g. *Alternaria* leaf spot. Bacterial black rot (*Xanthomonas campestris*) is controlled by the hot water treatment.

# CHAPTER 9

# PESTS AND DISEASES OF FRUIT AND HOPS

## LOW VOLUME SPRAYING

Chemical doses per acre for low volume spray applications are commonly quoted in this chapter as twice or $2\frac{1}{2}$ times the amount per 100 gal for a high volume spray. The general basis for such figures is the thesis that a low volume spray should apply about the same amount of chemical per acre as would be applied by a high volume spray by automatic machines, and that 200–250 gal/acre embraces common or 'average' rates. The effectiveness of doses arrived at in this way has been verified, or in some cases modified, by large orchard scale trials of simple design, and by experience. A few experiments have shown that with good spray distribution smaller amounts of toxicant can in fact give adequate control of certain pests, but in practice there is a tendency to advocate more than the minimum dose, on the grounds that the excess may help to compensate for inadequacies of spraying; the chemical deposit may later be redistributed by rain, or if it remains patchy, the more concentrated the toxicant in the patches the more likely they are to affect mobile insects.

It is unfortunately true to say that deposit requirements for most pests and diseases, in terms of size of droplets, their distribution and density on the tree, and concentration of toxicant, for each of the chemicals in use, are not known with any precision. Until these are known the optimum quality and quantity of mist spray for any given plant of fruit trees or bushes cannot be closely defined, nor criteria offered by which a grower can decide for himself how much to use. Because of this lack of information it has been difficult to present doses per acre in this chapter without some inconsistencies, and manufacturers' recommendations should always be studied, especially as these are liable to be modified with increasing experience.

While experiments have shown that with a good standard of spraying, and appropriate doses of toxicant, good control of common pests and diseases on apple trees can be obtained when the same amount of insecticide or fungicide is applied in low, medium or high volumes of carrier, they have also shown that the degree of control tends to improve as the volume is increased. This is because of a more uniform distribution of spray deposit, and is particularly noticeable with summer sprays against red spider, codling moth and apple mildew. It should be noted that if the spray volume is increased to obtain better cover, coalescence of droplets begins to occur to a marked extent on the foliage and fruit in the volume range of about 80–120 gal/acre. If a compound spray of several insecticides and fungicides is being applied this can mean an increased risk of phytotoxicity. Unless improved spraying techniques or more frequent applications at shorter intervals show that less toxicant per acre than that recommended for low volume spraying can be effective, it would be safer, at least with some mixtures of insecticides and fungicides, to increase the rate to a full high volume spray.

## 9.1 APPLE

### 9.1A PESTS

Pests usually calling for annual routine spray are aphids, winter moth and apple sucker (at green cluster); apple sawfly (at petal fall); fruit tree red spider mite (petal fall and fruitlet stages); and codling and tortrix moths (fruitlet stages). For most of these pests insecticides can be included with routine applications of fungicides for scab and mildew. Sprays may not be necessary every year for winter moth except where orchards adjoin woodland, nor are special sprays needed for apple sucker which is kept in check by insecticides commonly used against aphids. Of the other pests mentioned below many are controlled by insecticides regularly used against other pests, while others require treatment only on special occasions when they occur. A few, such as fruitlet mining tortrix, may need annual sprays in certain localities.

In the following recommendations all the insecticides of known effectiveness are mentioned, but whenever possible preference should be given to insecticides other than the more toxic and persistent organochlorine compounds such as aldrin, DDT, dieldrin, gamma-BHC, endosulfan, endrin and TDE.

*****Apple aphids** (A.L. 106). Apart from woolly aphid, which is described separately, there are four principal species: apple-grass aphid (*Rhopalosiphum insertum*), green apple aphid (*Aphis pomi*), rosy apple aphid (*Dysaphis mali*), and rosy leaf-curling or red-leaf aphid (*D. devecta*). These begin hatching from winter eggs on apple trees at about the time of bud break of Cox's Orange Pippin or Bramley's Seedling. Hatching is virtually complete by the green cluster stage of these varieties, but with the rosy leaf-curling and green apple aphids it may be a little later.

Apple-grass aphid causes slight curl of rosette leaves and departs soon after petal fall for grasses, especially annual meadow grass. Heavy infestations are more likely to follow summers with sufficient rainfall to maintain continuous growth of the grass. An indication of the degree of infestation to be expected in the following spring can be obtained by counting the small, wingless, yellow-green aphids (the egg-laying females) on the undersides of the leaves in late October. An average of one or fewer aphids per leaf from a sample of twenty leaves from each of about six trees across the orchard indicates a light infestation.

Rosy apple aphid causes severe leaf curl, and the leaves may turn yellow but never red; it disperses to plantains in June and July but some colonies may persist on apple into August.

Rosy leaf-curling aphid, a local pest tending to appear unless controlled on the same trees year after year, causes severe leaf curl with conspicuous red areas; it lays its winter eggs deep in crevices under the bark in June. It is restricted to apple and spreads very slowly from tree to tree.

Green apple aphid is less common in the spring in orchards than the apple-grass aphid. In late May, June and July it disperses to other apple and related trees such as pear, hawthorn and rowan. It infests mainly young extension growth and is more a pest of young trees which may become re-infested in the summer.

Either winter or delayed dormancy washes as ovicides, or spring sprays against the newly-hatched aphids give good control.

(1) Winter washes. Except with rosy leaf-curling aphid these are effective if a good cover of the eggs is obtained; H.V. sprays are normally recommended as a good cover is difficult with L.V. sprays. Washes are: tar oil, 5 gal (e.c.) or 5–6 gal (stock emulsion)/100 gal (H.V.), *or* tar-petroleum oil 10 gal e.c. or stock emulsion/100 gal (H.V.), when buds dormant (December–February), *or* DNOC-petroleum oil, at 0·1% DNOC and 3% oil ($=$ 4·5 gal conc.)/100 gal at the delayed dormant stage when the buds begin to break (March).

The tar oil spray also kills scale insects, hibernating caterpillars of small ermine moth, eggs of apple sucker and some of those of winter moths. Stock emulsion types are preferable to e.c. types with very hard or brackish waters.

DNOC-petroleum oil also controls scale insects, and eggs of winter moths, apple capsid (at $7\frac{1}{2}$%, see page 163), and gives some control of winter eggs of fruit tree red spider mite.

Do not winter wash in windy or frosty weather or when the trees are wet, nor use DNOC-petroleum oil after the buds have reached the breaking stage. Some apple varieties, e.g. Beauty of Bath, Gladstone, Cox's Orange Pippin and Lane's Prince Albert, are more susceptible to oil damage to buds and should be sprayed first.

(2) Spring sprays. (*a*) The most reliable control is given by the systemic or penetrant organophosphorus insecticides applied at the time when Cox or Bramley have reached the green cluster stage, when egg hatch is virtually complete. It is also desirable that these insecticides should be used in preference to organochlorine compounds. The following are

M

effective (doses as a.i./100 gal for H.V. and a.i./acre for L.V. spraying): azinphos-methyl + demeton-S-methyl sulphone, H.V. 1 lb 25%/7·5% w.p., L.V. 2 lb w.p.; *or* demephion, H.V. 1·2 g (4 fl. oz 30% e.c.), L.V. 2·4 oz; *or* demeton-S-methyl, 1·16 oz (2 fl. oz 58% e.c.), L.V. 2·32 oz; *or* dimethoate, H.V. 2·4 oz (6 fl. oz 40% e.c.) L.V. 4·8 oz; *or* ethoate-methyl, H.V. 2·4 oz (12 fl. oz 20% e.c.), L.V. 4·8 oz; *or* formothion, H.V. 3·5 oz (8 fl. oz 43% e.c.), L.V. 7 oz; *or* malathion, H.V. 9 oz (¾ pint 60% e.c.), L.V. 18–24 oz; mecarbam, 5·8 oz (8 fl. oz 80% e.c.), (L.V. not recommended); *or* mevinphos, H.V. 1–1½ oz (1–1½ fl. oz 99% conc.), L.V. 2–3 oz (use the higher rate in cool weather and increase to H.V. 2 oz, L.V. 4 oz if caterpillar control is required); *or* oxydemeton-methyl, H.V. 1·14 oz (2 fl. oz 56% e.c.), L.V. 2·28 oz; *or* phosphamidon, H.V. 4·8 oz (16 fl. oz 30% s.c.), L.V. 9·6–12 oz; *or* vamidothion, H.V. 3·2 oz (8 fl. oz 40% e.c.).

Azinphos-methyl, carbaryl or phosalone at green cluster will also reduce aphid numbers, and where used against winter moth caterpillars (see p. 176) will probably give adequate aphid control in years when infestations are light.

With rosy leaf-curling aphid good cover is important; a second spray at petal fall may be needed and can be combined with control of red spider by organophosphorus insecticides (see p. 170). Vamidothion appears to be the most effective, and one spray at green cluster or petal fall may virtually eliminate the pest.

With young trees subject to infestation by green apple aphid in the summer, spray in late May and again later if necessary.

The alternative organochlorine insecticides DDT and gamma-BHC are not so effective as the systemic organophosphorus compounds and, for aphid control, are best applied at late bud burst which is too early for a combined control of other pests such as caterpillars and apple sucker. Their use should be avoided unless control of some other pest is required for which there is no alternative.

Nicotine at 8 oz (95–98%)/100 gal (H.V.) or 8–10 oz/acre (L.V.) can also be effective against aphids if temperatures are 60°F or over, but is non-persistent and will not control caterpillars.

To reduce hazards to bees, do not spray open blossoms; cut down flowering weeds beneath the trees before spraying. Avoid contamination of ponds and streams. Because of the risk of causing taint, do not allow gamma-BHC to fall or drift on currants at any time, or on other soft fruit between flowering and picking, or on cauliflowers before curd forming, or on beetroot, onion, carrot and peas. See pp. 28–61 for minimum intervals between application and harvest.

****Apple blossom weevil** (*Anthonomus pomorum*) (A.L. 28). Adult weevils winter under loose bark, in leaf litter, etc., emerging in early spring when they feed on young apple foliage. From bud burst onwards the female bores holes into blossom buds, laying one egg in each blossom, and the larva feeds on the stamens and base of the flower. The petals are prevented from expanding and turn brown ('capped blossom'). At one time a serious pest but now rare in commecial orchards owing to extensive use of DDT. Pear, quince and medlar may also be attacked.

The adult weevils are readily controlled by DDT at bud break or bud burst at $\frac{3}{4}$ lb actual DDT/100 gal. (H.V.). Low volume sprays at $1\frac{1}{2}$ lb/acre also give good control, but the degree of control appears to improve as the volume applied is increased. DNOC-petroleum oil with DDT added at 1 lb/100 gal, at bud break in March, is also effective. Gamma-BHC gives good results though it is less effective than DDT.

Azinphos-methyl *or* demeton-S-methyl sulphone + azinphos-methyl, at rates as for winter moth control at green cluster, will also reduce damage.

Annual sprays against this pest are not necessary.

Follow the precautions against poisoning bees, and for avoiding taint of nearby crops with gamma-BHC, given under apple aphids on p. 162.

****Apple capsid** (*Plesiocoris rugicollis*) (A.L. 154). This pest overwinters as eggs inserted in the bark, which have started hatching by green cluster of Cox or Bramley. The young bugs puncture leaves, shoots and fruit. Reddish spots form on the leaves which may become distorted in shape, shoots are scarred and stunted, and rough russeted areas with scattered pits and pimples appear on the fruit. The bugs mature and lay their winter eggs from mid June to about mid July. This pest has virtually disappeared since the advent of DDT but, together with common green capsid, should be watched for where annual pre-blossom DDT sprays against caterpillars are omitted.

The eggs may be killed by petroleum oil washes, containing not less than 5% petroleum oil on dilution, if thoroughly applied, but spring sprays against the newly hatched bugs are more efficient. For winter washing, use a DNOC-petroleum oil, $7\frac{1}{2}$% (H.V.), at bud break (March), *or* tar-petroleum oil, 10% (H.V.) at delayed dormant (before end February), *or* a winter petroleum oil, $7\frac{1}{2}$–10% (H.V.), at bud burst (late February or early March). The precautions against oil damage given under apple aphids (p. 161) should be observed.

For spring spraying, use if necessary one of the following at green cluster (doses refer to amount of actual insecticide): gamma-BHC, 2–4 oz/100 gal (H.V.) or 4–8 oz/acre (L.V.), *or* DDT, 1 lb/100 gal (H.V.) or 2 lb/acre (L.V.), *or* nicotine 8 oz (95–98%)/100 gal (H.V.).

With nicotine, which is non-persistent, a second spray may be needed at pink bud. DDT also controls caterpillars and gamma-BHC apple sucker, and formulations of the two insecticides together are available for multipurpose sprays. DDT may also be included at 1 lb a.i./100 gal with

DNOC-petroleum oil at bud break for capsid control. Recent evidence suggests that aginphos-methyl at green cluster at 2·64 oz (12 fl. oz 22% e.c.)/100 gal H.V. can give control of apple capsid equal to or better than that by DDT.

Follow the precautions against poisoning bees, and for avoiding taint of nearby crops with gamma-BHC, given under apple aphids on p. 162.

***Apple fruit miner** (*Argyresthia conjugella*). These moths emerge in June and lay eggs on the fruit, and the caterpillars form a maze of small tunnels in the flesh. When full grown they leave the fruit, spin a cocoon on the trunk or among dead leaves, entering the chrysalis stage in October. The insect also occurs on the rowan or mountain ash.

Measures used against codling moth (see p. 167) should also control this pest.

***Apple fruit rhynchites** (*Caenorhinus aequatus*), is a small brownish red weevil occurring on the trees from May to July, but now scarce as a pest, probably due to the widespread use of insecticides. Cylindrical holes are drilled in the side of the fruit, resembling those made by pushing in a pencil point.

If required, spray at pink bud with DDT at 1 lb/100 gal (H.V.) and to avoid danger to bees, do not spray open blossoms; cut flowering weeds before spraying.

***Apple leaf midge** (*Dasyneura mali*). Although generally a minor pest, several conspicuous infestations have occurred in recent years. Eggs are laid in the unopened or partially uncurled leaves and, as with pear leaf midge, the leaf margins become tightly rolled inwards. Affected leaves turn reddish and finally black, falling from the tree. The larvae are bright pink and feed within the rolled leaves, most of them dropping to the ground when full-fed to pupate. There are three overlapping generations in the year from May to August. Miller's Seedling seems to be particularly susceptible.

There is little information on control though vamidothion appears to be effective. Summer sprays of azinphos-methyl or demeton-S-methyl have little effect.

*****Apple sawfly** (*Hoplocampa testudinea*) (A.L. 13). Adult sawflies emerge about the time when mid-season varieties are flowering; they are active in warm sunny weather and are only attracted to trees in blossom. Eggs are laid, usually one per flower, in a slit cut just below the calyx. Hatching normally begins 4 or 5 days after 80% petal fall and is complete within 14–15 days. The larva mines under the skin, then penetrates to the core. It later leaves to bore straight into another fruitlet, making a large entry hole at which sticky frass accumulates. When fully fed the larva drops to the ground, and builds a cocoon in the soil; adults emerge the following spring, but some not until the second spring.

Worcester Pearmain, Charles Ross, James Grieve, Ellison's Orange and Early Victoria are particularly susceptible, while culinary varieties, except Edward VII, are generally only lightly attacked.

One of the most effective insecticides is gamma-BHC, 2 oz a.i./100 gal (H.V.) or 4 oz/acre (L.V.).

Spray within 7 days after 80% petal fall, using Worcester Pearmain or Bramley's Seedling as a general guide for timing. Late varieties, e.g. King Edward VII, should be sprayed separately. With heavy infestations repeat this spray for at least two seasons.

In regularly sprayed orchards infestations are kept in check by organo-phosphorus insecticides used at petal fall for red spider, e.g. demephion, demeton-S-methyl, dimethoate, formothion, oxydemeton-methyl, phos-phamidon, and vamidothion. Nicotine, 8 fl. oz (95–98%)/100 gal (H.V.) gives variable results.

Follow the precautions against poisoning bees, and for avoiding taint of nearby crops with gamma-BHC, given under apple aphids on p. 162. See pp. 28–61 for minimum interval between last application of insecticides and harvest.

**Apple sucker** (*Psylla mali*) (A.L. 96). This insect overwinters as eggs on the bark, and these hatch over a fairly long period which may extend through April into May. The young suckers feed on the leaves and flowers and blossom trusses may turn brown as if killed by frost. The insects remain on the tree and eggs are laid in the autumn.

Either winter washes against the eggs or spring sprays against the newly hatched suckers may be used for control. For winter washing use either a tar oil, 5% (H.V.) *or* tar-petroleum 10% (H.V.) when the buds are dormant (December–February). The precautions to be taken are those given under apple aphids, p. 161.

Apple sucker is no longer a common pest in regularly sprayed orchards where it is probably kept in check by the organophosphorus insecticides used against aphids, sawfly and fruit tree red spider mite. Experimental evidence is not available for all these compounds but the following are known to be effective at green cluster (doses as a.i./100 gal H.V. and a.i./acre L.V.): azinphos-methyl, H.V. 2·64 oz (12 fl. oz 22% e.c.), L.V. 5·28 oz; *or* azinphos-methyl + demeton-S-methyl sulphone, H.V. 1 lb 25%/7·5% w.p., L.V. 2 lb w.p., *or* dimethoate, H.V. 4·8 oz (12 fl. oz 40% e.c.), L.V. 9·6–12 oz; *or* formothion, H.V. 7 oz (16 fl. oz 43% e.c.), L.V. 14 oz; *or* malathion, H.V. 18 oz (1½ pints 60% e.c.), L.V. 36–45 oz; *or* mecarbam, H.V. 5·8 oz (8 fl. oz 80% e.c.), L.V. not recommended; *or* mevinphos, H.V. 1–1½ oz (1–1½ fl. oz 99% conc.), *or* phosphamidon, H.V. 4·8 oz (16 fl. oz 30% s.c.) L.V. 9·6–12·0 oz; *or* vamidothion, H.V. 6·4 oz (16 fl. oz 40% e.c.), L.V. 12·8 oz.

Gamma-BHC, H.V. 2–4 oz, L.V. 4–8 oz, is also very effective. For heavy infestations, gamma-BHC or mevinphos at late green cluster are the most effective.

Carbaryl and phosalone at rates recommended for pre-blossom caterpillar control will reduce infestations of apple sucker, but in trials the degree of control has been variable. Nicotine, 8 oz (95–98%) H.V. or 8–10 oz L.V. can also be effective but requires temperatures of 60°F and over for best results and is not persistent. DDT is not effective.

To reduce hazards to bees and the tainting of crops by gamma-BHC, observe the precautions given under apple aphids (p. 162).

***Apple twig cutter** (*Rhynchites coeruleus*) is a small, deep blue weevil appearing in May and feeding on foliage, laying eggs in early June in the young shoots, which are then cut off just below the oviposition point. These shoots wither and drop off or remain hanging on the tree. A local pest, more important on young trees, which may be controlled by DDT 1 lb a.i./100 gal (H.V.) *or* 2–2½ lb/acre (L.V.) applied at pink bud.

To reduce hazard to bees, do not spray open blossoms and cut down flowering weeds beneath the trees before spraying.

***Bark beetles and Shot-hole borers.** The shot-hole borers (*Anisandrus dispar* and *Xyleborus saxeseni*) bore into the heart wood of the trunk and main branches. The females excavate galleries in which the eggs are laid, and the grubs feed largely on fungi which grow on the tunnel walls. Grubs may be found at all times of the year, while adults are more common from January to June. The bark beetles (*Scolytus rugulosus* and *S. mali*) tunnel under the bark where the grubs excavate a maze of galleries. As a rule only trees unhealthy due to root disorders or disease are attacked.

There are no standard recommendations for the control of these pests but, in addition to cultural measures to increase the health of the trees, the following uses of insecticide have been suggested: tar oil 10% (H.V.), applied to the trunks in March or April to drive out the beetles, taking care not to spray any foliage or buds, *or* DDT 5 lb a.i./100 gal (H.V.), sprayed or painted onto the trunk and main branches in April, to kill beetles after they emerge, taking care not to wet any foliage.

***Brown leaf weevil** (*Phyllobius oblongus*)—see Leaf weevils.

***Bryobia mite, Apple and Pear Bryobia** (*Bryobia rubrioculus*). The adults of this mite can be distinguished from those of fruit tree red spider by the greater length of their front legs, which are twice as long as the second pair and by the absence of long hairs on the body. It overwinters as eggs on the bark of the trunk and main branches. These begin hatching in March or April, about a fortnight earlier than those of fruit tree red spider.

The mites are active on the leaves in warm, sunny weather, retiring to crevices in the bark when the air temperature drops below about 56°F. They also return to the bark for moulting, and whitish clusters of cast skins are another aid to the recognition of their presence. There are three generations during the summer, the period egg to adult lasting some 6 weeks. Summer eggs are also laid on the bark, and the winter eggs are laid from about the third week of July onwards. The species also occurs on pear but is apparently less important in some areas than *B. cristata* (see p. 187).

The acaricides listed on p. 171 for the control of fruit tree red spider mite are probably all as satisfactory against bryobia mite. Where heavy infestations have arisen, spray at green cluster with an acaricide effective against the active stages; by choosing a suitable organophosphorus compound, aphids can be controlled at the same time. Follow with a second spray at petal fall. In subsequent years the pest should be kept under control by normal fruit tree red spider programmes.

For best results sprays should be applied in warm conditions (air temperature near 60°F if possible) when the mites are active on the leaves. On culinary varieties lime sulphur at green cluster can be effective.

Observe the minimum intervals between last application and harvest.

Some control of the winter eggs may be obtained by winter washes such as: DNOC-petroleum oil, $4\frac{1}{2}\%$ (H.V.), when the buds are beginning to break (March), *or* winter petroleum oil $4\frac{1}{2}\%$ (H.V.), at bud burst.

The precautions given under apple aphids (p. 161) should be observed to reduce hazard of oil damage.

****Chafer beetles** (A.L. 235). Adult cockchafer (*Melolontha melolontha*) and garden chafer beetles (*Phyllopertha horticola*), which may be abundant in May and June in occasional seasons, sometimes bite into fruitlets, causing damage resembling that by caterpillars. The large white larvae which live in the soil may also attack the roots of nursery trees. To prevent fruit damage, spray with DDT 1 lb a.i./100 gal, without delay if an attack commences. For control of the grubs see under strawberry, p. 215.

The use of DDT may encourage a build-up of red spider and an acaricide may have to be used (see p. 170).

***Cherry-bark tortrix** (*Enarmonia formosana*)—see under Cherry, p. 195.

***Clay-coloured weevil** (*Otiorhynchus singularis*)—see under Currant, p. 205.

*****Codling moth** (*Laspeyresia pomonella*) (A.L. 42). In an average season, these months emerge from the late May or early June until early August, with the main flight from late June to mid July. In some years a small second generation occurs in September. Eggs are generally laid in the evening, on leaves and fruit, and in most numbers when temperatures exceed 60°F. The caterpillars, which hatch in 10–14 days, bore into the fruit, feeding for the first few days in a cavity just beneath the skin, then tunnel to the core. The entry hole is small and covered with dry frass, in contrast with the large hole with a mass of wet frass typical of apple sawfly. About half of the early hatched caterpillars enter by the calyx. When fully fed, in about 4 weeks, the larva leaves the fruit and spins a cocoon under loose bark or other shelter. Second generation moths arise only from larvae which have spun up by early August. Later caterpillars overwinter in their cocoon to produce moths the following summer.

Since egg hatch extends over at least 2 months, the ideal control would be to maintain an insecticide deposit effective against eggs or young larvae throughout this time, and with the most persistent materials this would mean at least three sprays. Usually two sprays are given to destroy all except the latest hatched larvae, though on some farms a third is considered desirable. The first should be applied just before the earliest eggs hatch, which in southern and eastern England is about 4 weeks after petal fall of Cox, i.e. mid June in average seasons. Guidance on spray timing in any particular season is usually available from N.A.A.S. entomologists who operate light traps in the main fruit growing areas. In districts where codling moth infestations are light some growers begin spraying about a week later, so as to have the maximum insecticide deposits at the height of the emergence period. This may entail risking the escape of larvae hatching from the earliest laid eggs and consequently in a hot summer a bigger than usual second generation. In some areas only one spray has proved adequate and this should be applied early in July, preferably immediately after a spell of warm weather favouring egg laying. In the south-west, where tortrix caterpillars are more troublesome than codling moth, it is recommended that the first spray against both pests should be applied in late June.

Of the following recommended insecticides lead arsenate is a stomach poison, the remainder kill mainly by contact and have some effect on the moths as well as larvae and/or eggs (doses as a.i./100 gal, H.V., or a.i./acre, L.V.):

Azinphos-methyl, H.V.: 5·28 oz (24 fl. oz 22% a.c.), L.V. 10·56 oz, twice at 3 week intervals, or 3 sprays at 2 week intervals at 3·3 oz (15 fl. oz 22% e.c.) H.V., 6·6 oz L.V.; *or* carbaryl H.V. $\frac{3}{4}$–1 lb (1$\frac{1}{2}$–2 lb 50% w.p.), L.V. 1$\frac{1}{2}$–2 lb; *or* DDT, H.V. 1 lb, L.V. 2–2$\frac{1}{2}$ lb; *or* lead arsenate, H.V. 2–3 lb (powders) or 4 lb (pastes and colloidal susp.), L.V. 4–5 lb (powders) or 8–10 lb (pastes and colloidal susp.); *or* malathion, H.V. 18 oz (1$\frac{1}{2}$ pints 60% e.c.); *or* phosalone, H.V. 5 oz (15 fl. oz 33% e.c.), L.V. 10 oz (3 sprays 2 weeks apart) *or* phosphamidon, H.V. 4·8 oz (16 fl. oz 30% s.c.), L.V. 9·6–12·0 oz; *or* TDE, H.V. $\frac{3}{4}$ lb (1$\frac{1}{2}$ lb 50% w.p.), L.V. 1$\frac{1}{2}$ lb. High volume sprays, of at least 100 gal/acre, are generally more effective than low volume sprays.

Of these insecticides azinphos-methyl, carbaryl and phosalone are probably slightly more effective than DDT, phosphamidon and TDE. Lead arsenate is less effective than the other compounds; control can be improved by applying an additional spray at petal fall before the calyx closes, using a high volume spray with added spreader. Malathion has shorter persistence than the other chemicals and is useful if a late spray is required.

Moths from caterpillars wintering in packing sheds and box stacks may be responsible for heavy attacks on trees nearby. Shed windows and the outside of box stacks should be sprayed with DDT at the end of May. Keep doors closed as much as possible.

On a small scale corrugated paper bands can be tied round the tree trunks in mid July to trap caterpillars, which are then readily attacked by birds or can be removed and destroyed after the crop is picked.

Do not use lead arsenate with lime sulphur after blossom, or when temperatures are above 70–75°F.

If carbaryl is used between late pink bud and 3 weeks after petal fall it may lead to fruit thinning.

The insecticides mentioned, except lead arsenate, have adverse effects on predators of red spider. Azinphos-methyl, phosalone and phosphamidon are also acaricides, but DDT, carbaryl and TDE are not and may lead to serious infestations of this pest so they should be applied in combination with suitable acaricides (see under fruit tree red spider, p. 170).

Follow the precautions against poisoning bees given under apple aphids on p. 162. See pp. 28–61 for minimum interval between last application of insecticides and harvest.

**Common green capsid** (*Lygocoris pabulinus*). Damage by this pest to apples, and to a lesser extent to pears, appears to have become more common in recent years. Infestations almost certainly arise from eggs laid in the shoots in the previous autumn, and these hatch between early pink bud and petal fall of Bramley's Seedling. Unfortunately it is difficult to predict the risk of infestation in any particular orchard, and a watch should be kept where pre-blossom DDT sprays are omitted.

The pest is normally controlled by DDT, 1 lb a.i./100 gal H.V., or 2 lb/acre L.V., at green cluster as for caterpillar control. Experimental evidence on the effectiveness of alternatives to DDT is limited. The most effective organophosphorus insecticides are dimethoate, formothion and malathion, but applied at green cluster or pink bud would probably have insufficient persistence to be effective over the whole hatching period. Applied at petal fall dimethoate, H.V. 4·8 oz a.i. (12 fl. oz 40% e.c.), L.V. 9·6 oz a.i. *or* formothion, H.V. 7 oz a.i. (16 fl. oz 43% e.c.), L.V. 14 oz a.i. *or* malathion, H.V. 18 oz a.i. (1½ pints 60% e.c.), L.V. 36 oz, they would give a good control but it is uncertain to what extent fruitlet damage would be prevented.

**Common leaf weevil** (*Phyllobius pyri*)—see under Leaf weevils, p. 173.

***Dock sawfly** (*Ametastegia glabrata*). There are two and sometimes three generations of this sawfly in the year, those of the last brood laying eggs in August or September. The light green caterpillers feed on docks and fat-hen. They hibernate in hollow stems or suitable crevices, and sometimes tunnel into apples in search of winter quarters.

Insecticides used against codling moth should also control sawfly caterpillars on weeds under the trees. Where keeping the ground weed free is impracticable, spraying the weeds with malathion at 18 oz a.i. (1½ pints 60% e.c.)/100 gal *or* derris (0·002% rotenone/100 gal) in mid July and again in early August should be effective.

***Earwig** (*Forficula auricularia*). Earwigs lay their eggs in the soil from December to March, and a second period of egg laying occurs in May and June. Earwigs shelter by day in dark crevices, ascending plants at night to feed. Numbers appear to be highest from July to September, and in some orchards in grass damage to fruit occurs, in the form of deep rounded cavities with small entry hole, and also soiling by frass where the insects shelter between the stalk and the fruit or a touching leaf.

Where earwigs are a pest, apply a coarse spray to the base of the trunk and surrounding vegetation with gamma-BHC 4 oz a.i./100 gal in late July. Carbaryl 2 lb a.i. (4 lb 50% w.p.)/100 gal is also effective.

*****Fruit tree red spider mite** (*Panonychus ulmi*) (A.L. 10). This species winters as bright red spherical eggs on the bark. Hatching normally begins at the pink bud stage of Cox's Orange Pippin or Bramley's Seedling, in late April or early May; about half the eggs have hatched by petal fall, the remainder hatching in the next 3–4 weeks, i.e. up to about mid-June. Surveys have shown that there are strains of the mite with different hatching periods, the time of peak hatch of these ranging from 'normal' to up to 3–4 weeks later.

The mites feed on the undersides of the leaves causing a minute speckling, and in heavy attacks the leaves become dull green and then bronzed. Summer eggs are laid, mainly on the undersides of the leaves, and four or five generations occur during a warm summer. The period egg to adult normally takes about 4 weeks (egg stage 2–3 weeks, young stages 14 days). Breeding ends in the first half of September, when daylight is reduced to about 14 hours, and winter eggs are laid. Winter eggs may be laid earlier on severely bronzed trees.

For control in the summer, two sprays, applied 3 weeks apart, are usually required. With heavy infestations—clusters of winter eggs evident on the bark—the first should be at petal-fall or as soon afterwards as peak hatching is apparent, using an acaricide effective against the active stages. Where winter eggs are few this spray can be delayed until mid-June when an insecticide active against the summer eggs is an advantage. Persistent use of one acaricide is liable to select out a resistant strain of red spider, as happened in many orchards with 'summer ovicides' (chlorfenson, etc.) and more recently in some orchards with certain organophosphorus compounds.

Resistance seems likely to develop after 5 years of regular use of the same group of chemicals; a careful watch should be kept after 3 years' use for signs of resistance so that a change can be made to different chemical group of acaricides. Unnecessary routine sprays should be avoided where possible. The acaricides now in common use may be grouped as follows (doses as a.i./100 gal, H.V., or a.i./acre, L.V.; the stages against which they are effective are given in brackets):

(*a*) chlorbenside, H.V. or L.V. 6–10 oz ($1\frac{1}{2}$–$2\frac{1}{2}$ pints 20% conc. or 3 lb 20% w.p.); *or* chlorfenson, H.V. or L.V. 1 lb (2 lb 50% w.p.) (eggs and young stages);

(*b*) tetradifon, H.V. 2 oz ($1\frac{1}{4}$ pints 8% conc.); L.V. 4 oz (eggs and young stages; apply early in the hatching period, when the first adult mites are seen, usually late blossom or early petal fall, and repeat in early July and again mid August); *or* tetrasul, H.V. 6·4 oz a.i. (2 lb 20% w.p.), L.V. 12·8 oz (apply when 20–40% of winter eggs have hatched, usually mid to late blossom; or use at this time at 3·2 oz a.i. and repeat 4 weeks later).

(*c*) dicofol, H.V. 6·4 oz (32 fl. oz 20% e.c.), L.V. 16 oz (4 pints 20% e.c.) (all stages);

(*d*) organophosphorus compounds: azinphos-methyl, H.V. 5·3 oz (25 fl. oz 22% e.c.), L.V. 11·0 oz (active stages); *or* demeton-S-methyl, H.V. 3·5 oz (6 fl. oz 58% e.c.), L.V. 7·0 oz (active stages); *or* dimethoate, H.V. 3·6 oz (9 fl. oz 40% e.c.), L.V. 7·2–9·0 oz (18–22 fl. oz 40% e.c. (active stages); *or* ethoate-methyl, H.V. 4·0 oz (1 pint 20% e.c.), L.V. 8 oz (active stages); *or* formothion, H.V. 5 oz (12 fl. oz 43% e.c.), L.V. 10 oz (active stages); *or* malathion, H.V. 18 oz ($1\frac{1}{2}$ pints 60% e.c.), L.V. 36–45 oz active stages); *or* oxydemeton-methyl, H.V. 3·4 oz (6 fl. oz 56% e.c.), L.V. 6·8 oz (active stages); *or* phenkapton, H.V. 3·2 oz (16 fl. oz 20% e.c.), L.V. 6·4–8·0 oz (active stages); *or* phosalone, H.V. 5·0 oz (15 fl. oz 33% e.c.), L.V. 10·0 oz (active stages) (3 sprays timed as for codling moth); *or* phosphamidon, H.V. 4·8 oz (16 fl. oz 30% s.c.), L.V. 9·6–12·0 oz (active stages); *or* vamidothion, H.V. 6·4 oz (16 fl. oz 40% e.c.), L.V. 12·8–16·0 oz (active stages) (only two post blossom applications permitted). Demephion, H.V. 3·6 oz (12 fl. oz 30% e.c.), L.V. 7·2 oz, after petal fall will give some control.

(*e*) quinomethionate, 4 oz (1 lb 25% w.p.) is at least 150 gal/acre (active stages), at 80% winter egg hatch with a second spray 10–14 days later and a third if necessary;

(*f*) binapacryl *or* dinocap, used regularly against powdery mildew (see p. 179) will also control red spider (active stages, and partially ovicidal, at least until the end of July, provided the season does not start with heavy infestation of winter eggs. Dinobuton, not currently included in the

A.C.A.S., H.V. 1 lb a.i. (2 lb 50% w.p.), L.V. 2 lb a.i., appears to be similar in activity against red spider to binapacryl and dinocap.

Malathion gives only a short term effect as it is less persistent than the other organophosphorus compounds listed.

Chlorfenson may damage D'Arcy Spice; dicofol should not be used for 3 weeks after petal fall, as severe russeting and cracking may occur on Cox. Quinomethionate should not be used at more than 4 oz/acre, or as L.V. sprays, or when temperatures exceed 80°F. It may be mixed only with captan or gamma-BHC w.p.

Follow the precautions against poisoning bees given under apple aphids on p. 162. See pp. 28–61 for minimum interval between last application of insecticides and harvest.

For use in winter, washes containing 3% petroleum oil when diluted give some control of winter eggs with thorough high volume spraying though are less efficient than summer sprays. Use either DNOC-petroleum oil, $4\frac{1}{2}$% (H.V.), when the buds are breaking in March, *or* winter petroleum oil, $4\frac{1}{2}$%, at bud burst.

The precautions given under apple aphids (p. 161) should be taken to reduce the hazard of oil damage.

****Leafhoppers.** Two common species are *Erythroneura alneti* and *Typhlocyba froggatti*, which overwinter as eggs in young shoots; those of *T. froggatti* hatch between mid April and early May. This species has two generations a year, eggs of the second generation being laid in or near the midrib of the leaf from June onwards, and adults occur from about mid June into November. *E. alneti* has one generation, its eggs hatching in May, and adults occur from late June into October. *T. rosae* overwinters on rose and adults may fly to apple for the second generation returning to rose in the autumn. Several other species occur in small numbers.

The following insecticides are effective (doses as a.i./100 gal, H.V., or a.i./acre, L.V.): DDT, H.V. 1 lb, L.V. 2 lb; *or* demephion, H.V. 3·6 oz (12 fl. oz 3% e.c.), L.V. 7·2 oz; *or* demeton-S-methyl, H.V. 3·5 oz 6 fl. oz 58% e.c.), L.V. 7·0 oz; *or* malathion, H.V. 18 oz ($1\frac{1}{2}$ pints 60% e.c.), L.V. 36 oz; *or* oxydemeton-methyl, H.V. 3·4 oz (6 fl. oz 57% e.c.), L.V. 6·8 oz; *or* at H.V., azinphos-methyl, 6·6 oz ($1\frac{1}{2}$ pints 22% e.c.); *or* carbaryl, 1 lb (2 lb 50% w.p.); *or* dimethoate, 4·8 oz (12 fl. oz 40% e.c.); *or* formothion, 7 oz (16 fl. oz 43% e.c.). Low volume sprays of the four latter insecticides are probably also effective.

Good control has been obtained in trials by sprays at the end of June with a second 3 weeks later. In most seasons the above insecticides, when used against codling moth, are also adequate against leafhoppers.

For minimum intervals between last application and harvest see pp. 28–61.

****Leaf weevils** (*Phyllobius* spp.). The brown leaf weevil (*P. oblongus*), common leaf weevil (*P. pyri*) and silver-green leaf weevil (*P. argentatus*) occasionally damage leaves and flowers. The larvae feed at the roots of grasses and adults emerge about the blossom time of apples, sometimes appearing in large numbers but often disappearing after a week or so to feed elsewhere. Damage is only likely to be important on very young trees, where it may be reduced by DDT, 1 lb a.i./100 gal (H.V.) or 2 lb/ acre (L.V.).

To reduce hazard to bees, do not spray open blossom, and cut down flowering weeds beneath the trees before spraying. The use of DDT after the green cluster stage may lead to an increase in red spider infestations (see page 170).

***March moth** (*Alsophila aescularia*) (A.L. 11)—see winter moths.

***Mottled umber moth** (*Erannis defoliaria*) (A.L. 11)—see winter moths.

***Mussel scale** (*Mytilococcus ulmi*) (A.L. 36). The adult is about $\frac{1}{8}$ in long, shaped like a mussel shell, and lies flat on the bark. Eggs are laid beneath the scale in late summer, and these hatch in May, the young insects settling in a suitable place and gradually developing the waxy covering. Winter washes thoroughly applied are very effective: tar oil, 5% (H.V.) when the buds are dormant (December–February), *or* DNOC-petroleum oil, $4\frac{1}{2}$% (H.V.) at bud break (March), *or* winter petroleum oil, $4\frac{1}{2}$% at bud burst. The precautions to be taken in the use of winter washes are given under apple aphids (see p. 161).

If control is required out of the dormant season spray with diazinon, 5 oz a.i. (25 fl. oz 20% e.c.)/100 gal H.V., *or* malathion, 30 oz a.i. ($2\frac{1}{2}$ pints 60% e.c.)/100 gal H.V. in late May or early June. This spray is directed against the young stages or 'crawlers' and good cover of the wood and undersides of the leaves is essential. With heavy infestations a second spray 14 days later may be necessary. This pest has become uncommon in regularly sprayed orchards, probably kept in check by insecticides used in spring and summer against other insects.

***Oyster-shell scale** (*Quadraspidiotus ostreaeformis*). The scale is brown or grey, variable in shape and resembling an oyster shell. It lives on the bark, and the eggs which are laid beneath the parent scale hatch in June. The pest is now uncommon in sprayed orchards, but if present, may be controlled by the methods given above under mussel scale.

****Red bud borer** (*Thomasiniana oculiperda*). Midges appear from June to August. Eggs are laid in the cuts made when budding nursery stock and the red maggots feed under the bark. The tying material should be coated liberally with petroleum jelly immediately after tying in.

****Small ermine moths** (*Yponomeuta* spp.) (A.L. 40). Moths appear in late July and in August, and lay their eggs in flat scale-like masses on the twigs. Caterpillars hatch in September but remain under the egg scales during the winter, crawling out in the spring when they first mine the young leaves and later leaves are spun together to form a 'nest'. By late May and in June these nests are large and conspicuous, with numerous caterpillars and defoliation of the enveloped twigs. The caterpillars, about $\frac{1}{2}$ in long when full grown, are grey with black spots, and build cocoons for pupation in July. *Y. padella* feeds on plum, hawthorn and related trees, *Y. malinella* on apple. Moths infesting spindle and willow are different species.

The pest may be controlled either by winter or spring treatment:

(*a*) Winter washes. 5% tar oil when the buds are dormant in December–February, *or* $4\frac{1}{2}$% DNOC-petroleum in March at bud break as for aphid control will kill overwintering caterpillars.

(*b*) Spring sprays. DDT sprays applied at green cluster for winter moth caterpillars are too early for small ermine caterpillars. Spraying at pink bud or petal fall, before 'nests' have been formed, would be more effective, using DDT at 1 lb a.i./100 gal (or $2–2\frac{1}{2}$ lb/acre L.V.) *or* lead arsenate powder 2–3 lb/100 gal, or pastes and suspensions 4 lb (2 pints)/100 gal. Alternatives to DDT cannot yet be recommended; trichlorphon at 12·8 oz (1 lb 80% s.p.)/100 gal is very effective against small ermine caterpillars in hedgerows but there may be phytotoxicity risks with apples.

Do not spray open blossoms; cut down flowering weeds under the trees before spraying. DDT used after green cluster may lead to increases in red spider populations.

***Silver-green leaf weevil** (*Phyllobius argentatus*)—See under leaf weevils.

*****Tortrix caterpillars.** Some 16 species can be found on apple trees, but the larvae of only three are known to cause damage to the fruit. These are the fruit tree tortrix (*Archips podana*), the summer fruit tortrix (*Adoxophyes orana*) and the fruitlet mining tortrix (*Pammene rhediella*).

The fruit tree tortrix hibernates as young caterpillars in cocoons fixed to twigs or buds. These emerge over a fairly long period from late March to green cluster or pink bud stages, first boring into fruit buds and then feeding on the young leaves, which are frequently spun together. Pupation occurs between leaves spun together in late May to early June. Moths begin emerging 1–2 weeks later than codling moth and they occur until mid August with greatest numbers in late June and the first half of July. Scale-like eggs, in flat green batches, are laid on the leaves. Caterpillars hatch, feeding at first on the leaves, then on the fruit, eating out irregular shallow areas under the protection of a leaf attached with silk. These caterpillars go into hibernation in the autumn, but some, especially in fine summers, mature earlier and produce moths from late August to October. These lay eggs and the caterpillars may cause further damage to the apples before hibernating, though the most important damage is usually caused by caterpillars from the moths of the first brood. Caterpillars taken into apple stores continue to feed on the fruit.

The summer fruit tortrix mainly occurs in Kent, and has two complete generations a year. It overwinters as young caterpillars under leaf fragments fastened with silk to crevices in the bark, etc., and these emerge in late March or early April to feed on the leaves. The caterpillars pupate and produce a first brood of moths which occur over a similar period to that of the fruit tree tortrix. These lay eggs and the caterpillars feed on the leaves and also remove extensive areas of skin from the fruits. A second brood of moths occurs from late July to September, lays eggs, and the caterpillars feed a little before hibernating. This second brood is usually much larger than the first.

The fruitlet mining tortrix occurs locally in many of the fruit growing areas, the moths emerging in May and laying eggs on the undersides of leaves. The caterpillars attack the fruitlets, usually where two are touching, forming groups of round, black-rimmed holes. They are fully fed by early July and go into hibernation in cocoons under loose bark.

The bud moth (*Spilonota ocellana*) is a common species with a life cycle similar to that of the fruit tree tortrix, and the caterpillars emerging in the spring bore into the blossom buds.

Control of the fruit tree tortrix and summer fruit tree tortrix is best obtained by the sprays recommended for codling moth, and the most effective insecticides are azinphos-methyl, carbaryl, phosalone and TDE. The first spray can be applied about a week later than the date recommended for codling moth in areas where the latter is not an important pest. Spring sprays at bud burst or green cluster against the caterpillars emerging from hibernation give variable results and often only partial control, especially with DDT. If control in the spring is required, azinphos-methyl as for winter moth (see below) appears to be one of the most effective insecticides.

The fruitlet mining tortrix can be controlled, where required, by including DDT in petal fall sprays, 1 lb a.i./100 gal (H.V.) or $2-2\frac{1}{2}$ lb/acre (L.V.), *or* TDE $\frac{3}{4}$ lb a.i. ($1\frac{1}{2}$ lb 50% w.p.)/100 gal (H.V.) or $1\frac{1}{2}$ lb/acre (L.V.). Organophosphorus insecticides have proved successful on the continent against this pest and where used at petal fall against fruit tree red spider mite would probably give control. Those most effective against tortrix caterpillars are azinphos-methyl and phosalone; carbaryl is also effective.

Do not spray open blossoms and cut down flowering weeds before spraying. DDT used after green cluster may lead to increases in red spider populations.

**Wasps** (A.L. 451). The two commonest are *Vespula vulgaris* and *V.*

*germanica*, which may cause damage to ripening fruit. The best control measure is the systematic nest destruction by dusting insecticide into the entrance hole at dusk. DDT 20 or 25% w.p., *or* derris 2% dust, 1 tablespoon per nest, are effective. Another effective treatment is injection of the nest with $\frac{1}{4}$–$\frac{1}{2}$ pint carbon tetrachloride.

****Winter moths** (A.L. 11). The caterpillars of 3 species attack foliage, flowers and fruitlets—only the winter moth (*Operophtera brumata*) occurs in sufficient numbers to be a pest; the March moth (*Alsophila aescularia*) and the mottled umber moth (*Erannis defoliaria*) occasionally occur in small numbers. The females of all 3 have vestigial wings and cannot fly. Adults of the winter moth appear in October to December, those of the mottled umber October to December, but occasionally in January and February, and the March moth in March. Eggs are laid on the bark, those of the March moth in bands around twigs. Caterpillars hatch from about bud break or burst to green cluster, and when fully fed drop to the ground and pupate in the soil. Young winter moth larvae may be blown from one tree to another, and from woods or hedgerows into neighbouring fruit trees.

Partial control is obtained by winter washes; spring sprays are more effective. The DNOC-petroleum oil $4\frac{1}{2}$% applied H.V. in March at bud break for control of eggs of aphids, apple capsid and red spider, *or* winter petroleum $4\frac{1}{2}$% oil at burst for apple capsid and red spider, will also destroy eggs of winter moth and mottled umber moth.

For precautions against oil damage to buds, see under apple aphids, p. 161.

The following spring sprays used at green cluster are effective (doses as a.i./100 gal, H.V., or a.i./acre, L.V.): azinphos-methyl, H.V. 2·64 oz (12 fl. oz 22% e.c.), L.V. 5·28 oz; *or* azinphos-methyl + demeton-S-methyl sulphone, H.V. 1 lb 25%/7·5% w.p., L.V. 2 lb w.p.; *or* carbaryl, H.V. $\frac{3}{4}$–1 lb ($1\frac{1}{2}$–2 lb 50% w.p.), L.V. $1\frac{1}{2}$–2 lb; *or* DDT, H.V. $\frac{1}{2}$ lb, L.V. 1–$1\frac{1}{4}$ lb; *or* lead arsenate, H.V. 2–3 lb powders or 4 lb pastes or suspensions, L.V. 4–6 lb powders or 8–10 lb pastes or suspensions; *or* mevinphos, H.V. 2 fl. oz 99% conc., L.V. 4 oz; *or* phosalone, H.V. 5 oz (15 fl. oz 33% e.c.), L.V. 10 oz; *or* TDE, H.V. $\frac{3}{4}$ lb ($1\frac{1}{2}$ lb 50% w.p.), L.V. $1\frac{1}{2}$ lb.

In recent trials azinphos-methyl, carbaryl and phosalone have given a control of winter moth caterpillars equal to that by DDT and should preferably be used instead of DDT or TDE unless there are other reasons, such as a capsid problem, for using the latter insecticides. Good spray cover is important and better results will probably be obtained by spraying at late green cluster or early pink bud rather than earlier. Fenitrothion, 5·0 oz ($\frac{1}{2}$ pint 50% e.c.) (H.V.) *or* 'Supracide' 6·4 oz (1 lb 40% w.p.) (H.V.) (not currently included in the A.C.A.S.) at green cluster have also given good results in recent trials.

****Woolly aphid** or **American blight** (*Eriosoma lanigerum*) (A.L. 187). This aphid passes its whole life cycle on the tree. It winters as young aphids, devoid of wool, sheltering in cracks or under loose bark. These become active in March or April, secreting the typical waxy 'wool', and breeding colonies are present by May. Some winged forms occur in July but are not usually an important source of new infestations. Breeding continues through the summer. Eggs are laid in September but are usually sterile, and the adults die as winter approaches.

The best control is given by vamidothion, 6·4 oz a.i. (16 fl. oz 40% e.c.)/100 gal (H.V.) or 12·8–16 oz a.i./acre (L.V.) and one application in June can give adequate control (not more than two post blossom applications per season are permitted). H.V. sprays are recommended for very heavy infestations. The following can also give useful control as H.V. drenching sprays which are probably best applied in late June (doses as a.i./100 gal): demephion, 3·6 oz (12 fl. oz 30% e.c.); *or* demeton-S-methyl, 3·5 oz (6 fl. oz 58% e.c.); *or* dimethoate, 4·8 oz (12 fl. oz 40% e.c.); *or* ethoate-methyl, 6·0 oz (30 fl. oz 20% e.c.); *or* formothion, 7 oz (16 fl. oz 43% e.c.); *or* malathion, 18 oz (1½ pints 60% e.c.); or menazon,(8 oz (1 pint 40% e.c.)/acre in not less than 100 gal; *or* mevinphos, 2–3 oz 99% conc./acre (use the higher rate in cool weather); *or* phosphamidon, 4·8 oz (16 fl. oz 30% s.c.).

Petal fall sprays of gamma-BHC and organophosphorus insecticides, except vamidothion, applied against apple sawfly, are only partially effective for woolly aphid.


## 9.1B DISEASES

****Blossom wilt** (*Sclerotinia laxa* f. *mali*) (A.L. 155). Spores of the fungus are carried to the blossom where they germinate causing blossom wilt. The mycelium may continue along the flower stalk into the spur and even into the spur-bearing branch where it causes dieback. In moist weather, spore pustules develop on the flowers shortly after infection to continue the spread. Spurs and killed wood release spores the following spring, so renewing the disease cycle.

Where lime sulphur is used preblossom for scab control (p. 185), sources of infection are likely to be reduced. Routine spraying with tar oil winter washes in the late dormant period (see p. 161) is also effective. Spray treatments only partly suppress the disease so that infected trusses, cankers and dead wood should be removed and burnt. This can best be done in spring or early summer when infected parts can be recognized by the wilted blossoms. The varieties Lord Derby, Cox's Orange Pippin and James Grieve are very susceptible. Bramley's Seedling is resistant.

*****Canker** (*Nectria galligena*) (A.L. 100). *N. galligena* is responsible

N

for most cankers on apple trees, but it is not always easy to distinguish from those caused by *Gloeosporium* spp. and also by other agencies including mechanical damage and woolly aphid. Common canker is usually seen as sunken zones of bark around bud, leaf scar, base of small dead side shoot, or an open wound. Small branches are often encircled but on larger branches the canker is restricted to one side. Fruits are occasionally infected. Two types of spores are produced: summer spores ooze from cankers as white pustules; winter spores develop on the canker in red pear-shaped receptacles (perithecia), sometimes mistaken for red spider eggs; the spores may be shot out of the perithecia at any time of the year, particularly in winter.

Spores can infect only through breaks in the bark layer, including pruning or other wounds, leaf scars during autumn, woolly aphid, or scab injury. All badly infected small shoots should be removed and the cankered pockets in large boughs pared away. Large cut surfaces, whether from healthy or diseased zones should be protected immediately with soft grafting wax or a special preparation for tree sealing. On shoots not more than half girdled a proprietary paint containing 1·9–2·0% mercury in organic form in oil may be used during the dormant period instead of cutting out. Treatment should not extend beyond areas adjacent to the diseased wood. Mercury canker paints should not be applied to young wood of Bramley's Seedling, Grenadier, Lord Lambourne or Scarlet Pimpernel. Attention to control of woolly aphid and to scab will prevent entry through injuries made by these. Where disease is severe spray with Bordeaux mixture applied (1) just before leaf fall begins, (2) at about 50% leaf fall and (3) in spring when the buds start to swell. The spray should contain the equivalent of 2·5 lb metallic copper/100 gal (H.V.) *or* 5 lb metallic copper/acre (L.V.).

The apple varieties Cox's Orange Pippin, James Grieve and Worcester Pearmain are particularly liable to canker, whereas Bramley's Seedling, Lane's Prince Albert and Newton Wonder are resistant. Environmental and cultural conditions influence liability to canker (see A.L. 100).

**Collar Rot** (*Phytophthora cactorum* and *P. syringae*). Infected areas of bark occur at, or just above, soil level. They are best seen in spring and autumn, often with cracks at margin. Small oily or watersoaked patches may occur on the infected area and also an ooze of reddish brown droplets. If extension of the area ceases, the bark shrinks and appears smooth and shrunken and cracks away clearly from the surrounding healthy bark.

Severe outbreaks have been recorded only in mature orchards of Cox's Orange Pippin and less severe in several other commercial varieties. The common commercially used Malling and Malling-Merton rootstocks are resistant.

Rotted pockets of bark, together with about 2 in of the surrounding healthy bark, should be cut out and the chippings burnt. The wound should be covered with Bordeaux mixed slurry made from Bordeaux mixture w.p. and water to protect from further infection.

The disease is soil-borne and all trees in the vicinity should be protected as follows: (1) clear trunk bases of debris and weeds, keeping free with herbicides; (2) remove soil from graft unions where feasible; (3) avoid mechanical injury to base of tree; (4) remove fallen fruit; (5) paint trunk bases with slurry made from Bordeaux w.p. and water. (Undetected infection will often appear reddish brown after slurry has dried.)

In subsequent years, tree bases and 1 ft soil width around should be protected by spraying with copper sulphate, 20 lb/100 gal, *or* copper oxychloride w.p. or liquid formulation at 20 lb a.i./100 gal.

***Fire blight** (*Erwinia amylovora*). Although this disease is at present confined mainly to pears and to certain ornamental plants, a few trees of the apple varieties Crawley Beauty, Egremont Russet, Charles Ross and Miller's Seedling have been infected. Precautions recommended for pear (see p. 190) should be taken where the disease is known to exist.

*****Gloeosporium rot (Bitter rot) of apple** (*Gloeosporium album, G. perennans* and *G. fructigenum*). Most of the rotting of apples stored until January or later is due to one or more species of *Gloeosporium*. These fungi exist on small cankers which may cause dieback of shoots or may be insignificant. From the cankers, spores are produced all the year round, but especially in autumn, when pruning cuts and other wounds may be infected. During wet periods the spores are washed to the fruit surfaces where after entry of the lenticel chambers they remain dormant until the fruit reaches a stage of maturity in store which permits further penetration by the fungus.

The variety Cox's Orange Pippin is particularly liable to the disease.

At present this disease can only be reduced in severity; there is no complete control. Spray with captan at 1 lb a.i. (2 lb 50% w.p.)/100 gal (H.V.) *or* at 2·5 lb a.i./acre (L.V.) at about mid July and then at 2–3 week intervals until orchard conditions make further applications impracticable. The variety Worcester may be sensitive to thiram in some conditions.

Pruning should be delayed until January at least. Cankers, dead shoots and mummified apples should be removed and burnt.

*****Powdery mildew** (*Podosphaera leucotricha*) (A.L. 205). The fungus overwinters in buds and when an infected spur bud breaks the emerging growth appears white and mealy due to the presence of a large number of spores. Diseased blossoms and leaves wither and drop from the tree. A terminal bud in which the fungus has overwintered often produces a 'silvered' shoot with mildewed leaves. Primary outbreaks can cause secondary infections of unexpanded leaves, of new shoots of growing points, and also of young buds. The latter begin the cycle again during the following year.

The cutting-out of diseased parts is strongly recommended (A.L. 205) but, where it is not practicable, spray with DNOC-petroleum oil, at 0·1% DNOC and 2·9% oil (4½% e.c.)/100 gal (H.V.) once in December or January during the dormant period but not later than bud swelling. Lime sulphur at 1·5–2·0 gal/100 gal (H.V.) with a wetter applied at early pink bud stage will also help to reduce infection but may cause leaf damage. From blossom time, spray with binapacryl at 0·5 lb a.i./acre (7-day interval) *or* 0·75 lb a.i./acre (10-day interval) *or* 1 lb a.i./acre (14-day interval) *or* dinocap at 0·25 lb a.i./100 gal (H.V.) *or* 0·5 lb a.i./acre (L.V.) *or* lime sulphur at 1 gal/100 gal (H.V.) *or* wettable sulphur at 3 lb/100 gal (H.V.) at 7–10 day intervals for at least 6 weeks. The trees, especially the growing tips, must be thoroughly wetted. For mildew control, applications at high volume are considered more effective than those at low volume. With binapacryl, the quantities stated should be applied in not less than 50 gal water and the minimum interval between the last application and harvesting must be 4 weeks. It should not be used where trees are under-planted with soft fruit. Several apple varieties are sensitive to sulphur (see apple scab, Table 9.2, p. 185).

****Replant problem.** In some orchards, retarded development of trees planted on or next to grubbed trees of the same type (particularly apple after apple or pear, and cherry after cherry or plum) often occurs. The reason for this is not known but retardation of growth may be avoided in most 'problem' soils if the sites of the grubbed trees are first treated with chloropicrin.

Treatment of large areas is done with a machine (e.g. modified Egedal soil injector) which dribbles the fluid behind special tines attached to a 4 ft 6 in wide drawbar. The machine is drawn along each row to be replanted. A 5 gal drum of chloropicrin will treat 689 yds of row when applied at 25 gal/treated acre.

For gapping up a single grubbed tree and for other small areas, a hand-operated injector may be used. With this a square area with a side of 2 yds—tree site in centre—should be treated at 9 in staggered intervals and a depth of 6 in. Each area requires 3½ fl. oz.

Chloropicrin is the only chemical which has proved effective in solving the replant problem but because it is unpleasant to use, special precautions must be observed. These include the use of protective clothing and a respirator fitted with a type C or CC canister when the container is opened and when attaching to machine or filling hand injectors; also when washing out the injector and container.

Before treatment the soil must be worked to a fine tilth and have a slight moisture content. Excessive moisture or dryness reduces the effectiveness of the treatment.

*****Scab** (*Venturia inaequalis*) (A.L. 245). The scab fungus overwinters within the tissues of fallen apple leaves. Spore cases (perithecia) develop in early spring and, during wet periods, spores (ascospores) are ejected. When the weather is warm and humid the spores germinate on young leaves and fruitlets causing scab infections. Diseased parts soon produce summer spores (conidia) which themselves continue the spread. Infected fruits become spotted, distorted and unsaleable. In some varieties, including Cox's Orange Pippin, young extension shoots become infected and the following spring produce cushions of conidia beneath blister-like swellings.

Routine scab control by the application of fungicides is an essential part of apple growing. Two methods of determining the intervals between applications are recognized according to the type of programme (preventive or curative). With preventive spraying, the aim is to ensure that there is always a sufficient deposit of fungicide on the tree during the vulnerable period. This object is achieved by spraying at 10-day intervals from bud burst until late June, though the time intervals between sprays may be modified according to the maker's instructions. With a curative programme the aim is to apply a suitable fungicide immediately after weather conditions have been favourable for infection, that is, as soon as possible after an infection period assessed from a Mills table (Table 9.1) giving periods of leaf wetness needed at various temperatures. Fungicides containing dodine acetate or an organomercury compound are recommended for curative spraying.

Combinations of a fungicide with an insecticide or with another fungicide are commonly used. Before adopting one of these, it is necessary to ensure that the products to be mixed are compatible and guidance from the manufacturers should be sought. Suitable spray schedules are given in Table 9.2.

TABLE 9.1

*Showing hours of wetness needed for ascospores to infect leaves on the tree**

| Average temperature during wet period (degrees Fahrenheit) | Time wetness must persist for scab infection (hours) |
| --- | --- |
| 33 to 41 | 48 or more |
| 42 | 30 |
| 45 | 20 |
| 50 | 14 |
| 55 | 11 |
| 58 | 10 |
| 62 | 9 |

* From figures from Western New York State, U.S.A. (Mills, L.D. and Laplante, A.A., 1951, *Cornell Ext. Bull.* No. 711, pp. 21–7.)

TABLE 9.2

*Apple and Pear Scab—Fungicide (a.i.) Rates*

(Unless certain that a fungicide is safe to use on a particular variety in a particular orchard, an Advisory Officer should be consulted. Always read the label on the container.)

| | Bud burst | Green cluster | Petal fall and after | Remarks |
|---|---|---|---|---|
| Captan w.p. | 1 lb a.i./100 gal (H.V.)<br>2·5 lb a.i./acre (L.V.) | As at bud burst<br>As at bud burst | As at bud burst<br>As at bud burst | The following varieties are sensitive under some conditions: Bramley's Seedling, Monarch, King Edward, Kidd's Late Orange, Winston and also D'Anjou pears. |
| Copper oxide w.p.<br>Bordeaux mixture w.p. | 1·5–2 lb metallic copper/ 100 gal (H.V.)<br>5 lb metallic copper/acre (L.V.) | No | No | (1) At bud burst or early bud stage only.<br>(2) Russeting may be caused unless safeners are added.<br>(3) Some pear varieties, including Doyenne du Comice, susceptible to copper damage. |

TABLE 9.2.—*cont.*

| | Bud burst | Green cluster | Petal fall and after | Remarks |
|---|---|---|---|---|
| Dithianon liquid suspension | 9 oz a.i./100 gal (H.V.)<br>18 oz a.i./acre (L.V.) | As at bud burst<br>As at bud burst | As at bud burst<br>As at bud burst | Last treatment at least 8 weeks before harvest. |
| Dodine w.p. | *Protective*<br>*7-day interval*<br>3·9 oz a.i./100 gal (H.V.)<br>7·8 oz a.i./acre (L.V.)<br>*or*<br>*10-day interval*<br>5·2 oz a.i./100 gal (H.V.)<br>10·4 oz a.i./acre (L.V.)<br>*or*<br>*14-day interval*<br>7·8 oz a.i./100 gal (H.V.)<br>15·6 oz a.i./acre (L.V.)<br><br>*Curative*<br>7·8 oz a.i./100 gal (H.V.)<br>15·6 oz a.i./acre (L.V.) | <br><br>As at bud burst<br>As at bud burst<br><br><br>As at bud burst<br>As at bud burst<br><br><br>As at bud burst<br>As at bud burst<br><br><br>As at bud burst<br>As at bud burst | <br>*7-day interval*<br>2·6 oz a.i./100 gal<br>5·2 oz a.i./acre<br>*or*<br>*10-day interval*<br>3·9 oz a.i./100 gal<br>7·8 oz a.i./acre<br>*or*<br>*14-day interval*<br>5·2 oz a.i./100 gal<br>10·4 oz a.i./acre<br><br><br>As at bud burst | For pears, concentration as at bud burst until 2 weeks after petal fall |
| Dodine liquid | *Protective*<br>5·0 oz a.i./100 gal (H.V.)<br>10·0 oz a.i./acre (L.V.)<br><br>*Curative*<br>8·0 oz a.i./100 gal (H.V.)<br>16·0 oz a.i./acre (L.V.) | <br>As at bud burst<br>As at bud burst<br><br><br>As at bud burst<br>As at bud burst | <br>As at bud burst<br>As at bud burst<br><br><br>As at bud burst<br>As at bud burst | |

TABLE 9.2.—*cont.*

| | Bud burst | Green cluster | Petal fall and after | Remarks |
|---|---|---|---|---|
| Mancozeb with zineb w.p. | According to label | According to label | According to label | Last treatment at least 1 week before harvest. |
| Organomercury w.p. or liquid | *Protective and Curative* (apple)<br>0·24–0·48 oz mercury/100 gal (H.V.) | As at bud burst | As at bud burst | (1) Last treatment at least 42 days before harvest. |
| | 0·6–1·2 oz mercury/acre (L.V.) | As at bud burst | As at bud burst | (2) Some varieties of apples, particularly Cox's Orange Pippin, and some pear varieties, are sensitive and a lower rate must be used. Do not apply to Doyenne du Comice. |
| Organomercury with sulphur w.p. (severe mildew) | 0·25 oz mercury + 3 lb sulphur/100 gal (H.V.) | As at bud burst | As at bud burst | (3) Do not spray after end of July. |
| | | As at bud burst | As at bud burst | (4) Prevent access of animals and poultry to treated areas for at least 2 weeks. |
| (light mildew) | Organomercury alone | Organomercury alone | 0·15 oz metallic mercury + 2 lb sulphur/100 gal (H.V.) | (5) Harmful to bees. Do not apply at flowering stages. Keep down flowering weeds. |

For post harvest/pre bud sprays on apple see section below Table 9.1.

TABLE 9.2.—*cont.*

| | Bud burst | Green cluster | Petal fall and after | Remarks |
|---|---|---|---|---|
| Lime sulphur liquid | 3 gal/100 gal (H.V.)<br>3 gal/acre (L.V.) | 2 gal/100<br>2 gal/acre | 1 gal/100<br>1 gal/acre | (1) Lime sulphur may cause damage to several varieties, including Cox's Orange Pippin Lane's Prince Albert, Beauty of Bath and Newton Wonder.<br>(2) Apply lime sulphur to pears at pre blossom only.<br>(3) All types of sulphur cause damage to some apple varieties and to pear Doyenne du Comice. |
| Sulphur w.p. or equivalent as Colloidal sulphur | 4.55 lb sulphur/100 gal (H.V.)<br>7·5 lb sulphur/acre (L.V.) | 3 lb/100<br><br>7·5 lb/acre | 3 lb/100<br><br>7·5 lb/acre | |
| Thiram w.p. or liquid | 1·6 lb a.i./100 gal (H.V.)<br>3·2–4 lb a.i./acre (L.V.)<br>(pear only) | 1·6 lb/100 gal<br>3·2–4 lb/acre<br>(pear only) | 1·6 lb/100 gal<br>3·2–4 lb/acre<br>(pear only) | (1) Should not be used on fruit for canning or quick freeze.<br>(2) Last treatment at least 7 days before harvest. |

In apple orchards where traditional control measures have not been sufficiently effective, applications of phenylmercury chloride in autumn and again in spring have much reduced infection by ascospores and so made the normal routine sprays more effective. It should be applied (*a*) after picking but before leaf fall at 0·99 oz mercury (4 oz 40% conc.)/100 gal at not less than 200 gal/acre. This spray should be directed at the leaves on the tree and any fallen leaves; (*b*) before bud burst at 1·98 oz mercury (8 oz 40% conc.)/100 gal at not less than 100 gal/acre. This spray is directed mainly at the leaves on the ground but the trees must also be sprayed.

## 9·2 PEAR

### 9.2A PESTS

Annual sprays are usually applied for aphids, and pear sucker in areas where this is a pest. Control measures for codling moth and red spider mites may be required locally.

***Aphids.** The pear-bedstraw aphid (*Dysaphis pyri*) is the most important. It overwinters as eggs on the tree which have finished hatching by the white bud stage. The pinkish aphids cause severe leaf curling and may spread over the tree, persisting into July and then departing for bedstraw (*Galium* spp.), returning to pear in the autmn. The pear-coltsfoot aphid (*Anuraphis farfarae*) causes the leaves to fold upwards and turn red; adults are brown, the young yellow-green. Other aphids occurring include the apple-grass aphid—light infestations are often present—and green apple aphid (see under apple aphids, p. 160), and the pear-grass aphid (*Geoktapia pyraria*).

Control may be achieved either by winter washes against the eggs, or spring sprays against the newly hatched aphids.

(*a*) Winter washes. The following are effective if thoroughly applied: tar oil, 5% (miscibles) or 5–6% (stock emulsions) when the buds are dormant in December–mid February, *or* DNOC-petroleum oil 4½%, in February–March at bud break.

Do not winter wash in windy or frosty weather or when the trees are wet. Do not use DNOC-petroleum oil after the buds have reached the breaking stage.

(*b*) Spring sprays. Spray at green cluster with one of the following organophosphorus insecticides (doses as a.i./100 gal for H.V. and a.i./acre for L.V.: demeton-S-methyl, H.V. 1·16 oz (2 fl. oz 58% e.c.), L.V. 2·32 oz; *or* azinphos-methyl + demeton-S-methyl sulphone, H.V. 1 lb 25%/7·5% w.p., L.V. 2 lb w.p.; *or* dimethoate, H.V. 4·8 oz (12 fl. oz 40% e.c.), L.V. 9·6–12 oz; *or* formothion, H.V. 7 oz (16 fl. oz 43% e.c.), L.V. 14 oz; *or* malathion, H.V. 9 oz (¾ pint 60% e.c.), L.V. 18–24 oz; *or* mecarbam, H.V. 5·8 o (8 fl. oz 80% e.c.), (L.V. not recommended); *or* mevinphos,

H.V. 1–1½ oz (1–1½ fl. oz 99%), L.V. 2–3 oz (use the higher rate in cool weather and increase to 2 oz H.V. or 4 oz L.V. if caterpillar control required); *or* oxydemeton-methyl, H.V. 1·14 oz (2 fl. oz 56% e.c.), L.V. 2·28 oz; *or* phosphamidon, H.V. 4·8 oz (16 fl. oz 30% s.c.), L.V. 9·6–12·0 oz; *or* vamidothion, H.V. 6·4 oz (16 fl. oz 40% e.c.), L.V. 12·8–16 oz.

Alternatively spray at petal fall, when combined control of aphids, pear sucker (p. 189), winter moth (p. 176) and bryobia mite (see below) can be obtained if required by choosing the appropriate insecticides.

Also effective are: gamma-BHC, H.V. 2–4 oz, L.V. 4–8 oz; *or* DDT, H.V. ¾–1 lb, L.V. 1–2 lb; *or* nicotine, H.V. 8 oz (95–98%), L.V. 8–10 oz. With nicotine a minimum air temperature of 60°F is desirable. Preference should be given to the organophosphorus insecticides, which in any case are more reliable.

Follow the precautions against poisoning bees, and for avoiding taint of nearby crops with gamma-BHC, given under apple aphids on page 162. See pp. 28–61 for minimum interval between last application of insecticides and harvest.

***Apple blossom weevil** (*Anthonomus pomorum*) (A.L. 28). See under Apple, p. 162.

***Apple twig cutter** (*Rhynchites coeruleus*). See under Apple, p. 166.

****Bryobia mites** (*Bryobia rubrioculus, B. cristatas*). Although *B. rubrioculus* (see p. 166 for life history) will attack pear as well as apple, the grass-pear bryobia (*B. cristata*) is apparently more important and serious infestations have been reported, especially in east Kent. This mite breeds on grasses (including annual meadow grass and couch), clover, cinquefoil, mallow and other weeds, where 5 generations occur during the summer. It winters mainly as adults and young, but breeding also occurs in mild weather in winter. In the spring, usually May, dispersal occurs to trees including damson, bullace, hawthorn, rose, apple and pear. Conference pear seems to be particularly favoured. Two generations occur on the trees and the mites then return to grasses and weeds.

Bryobia mites are susceptible to most if not all the acaricides recommended for fruit tree red spider (see p. 170). With both *B. rubrioculus* and *B. cristata*, a single spray of an organophosphorus insecticide at the time of petal fall can be very effective. With heavy infestations a second spray may be needed 2–3 weeks later. Well timed and applied sprays may prevent the need for a subsequent annual routine. The post blossom timing is sufficiently close to that for pear sucker and aphids for all these pests to be controlled at the same time where necessary, using an appropriate organophosphorus insecticide (see under pear sucker, p. 189, and aphids, p. 186).

With *B. cristata* clearing grass and weeds for about 2 ft round the trunk will help to prevent infestations.

Acaricides suitable for the control of *Bryobia* on pear and effective against active stages are (doses as a.i.) 100 gal, H.V., or a.i./acre, L.V.): azinphos-methyl, 5·28 oz (24 fl. oz 22% e.c.), L.V. 10·56 oz; *or* azinphos-methyl + demeton-S-methyl sulphone, H.V. 1 lb 25%/7·5% w.p.), L.V. 2 lb; *or* demeton-S-methyl, H.V. 3·5 oz (6 fl. oz 58% e.c.), L.V. 7·0 oz; *or* dimethoate, H.V. 3·6 oz (9 fl. oz 40% e.c.), L.V. 7·2–9·0 oz; *or* formothion, H.V. 5 oz (12 fl. oz 43% e.c.), L.V. 10 oz; *or* malathion, H.V. 18 oz (1½ pints 60% e.c.), L.V. 36–45 oz; *or* oxydemeton-methyl, H.V. 3·4 oz (6 fl. oz 56% e.c.), L.V. 6·8 oz; *or* phenkapton, H.V. 3·2 oz (16 fl. oz 20% e.c.), L.V. 6·4–8·0 oz; *or* phosalone, H.V. 5·0 oz (15 fl. oz 33% e.c.), L.V. 10 oz; *or* phosphamidon, H.V. 4·8 oz (16 fl. oz 30% s.c.), L.V. 9·6–12·0 oz; *or* vamidothion (single application only permitted), H.V. 6·4 oz (16 fl. oz 40% e.c.), L.V. 12·8–16·0 oz. Also effective are tetradifon, H.V. 2·0 oz (1¼ pints 8% e.c.), L.V. 4·0 oz; *or* tetrasul, H.V. 6·4 oz (2 lb 20% w.p.), L.V. 12·8 oz, which chiefly affect eggs and young stages; these could be used where later sprays are required and both these stages are present.

***Clay-coloured weevil** (*Otiorhynchus singularis*) (A.L. 154). See under Currant, p. 205.

****Codling moth** (*Laspeyresia pomonella*) (A.L. 42) is less common as a pest of pear than apple; one spray of lead arsenate, at rates as for apple, at petal fall, can be effective, or if required follow the spray programme for apple, see p. 167.

***Common green capsid** (*Lygocoris pabulinus*) (A.L. 154). There is little information on which to base an alternative to the standard control of DDT (1 lb a.i./100 gal H.V. or 2 lb/acre L.V.) at white bud but dimethoate (12 fl. oz 40% e.c./100 gal H.V., 24 fl. oz/acre L.V.), *or* formothion (16 fl. oz 40% e.c./100 gal H.V., 32 fl. oz/acre L.V.), *or* malathion (1½ pints 60% e.c./100 gal H.V., 3 pints/acre L.V.) are the most effective of the organophosphorus compounds, and their use at petal fall is suggested.

***Fruit tree red spider mite** (*Panonychus ulmi*) (A.L. 10) is not common as a pest on pears but, where required, follow the summer spray programme for apple, see p. 170, except that binapacryl, dinocap or quinomethionate should not be used.

***Leaf weevils** (*Phyllobius* spp.). See under Apple, p. 173).

***Mussel scale** (*Mytilococcus ulmi*) (A.L. 36). See under Apple, p. 173.

***Oystershell scale** (*Quadraspidiotus ostreaeformis*). See under Apple, p. 173.

***Pear leaf blister mite** (*Eriophyes pyri*) (A.L. 35). Adult mites hibernate under bud scales, invading leaves at bud burst. Leaves become dotted with yellowish or reddish blisters, and reddish or brown pustules occur on fruitlets. The mite is mostly a pest of wall trees, and is controlled by 5% lime sulphur (H.V.) in March just before the buds open. In Canada good control has been given by carbaryl, ½ lb a.i./100 gal (H.V.).

**Pear leaf midge** (*Dasyneura pyri*). Midges lay eggs in folds of young leaves in May; larvae feed on leaves, the edges of which become rolled upwards. There are several overlapping generations a year. The midge is sometimes a pest of nursery trees, and may be controlled by DDT 1 lb a.i./100 gal (H.V.) in late May or in June, repeated once or twice at 10–14 day intervals. Vamidothion has given promising results with one spray in June.

**Pear midge** (*Contarinia pyrivora*) (A.L. 26). Midges lay their eggs in the flower buds, and the yellowish white larvae feed inside the fruitlets which become swollen and deformed. To control, apply DDT 1 lb a.i./100 gal (H.V.) or 2–2½ lb/acre (L.V.), at white bud. Repeat in following season. Recent trials suggest that carbaryl at 1 lb a.i. (2 lb 50% w.p.)/100 gal at white bud is an effective and probably a superior alternative to DDT.

To reduce hazard to bees the precautions given under apple aphids (p. 162) should be observed.

**Pear sawfly** (*Hoplocampa brevis*), is a local pest, similar to the apple sawfly in habits and damage caused. It is controlled by spraying with gamma-BHC 2 oz a.i./100 gal (H.V.) or 4–5 oz/acre L.V., within 7 days after petal fall, *or* nicotine 8 oz/100 gal at petal fall. Organophosphorus insecticides as for apple sawfly are probably also effective.

Follow the precautions against poisoning bees, and for avoiding taint of nearby crops with gamma-BHC, given under apple aphids on p. 162. See pp. 28–61 for minimum interval between last application of insecticides and harvest.

**Pear slug sawfly** (*Caliroa cerasi*) (A.L. 84). The adults appear in May and June, eggs are laid in slits cut in the leaf. The caterpillars are yellowish white at first but turn dark greenish or black and are slug-like in appearance. They feed on the upper leaf surface and leaves may become skeletonized. There are 2–3 generations a year.

To control the pest, spray in June if caterpillars are seen with one of the following:

Derris (0·0035% rotenone in diluted wash), *or* lead arsenate powder, 2–3 lb/100 gal (H.V.) or 4–5 lb/acre (L.V.), or pastes and suspensions 4 lb (2 pints)/100 gal (H.V.) or 8–10 lb/acre (L.V.) *or* gamma-BHC 2 oz a.i./100 gal (H.V.), or 4–6 oz a.i./acre (L.V.), *or* nicotine, 8 oz/100 gal (H.V.).

Follow the precautions against poisoning bees, and for avoiding taint of nearby crops with gamma-BHC, given under apple aphids on page 162. See pages 28–61 for minimum interval between last application of insecticides and harvest.

**Pear sucker** (*Psylla simulans*). Adults hibernate at rest on the bark, or among dead leaves or other shelter. Eggs are laid on the shoots and spurs from March to petal fall, and the young suckers feed on the buds and blossom trusses. Two further generations follow and if the insects become numerous they render the leaves sticky with honeydew on which sooty moulds grow.

In some orchards one control spray is sufficient and the best time is probably about 3 weeks after petal fall when eggs are fewest. Where heavy infestations occur two sprays are probably necessary for full control, one at petal fall (which can be combined with aphid control) and a second 3 weeks later. The following are effective (doses as a.i./100 gal, H.V., or a.i./acre, L.V.): azinphos-methyl, H.V. 5·28 oz (24 fl. oz 22% e.c.), L.V. 10·5 oz; *or* azinphos-methyl + demeton-S-methyl sulphone, H.V. 1 lb 25%/7·5% w.p.; *or* dimethoate, H.V. 4·8 oz (12 fl. oz 40% e.c.), L.V. 9·6–12·0 oz; *or* formothion, H.V. 7 oz (16 fl. oz 43% e.c.), L.V. 14 oz; *or* malathion, H.V. 18 oz (1½ pints 60% e.c.), L.V. 36–45 oz; *or* mecarbam, H.V. 5·8 oz (8 fl. oz 80% e.c.) (L.V. not recommended); *or* phosphamidon, H.V. 4·8 oz (16 fl. oz 30% s.c.), L.V. 9·6–12·0 oz; *or* vamidothion, H.V. 6·4 oz (16 fl. oz 40% e.c.). Nicotine, H.V. 8 oz (96–98%), L.V. 16–10 oz, can also be effective but requires temperatures of 60°F or over and is not persistent.

The timing of sprays for pear sucker, aphids and bryobia mite are sufficiently similar for the three pests to be controlled at the same time if required and using the same insecticide.

5% tar oil in December–mid February will give some control by killing suckers hibernating on the trees.

See under apple aphids (p. 160) for precautions against poisoning bees, and risk of oil damage.

***Pear thrips** (*Taeniothrips inconsequens*). The insects feed on the flowers and young foliage, and may cause russeting of the fruit of apple, pear and plum. They are controlled by DDT ½ lb a.i./100 gal (H.V.) at white bud.

****Red bud borer.** See under Apple, p. 173.

***Tortrix caterpillars.** Several species occur on pear, including *Archips podana* (see under Apple, p. 174) but are rarely important.

****Wasps** (*Vespula* spp.). See under Apple, p. 176.

****Winter moth caterpillars.** See under Apple, p. 176.


## 9.2B DISEASES

****Canker** (*Nectria galligena*) (A.L. 100). Common canker of the pear is caused by the same fungus as that causing apple canker. Biology and control measures are the same as for apple (p. 178).

****Fire blight** (*Erwinia amylovora*) First recorded in Great Britain in

1957 on pears, this disease has spread rapidly and has since infected a number of apple varieties and also other members of the sub-family Pomoideae. These include hawthorn, cotoneaster, pyracantha, stransvaesia, white beam and mountain ash. In pears, infection by the causative bacterium generally occurs through late summer (secondary) blossom, killing this and progressing via the stalk into the twig, branch and finally the trunk. Because the pear variety Laxton's Superb is particularly prone to produce secondary blossoms over a long period, it is very liable to attack.

In summer and autumn, parts of the bark containing active bacteria usually show a red discoloration on cutting but during winter months the discoloration may be dark brown. In a mild winter the bacteria may continue to advance along the affected part but in certain circumstances a limited canker is formed and this may crack at the margin thus becoming isolated from healthy tissue external to it. The bacteria in some of these 'holdover cankers' may survive until the following spring when they can initiate new outbreaks.

Although infection is usually through the blossom, shoots and leaves may also be attacked.

Statutory regulations under the Fire Blight Disease Order require that where fire blight occurs or is believed to occur, it should be immediately reported to the Plant Health Branch of the M.A.F.F.

The grubbing of badly infected trees may be required, but where cutting out of branches is permissible, this should be done at not less than 2 ft below visible signs of the disease within the bark. To reduce the chance of reinfection, the cut surface should be painted immediately with white lead finishing paint conforming to B.S. 2526/7.

The cutting parts of all pruning tools should be immersed in 3% lysol between making cuts on both affected and healthy trees.

***Scab** (*Venturia pirina*) (A.L. 245). Pear scab is caused by a fungus different from that causing apple scab. Nevertheless, the life histories of the two fungi are almost identical and the reader is referred to the description of apple scab (p. 180). In contrast to apple, pustules on the wood are of general occurrence and may produce spores after the first year, so are more important.

Spray protectively at intervals of about 10 days from bud burst until at least the end of May with captan, dodine or wettable sulphur as for apples; later sprays at 14-day intervals may be necessary. Thiram is also suitable. Mercury- or copper-containing sprays are not recommended after green cluster because of possible damage. In general the same strengths of fungicide as recommended for apples are suitable for most pear varieties. Dodine has been recorded as damaging in cold spring weather. Doyenne du Comice must not be sprayed with copper, sulphur or mercury.

## 9.3 QUINCE

### 9.3A DISEASE

****Leaf blight and fruit spot** (*Fabraea maculata = Entomosporium maculatum*). The fungus infects foliage, fruit and young twigs. On the leaves, the spots are at first reddish becoming nearly black; they are irregular in shape and usually up to $\frac{1}{10}$ in across. Spores are produced from these spots. If spots run together the leaves turn yellow and fall early. On fruits, the spots are very dark brown, slightly sunken and produce spores; in severe cases deformity may occur. The disease is common on fruiting trees and may occur in nursery rows where quinces are propagated as rootstocks for pears. Malling Quince A is fairly resistant.

On fruiting trees infected twigs should be removed and burnt while the foliage should be sprayed with Bordeaux mixture or other copper-containing fungicide recommended for this purpose by the makers.

## 9.4 APRICOT, NECTARINE AND PEACH

### 9.4A PESTS

*****Aphids.** Several species overwinter as eggs on these trees. The peach-potato aphid (*Myzus persicae*) is green, causing severe leaf curl in the spring, and dispersal to summer host plants occurs in May and June. The peach aphid (*Appelia schwartzi*) is dark brown and infests the shoots, causing severe leaf curl; it disperses to other peach trees in the summer. The black peach aphid (*Brachycaudus persicaecola*) overwinters as aphids on the roots, and infests the shoots during the summer. The mealy peach aphid (*Hyalopterus amygdali*) also occurs, dispersing in the summer to reeds.

Either winter washes against the eggs or spring sprays against the aphids may be used to control these pests. In winter, tar oil 5% (H.V.), in first half of December for outdoor trees, is effective if thoroughly applied.

In spring, use one of the following at the end of the blossom period (doses as a.i./100 gal, H.V., or a.i./acre, L.V.): demeton-S-methyl, H.V. 3·5 oz (6 fl. oz 58% e.c.), L.V. 7·0 oz; *or* dimethoate, 4·8 oz (12 fl. oz 40% e.c.) H.V., 9·6–12·0 oz L.V.; *or* malathion, H.V. 18 oz (1½ pints 60% e.c.), L.V. 36 oz; *or* oxydemeton-methyl, H.V. 3·4 oz (6 fl. oz 56% e.c.), L.V. 6·8 oz.

Follow the precautions against poisoning bees given under Apple aphids on p. 162. See pp. 28–61 for minimum interval between last application of insecticides and harvest.

***Common green capsid** (*Lygocoris pabulinus*) (A.L. 164). See under Currant, p. 205.

***Fruit tree red spide mite** (*Panonychus ulmi*) (A.L. 10). This mite may occur on outdoor trees and its life history is as described under Apple (p. 170).

The timing of control sprays is as recommended for apple. With heavy infestations spray about the time of petal fall of Cox's Orange Pippin, and again 3 weeks later if necessary, using an acaricide effective against the active stages, e.g. (doses as a.i./100 gal, H.V., or a.i./acre, L.V.): demeton-S-methyl, H.V. 3·5 oz (6 fl. oz 58% e.c.), L.V. 7·0 oz; *or* dimethoate, 3·6 oz (9 fl. oz 40% e.c.) H.V., 7·2–9·0 oz L.V.; *or* malathion, H.V. 18 oz (1½ pints 60% e.c.), L.V. 36–45 oz; *or* oxydemeton-methyl, H.V. 3·4 oz (6 fl. oz 56% e.c.), L.V. 6·8 oz.

With light infestations spraying can be delayed until mid June, repeated 3 weeks later if necessary, when it is an advantage to use an acaricide effective against eggs and young stages, e.g. (doses as a.i./100 gal, H.V., or a.i./acre, L.V.): chlorbenside, H.V. or L.V. 6–10 oz (1½–2½ pints 20% e.c. or 3 lb 20% w.p.); *or* chlorfenson, H.V. 1 lb (2 lb 50% w.p.), L.V. 2 lb; *or* fenson, H.V. 6·4 oz (2 lb 20% w.p.); *or* tetradifon, H.V. 2 oz (1¼ pints 8% e.c.), L.V. 4 oz.

Where only a single spray is given, the best time is about 10 days after petal fall of the apple Cox's Orange Pippin and an acaricide effective against active stages should be used (see p. 171).

Follow the precautions against poisoning bees given under Apple aphids on p. 162. See pp. 28–61 for minimum interval between last application of insecticides and harvest.

***Red spider mite** (*Tetranychus urticae*) (A.L. 226). This is the red spider mite common on many crops outdoors and under glass, sometimes infesting top and soft fruit. It overwinters as adult females in the soil and other shelter, and these emerge in spring to feed and breed on the leaves.

Apply one of the following about mid May and again 2–3 weeks later if required (doses as a.i./100 gal, H.V., or a.i./acre, L.V.): chlorbenside, H.V. or L.V., 6–10 oz (1½–2½ pints 20% e.c. or 3 lb 20% w.p.); *or* chlorfenson, H.V. 1 lb (2 lb 50% w.p.), L.V. 2 lb; *or* demeton-S-methyl, H.V. 3·5 oz (6 fl. oz 58% e.c.), L.V. 7·0 oz; *or* fenson, 6·4 oz (2 lb 20% w.p.) H.V.; *or* oxydemeton-methyl, H.V. 3·4 oz (6 fl. oz 56% e.c.), L.V. 6·6 oz; *or* tetradifon, H.V. 2 oz (1¼ pints 8% e.c.), L.V. 4 oz.

Follow the precautions against poisoning bees given under apple aphids on p. 162. See pp. 28–61 for minimum interval between last application of insecticides and harvest.

***Scale insects.** The brown or peach scale (*Parthenolecanium corni*) (A.L. 88), mussel scale (*Mytilococcus ulmi*) (A.L. 36) and oyster-shell scale (*Quadraspidiotus ostreaeformis*) may occur. Tar oil 5% H.V. in first half December is effective. Where control during the growing season is required, malathion can be effective—see under mussel scale, p. 173, and brown scale, p. 204. With mussel scale under glass spray 3–4 weeks earlier than for outdoor trees.

****Wasps** (*Vespula* spp.) (A.L. 457)—see under Apple, p. 176.

***Winter moths** (A.L. 11). See under Apple, p. 176. Caterpillars of this group may occur on outdoor trees and are controlled by applying DDT $\frac{1}{2}$ lb a.i./100 gal (H.V.) or 1–1$\frac{1}{4}$ lb a.i./acre (L.V.) in April when caterpillars are first seen. There is no information on alternative insecticides for use on peaches.

### 9.4B DISEASES

****Peach leaf curl** (*Taphrina deformans*) (A.L. 81). Young diseased leaves are thick and yellow, tinged with red. Older leaves are crumpled and redder. Premature defoliation weakens growth and may render nursery stock uselss. The causative fungus overwinters in the bark and between the bud scales.

Apply a copper compound or lime sulphur before the buds begin to swell, usually in late February or early March and repeat immediately before leaf fall in the autumn. Suitable copper sprays are: Bordeaux mixture (w.p.) *or* copper oxychloride (w.p.) *or* copper oxide (w.p.) *or* cuprammonium compound (soluble), each applied at 1–1·5 lb metallic copper/100 gal (H.V.). The lime sulphur wash should contain 3 gal lime sulphur/100 gal (H.V.).

The removal of affected leaves before a bloom (spores) appears on them will help control.

****Peach powdery mildew** (*Sphaerotheca pannosa* var. *persicae*). Powdery (sporing) patches appear on infected leaves and young shoots. The fungus overwinters in the buds and these produce stunted shoots which bear narrow leaves. To control the disease apply, when mildew first seen and at 14 day intervals, sulphur sprays at 3 lb sulphur content/100 gal (H.V.).

## 9.5 CHERRY

### 9.5A PESTS

Routine sprays are required for cherry black fly, and DDT is included in spring sprays for winter moth caterpillars where necessary.

*****Aphids.** The cherry black fly (*Myzus cerasi*) is the only species on cherry. It overwinters as eggs on the bark and these hatch in March and April, hatching being complete by the white bud stage. Successive generations of black aphids are produced, then winged forms which disperse in June and July to bedstraws. Leaves are severely curled and new growth checked.

Either winter washes against the eggs or spring sprays against the aphids may be used:

(*a*) Winter washes. Tar oil, 5% (miscible) or 5–6% (stock emulsion) H.V. when the buds are dormant (December–January), *or* DNOC-petroleum oil, $4\frac{1}{2}$%, up to bud break.

(*b*) Spring sprays. Any of the following are effective at the white bud stage (doses as a.i./100 gal, H.V., or a.i./acre, L.V.): gamma-BHC, H.V. 2–4 oz, L.V. 4–8 oz; *or* DDT (as e.c.), H.V. $\frac{1}{2}$–1 lb, L.V. 1–2 lb; *or* demeton-S-methyl, H.V. 3·5 oz (6 fl. oz 58% e.c.), L.V. 7·0 oz; *or* dimethoate, H.V. 4·8 oz (12 fl. oz 40% e.c.), L.V. 9·6 oz; *or* formothion, H.V. 7 oz (16 fl. oz 43% e.c.), L.V. 14 oz; *or* malathion, H.V. 18 oz ($1\frac{1}{2}$ pints 60% e.c.), L.V. 36–45 oz; *or* mevinphos, H.V. $1$–$1\frac{1}{2}$ oz ($1$–$1\frac{1}{2}$ fl. oz 99%), L.V. 2–3 oz (use the higher rate in cool weather; increase to 2 oz H.V. or 4 oz L.V. if caterpillar control is also required); *or* oxydemeton- methyl, H.V. 3·4 oz (6 fl. oz 56% e.c.), L.V. 6·8 oz; *or* vamidothion, H.V. 6·4 oz (16 fl. oz 40% e.c.) (one application at petal fall). Nicotine, H.V. 8 oz (95–98%) L.V. 8–10 oz, will also kill aphids but temperatures of 60°F or over are required and it is not persistent.

Where caterpillars are not troublesome an organophosphorus compounds should be used and this would give better aphid control in years when black fly is abundant.

Follow the precautions against poisoning bees, and for avoiding taint of nearby crops with gamma-BHC, given under apple aphids on page 162. See pp. 28–61 for minimum interval between last application of insecticides and harvest.

***Cherry-bark tortrix** (*Enarmonia formosana*). The pinkish white caterpillar tunnels under the bark of the trunk, often just below the crotch; successive generations use the same galleries which may become extensive.

Heavy infestations sometimes occur in old orchards, and there is some evidence that extensive bark injury can kill large branches or even whole trees. Apple, pear, plum and cherry may be attacked. The moths appear from mid May until early September.

Control measures have been studied on apple. Drenching sprays of the trunk and main branches in May, when the caterpillars are near the surface for pupation, of trichlorphon at 1 lb a.i./100 gal have given good results, through such sprays would probably have to be repeated for several seasons. Rather more effective is liberal brushing of the trunk and main branches with creosote, or drenching with undiluted tar oil, in March while the trees are still dormant.

****Cherry fruit moth** (*Argyresthia curvella*). Moths appear in late June and in July, and lay eggs under the bud scales, in crevices in the bark, etc., especially towards the tips of branches. Most of the eggs hatch in the autumn, the caterpillars hibernating in a silk cocoon in bark crevices, though some do not hatch until spring. The caterpillars bore into flower buds in the spring, feeding on the flowers and later the young fruitlets, dropping to the ground when fully fed.

Winter washes are toxic to the eggs but will not reach hibernating larvae, and these are best controlled by spring sprays. Tar oil, 7–8% H.V. in December to January, thoroughly sprayed to wet all the shoots is the best winter treatment. In spring use DDT (1 lb a.i./100 gal H.V.) in late March when the buds are breaking, repeated at white bud with heavy infestations. It is important to wet the tops of the trees and to take precautions to reduce hazard to bees (see under apple aphids, p. 162).

Information is lacking on alternatives to DDT for the control of this pest.

****Fruit tree red spider mite** (*Panonychus ulmi*) (A.L. 10) is not usually a pest on cherry. See under Plum, p. 198, for control.

***Leaf weevils** (*Phyllobius* spp.). See under Apple, p. 173.

***Pear slug sawfly** (*Caliroa cerasi*) (A.L. 84). See under Pear, p. 189.

****Wasps** (*Vespula* spp.) (A.L. 457). See under Apple, p. 176.

***Winter moths** (A.L. 11). The standard recommendation is DDT (1 lb a.i./100 gal H.V.) at white bud; information is lacking on possible alternative insecticides, except for mevinphos, 2 oz a.i. (2 fl. oz 99% conc.) H.V., 4 oz L.V.


## 9.5B DISEASES

*****Cherry bacterial canker** (*Pseudomonas mors-prunorum*). During the autumn, the bacteria enter wounds and fresh leaf scars. Until spring the disease develops from the points of entry causing death of bark and young wood. In summer the bacteria become almost inactive but stems, branches or shoots which have been girdled by the 'cankers' die, often after producing pale green or yellow leaves. Where girdling has not occurred, a limited flat canker remains and the margins of this may produce callus. In some years, buds may be killed and sometimes small holes (shot-holes) appear in leaves near to cankers.

In order to reduce the numbers of bacteria during the most vulnerable period, a drenching spray of Bordeaux mixture should be applied three times at intervals of about 3 weeks from the end of August. An extra application at petal fall may be advisable on young trees of susceptible varieties in order to build a canker-free framework. The petal fall application may cause damage.

Bordeaux mixture: (1) August: 4 lb copper sulphate, 6 lb hydrated lime, 6 pints cotton seed oil/100 gal water. (2) September: 6 lb copper sulphate, 9 lb hydrated lime, 6 pints cotton seed oil/100 gal water. (3) October: 10 lb copper sulphate, 15 lb hydrated lime, 100 gal water. (4) Petal fall: as August application (see above).

Ready-prepared forms of copper may be used but at equivalent copper content they are likely to cause damage and should be used according to the makers' recommendations. Pruning or cutting back of trees should be done between May and the end of August.

Varieties differ markedly in their susceptibility to bacterial canker.

The risk of crotch infection is reduced by using seedling mazards or 12/1 rootstocks for frameworking.

****Replant problem.** Where cherry is to be replanted after cherry, retardation of growth is likely to occur and the pre-planting treatment of the soil with chloropicrin is recommended. The method of treatment is similar to that described under Apple, p. 180.

## 9.6 PLUM AND DAMSON

### 9.6A PESTS

Annual treatments, as winter washes or spring sprays, for control of aphids are usually required, with occasional sprays against caterpillars. Regular control measures for plum sawfly and red spider may be required locally.

*****Aphids** (A.L. 34). The three common species are leaf-curling plum aphid (*Brachycaudus helichrysi*), mealy plum aphid (*Hyalopterus pruni*), and damson-hop aphid (*Phorodon humuli*) which is mainly harmful to damsons. All overwinter as eggs laid in autumn on the twigs.

Leaf-curling plum aphid hatches early, usually by January, the young aphids feeding on the dormant buds; in the spring successive generations feed on the foliage and produce severe leaf curl. From May to July winged forms disperse to various summer host plants. Mealy plum aphids are later, but have hatched by the white bud stage; from the end of June winged forms disperse to grasses or reeds. Damson-hop aphid hatches in early spring, dispersing to hop from mid May until late July.

Either winter washes or spring sprays may be used for control. With winter washes, a thorough H.V. spraying gives a good kill of winter eggs or young leaf-curling plum aphids: tar oil, 5% (miscible) or 5–6% (stock emulsion), when buds dormant in December to early January for early varieties, to end January for main crop, *or* DNOC-petroleum, $4\frac{1}{2}$%, as delayed dormant wash when buds breaking, in February or March. This wash will also give some control of red spider and winter moth.

The buds of some plum varieties are very subject to tar oil damage. Tar oil should not be used at all on Myrobalan, and not after about mid January on Belle de Louvain, Victoria, Yellow Egg and the Gages. Do not spray in frosty weather or when the trees are wet.

In spring any of the following are effective at the white bud stage (doses as a.i./100 gal, H.V., or a.i./acre, L.V.): gamma-BHC, H.V. 2–4 oz, L.V. 4–8 oz; *or* DDT (as e.c.), H.V. $\frac{1}{2}$–1 lb, L.V. 1–2 lb; *or* demeton-S-methyl, H.V. 3·5 oz (6 fl. oz 58% e.c.), L.V. 7 oz; dimethoate, H.V. 4·8 oz (12 fl. oz 40% e.c.), L.V. 9·6–12·0 oz; *or* formothion, H.V. 7 oz (16 fl. oz 43% e.c.), L.V. 14 oz; *or* malathion, H.V. 18 oz (1½ pints 60% e.c.), L.V. 36–45 oz; *or* mevinphos, H.V. 1–1½ oz (1–1½ fl. oz 99%), L.V. 2–3 oz (use the higher rate in cool weather; increase to 2 oz H.V. or 4 oz L.V. if caterpillar control is also necessary); *or* oxydemeton-methyl, H.V. 3·4 oz (6 fl. oz 56% e.c.), L.V. 6·8 oz. Nicotine, H.V. 8 oz (95–98%), L.V. 8–10 oz, will also kill aphids but temperatures of 60°F or over are required, and it is not persistent.

Vamidothion, H.V. 6·4 oz (16 fl. oz 40% e.c.), L.V. 12·8 oz, may also be used between cot split and mid-June for the combined control of late aphid attacks and fruit tree red spider.

Gamma-BHC or DDT must be applied by white bud. Organophosphorus insecticides should preferably be used unless DDT is required for other pests, and should in any case be chosen if spraying is delayed until leaf growth has occurred. Mealy plum aphid infestations are often overlooked and, as they can persist until August, may require additional sprays of organophosphorus insecticides in May or June if not properly controlled in the spring.

Follow the precautions against poisoning bees, and for avoiding taint of nearby crops with gamma-BHC, given under apple aphids on p. 162. See pp. 28–61 for minimum interval between last application of insecticides and harvest.

***Common green capsid** (*Lygocoris pabulinus*) (A.L. 154). There is little information on alternatives to the standard DDT, 1 lb a.i./100 gal H.V. or 2 lb a.i./acre L.V. at white bud, but dimethoate at cot split for plum sawfly should also control common green capsid.

*****Fruit tree red spider mite** (*Panonychus ulmi*) (A.L. 10). For life history see under Apple, p. 170.

Winter washes based on petroleum oil (see under Apple, p. 172) give some control of winter eggs, but it is usually better to restrict winter washing to tar oil for aphids, and use summer sprays for red spider as these are more efficient.

The timing of summer sprays is as recommended for apple. With heavy infestations spray about the time of petal fall of Cox's Orange Pippin, and again 3 weeks later if necessary, using one of the following acaricides effective against the active stages: (doses as a.i./100 gal, H.V., or a.i./acre, L.V.: demeton-S-methyl, H.V. 3·5 oz (6 fl. oz 58% e.c.), L.V. 7·0 oz; *or* dimethoate, H.V. 3·6 oz (9 fl. oz 40% e.c.), L.V. 7·2–9·0 oz; *or* formothion, H.V. 5 oz (12 fl. oz 43% e.c.), L.V. 10 oz; *or* malathion, H.V. 18 oz (1½ pints 60% e.c.), L.V. 36–45 oz; *or* oxydemeton-methyl, H.V. 3·4 oz (6 fl. oz 56% e.c.), L.V. 6·8 oz; *or* vamidothion, H.V. 6·4 oz (16 fl. oz 40% e.c.), L.V. 12·8 oz (one application between cot split and mid June).

Where only a single spray is given the best time is about 10 days after petal fall of the apple Cox's Orange Pippin, when an acaricide effective against active stages should be used. Also effective are dicofol, 6·4 oz a.i. (32 fl. oz 20% conc.)/100 gal H.V., phenkapton, 3·2 oz a.i./100 gal H.V., and fenson 6 oz a.i./100 gal H.V.

With light infestations spraying can be delayed until mid June, and an acaricide effective against eggs and young stages used, e.g. (doses as a.i./100 gal, H.V., or a.i./acre, L.V.): chlorbenside, H.V. or L.V. 6–10 oz (1½–2½ pints 20% conc. or 3 lb 20% w.p.); *or* chlorfenson, H.V. 1 lb (2 lb 50% w.p.); *or* tetradifon, H.V. 2 oz (1¼ pints 8% conc.), L.V. 4 oz; *or* tetrasul, H.V. 6·4 oz (2 lb 20% w.p.), L.V. 12·8 oz.

Do not spray open blossom; cut down flowering weeds under the trees before spraying. Observe the minimum intervals between last application and harvest.

*Pear thrips (*Taeniothrips inconsequens*). See under Pear, p. 190.

**Plum fruit moth (*Laspeyresia funebrana*). This is a local pest affecting late varieties. There are two generations a year, and damage is probably done by caterpillars of the second generation hatching from eggs laid at the base of the fruit stalk towards the end of July and in early August. The red coloured caterpillar ('red plum maggot') bores into the fruit towards the stone; when fully fed in late August and September it leaves the fruit and builds a cocoon for the winter in crevices in the bark, etc. For control apply azinphos-methyl, H.V. 5·3 oz (24 fl. oz 22% e.c.)/100 gal, L.V. 10·6 oz, *or* DDT 1 lb a.i./100 gal (H.V.), or 2–2½ lb/acre (L.V.), in late July and again 2–3 weeks later. The use of DDT may encourage an increase in red spider populations and it may be necessary to use an acaricide. Do not use DDT within 14 days of picking.

***Plum sawfly (*Hoplocampa flava*). The adults appear at blossom time, the female laying eggs in the flowers. The creamy white caterpillars bore into the fruitlets, one caterpillar attacking as many as four before it is fully fed and drops to the ground, where it builds a cocoon in the soil for the winter. Marked varietal preference is shown, Czar being particularly susceptible.

To control, spray at the cot-split stage (7–10 days after petal fall) with one of the following insecticides (doses as a.i./100 gal, H.V., or a.i./acre, L.V.): demeton-S-methyl, H.V. 3·5 oz (6 fl. oz 58% e.c.), L.V. 7·0 oz; *or* dimethoate, H.V. 4·8 oz (12 fl. oz 40% e.c.), L.V. 9·6–12·0 oz; *or* formothion, H.V. 7 oz (16 fl. oz 43% e.c.), L.V. 14 oz; *or* gamma-BHC, H.V. 2–4 oz, L.V. 4–8 oz; *or* oxydemeton-methyl, H.V. 3·4 oz (6 fl. oz 56% e.c.), L.V. 6·8 oz. The organophosphorus insecticides appear to be the more effective. These will also control early hatching fruit tree red spider mites, but for full control of this pest the timing should be as given on p. 198.

Follow the precautions against poisoning bees, and for avoiding taint of nearby crops with gamma-BHC, given under apple aphids on p. 162. See pp. 28–61 for minimum interval between last application of insecticides and harvest.

***Red-legged weevil** (*Otiorhynchus clavipes*) (A.L. 57). This wingless weevil is about ½ in long, shining black with reddish legs. It appears in late April or May and shelters by day under clods and in rough vegetation. It feeds at night on foliage, flowers and fruitlets, and gnaws at bark often causing young shoots to break off. Peach, apricot, nectarine, raspberry, gooseberry, etc., are also attacked.

The control measures recommended for clay-coloured weevil can be used (see p. 205).

***Scale insects.** The brown or peach scale (A.L. 88), mussel scale (A.L. 36) and oystershell scale may occur on plum. Control is effected by tar oil 5% H.V. applied when the buds are dormant in December to early January for early varieties, to end January for main crop, *or* by DNOC-petroleum 4½% H.V., up to bud break in late January or early February.

The buds of some plum varieties are very subject to tar oil damage. Tar oil should not be used at all on Myrobalan, and not after about mid January on Belle de Louvain, Victoria, Yellow Egg and the Gages. Do not spray in frosty weather or when the trees are wet.

****Tortrix moths.** Caterpillars of the plum tortrix (*Hedya pruniana*) feed on foliage and tunnel in the shoots from April to June. Several of the tortrix caterpillars found on apple (p. 174) also occur on plum. The caterpillars are kept in check by spring sprays against winter moth caterpillars.

****Wasps** (*Vespula* spp.). See under Apple, p. 176.

****Winter moths** (A.L. 11). See under Apple, p. 176, for life histories. For control, use DDT ½–1 lb a.i./100 gal (H.V.), *or* 1–2 lb/acre (L.V.), at the white bud stage, *or* mevinphos, 2 oz a.i. (2 fl. oz 99%)/100 gal H.V., 4 oz L.V. Information is lacking on the use of other alternative compounds. To reduce danger to bees, do not spray open blossom; cut down flowering weeds beneath the trees before spraying.

## 9.6B DISEASES

*****Blossom wilt, Brown rot and allied diseases of plum** (*Sclerotinia fructigena* and *S. laxa*) (A.L. 248). *Sclerotinia fructigena* and *S. laxa* cause brown rot of plum fruits, while the latter species also causes blossom wilt sometimes followed by spur blight and canker and wither tip of shoots. Both species overwinter in mummified fruits and cankers which produce spores in spring so continuing the cycle. *S. laxa* can enter shoots and spurs via damaged leaves.

Diseased trusses, shoots and cankered spurs and branches should be cut out and burnt, preferably in the current spring or summer when their presence can more easily be recognized. Mummified fruits should also be collected and burnt. Tar oil spraying as for aphids (see p. 197) late in the following dormant period gives partial control by destroying spore cushions that arise on any diseased parts overlooked in the earlier cutting out.

## 9.7 HAZELNUT, COBNUT AND FILBERT

### 9.7A PESTS

**Nut weevil** (*Balaninus nucum*). The adult weevils emerge in May. The female bores a hole in the side of the young nut and deposits an egg, and the white grub feeds on the kernel. When full grown in late July or August the grub drops to the ground, and builds a cocoon in the soil for the winter, pupating in the spring. Holes drilled by the female weevil also permit the entry of the brown rot fungus (nut drop). Control is effected by DDT, 10–16 oz a.i./100 gal (H.V.), or 5% DDT dust $\frac{1}{2}$ cwt/acre, in late May or early June and again 2–3 weeks later.

**Winter moths** (A.L. 11). Caterpillars of the winter moth group (see under Apple, p. 176) may also infest nut trees, feeding on the foliage. Apply DDT $\frac{1}{2}$ lb a.i./100 gal (H.V.), about the middle of April, when caterpillars are seen.

### 9.7B DISEASES

**Gloeosporium bud rot and Twig canker of cobnut and filbert** (*Gloeosporium* spp.). In early spring, infected buds become brown and die. The causative fungus penetrates the twig causing a canker in which the fungus may overwinter. The disease is present in most plantations and is a common cause of crop loss.

Spray after harvest but before leaf fall with an organomercury compound, using w.p. formulations at rates equivalent to 0·5 oz metallic mercury/100 gal (H.V.) or 1·0–1·25 oz mercury/acre (L.V.).

## 9.8 CURRANT AND GOOSEBERRY

### 9.8A PESTS

Annual control measures are required for aphids and blackcurrant gall mite, and locally for sawflies and capsid.

**Aphids** (A.L. 176). Several species overwinter as eggs on currant and gooseberry. The eggs hatch in spring, the aphids breed and in the

summer disperse to various other host plants, returning in the autumn
to lay eggs.

The currant-sowthistle aphid (*Hyperomyzus lactucae*) causes the leaves
of red and black currants to curl downwards and stunts young growth;
it disperses to sowthistle. The lettuce aphid (*Nasonovia ribis-nigri*), a
darker green species, normally infests gooseberry and disperses to lettuce.
*Cryptomyzus* are pale yellow green delicate-looking species; the red currant
blister aphid (*C. ribis*) causes leaf blisters which are red on red and white
currant and yellowish green on black currant; the currant aphid (*C.
galeopsidis*), which causes no obvious injury, has 3 forms, two of which
remain respectively on black and red currants. The third, and red currant
blister aphid, disperse to hedge woundwort and hemp nettle. The gooseberry aphid (*Aphis grossulariae*), deep green in colour, causes severe curling
and distortion of young leaves of currant and gooseberry; some of the
aphids remain on the bushes all summer. The permanent currant aphid
(*A. schneideri*), blue green in colour, is responsible for similar damage on
red and black currant. Other species occur less commonly.

The currant root aphid (*Schizoneura ulmi*), covered with white waxy
material like the woolly aphid, infests the roots of currant and gooseberry.
It winters as eggs on elm, from which aphids disperse to currant and
gooseberry in the summer.

Except with the currant root aphid either winter washes against the
eggs or spring sprays against the aphids can be used.

Winter washes are effective if thoroughly applied. Use tar oil, 5%
(miscible) or 5–6% (stock emulsion) H.V., when buds dormant, in
December to January, not later, *or* DNOC-petroleum oil, $4\frac{1}{2}$% H.V., up
to bud break, January or early February.

Do not spray in windy or frosty weather or when the bushes are wet.
The variety Raby Castle is liable to oil injury and should not be winter
washed.

Alternatively apply one of the following in the spring, just before the
first flowers open (doses as a.i./100 gal, H.V., or a.i./acre, L.V.): demeton-
S-methyl, H.V. 3·5 oz (6 fl. oz 58% e.c.), L.V. 7·0 oz; *or* dimethoate,
H.V. 4·8 oz (12 fl. oz 40% e.c.), L.V. 9·6 oz; *or* formothion, H.V. 7 oz
(16 fl. oz 43% e.c.), L.V. 14 oz; *or* malathion, H.V. 18 oz ($1\frac{1}{2}$ pints 60%
e.c.); *or* oxydemeton-methyl, H.V. 3·4 oz (6 fl. oz 56% e.c.), L.V. 6·8 oz.

If endrin or endosulfan are used just before flowering for blackcurrant
gall mite (see below) these will also give some control of leaf midge and
aphids. Little is known of control measures for currant root aphid.

Follow the precautions against poisoning bees given under apple
aphids on p. 162. See pp. 28–61 for minimum interval between last
application of insecticides and harvest.

***Blackcurrant gall mite (Big bud mite)** (*Cecidophyopsis ribis*)

(A.L. 27). The microscopic colourless mites feed and multiply inside the buds of blackcurrant, which become swollen and in the following spring many fail to open, or produce much reduced flower trusses. Mites disperse from the swollen buds from March until the buds have dried up which may be late in July, the main emergence period being from early April to the end of June, with a peak usually in May. Emergence is accelerated by rising temperatures, and mites may be dispersed to fresh sites by air currents, or transporation by insects or rain.

The difficulty of chemical control is to protect the young buds against mite infestation for a sufficiently long period without leaving insecticide residues on the fruit. Endrin and endosulfan are effective, but cannot be used late enough in the season, except on non-fruiting bushes, for maximum control; hence, when used at the latest permissible time, they can only keep infestations in check. In trials in the west country three sprays of 1% lime sulphur with 4 fl. oz succinate type spreader/100 gal, at first open flower and repeated when the last flowers are opening and again 2 weeks later, have given good control. Sulphur sprays, however, need to be used with caution as other trials in south and east England with lime sulphur and colloidal sulphur have resulted in leaf scorch and reduced yield. The most effective acaricides are:

(a) On fruiting bushes: endosulfan, 8 oz a.i. (2 pints 20% e.c.)/100 gal (H.V.), at first open flower and again 3 weeks later (any later application is not permissible); endrin (6 oz a.i./100 gal H.V.) may also be used immediately before flowering (not later), but if the crop is grown under contract only with the purchaser's approval.

Thorough H.V. spraying is essential to give adequate cover, with at least 200 gal/acre on large bushes. Endosulfan will also control aphids and, to some extent, capsid bugs.

(b) On nursery stock and non-fruiting bushes: endrin, 6 oz a.i. ($1\frac{1}{2}$ pint 20% e.c.)/100 gal (H.V.), or endosulfan, 8 oz a.i. (2 pints 20% e.c.)/ 100 gal (H.V.).

Apply in latter half of April and repeat at least once more with an interval of 2–3 weeks. Thorough spraying is essential. Endrin also controls capsid bugs.

Do not use lime sulphur on sulphur shy varieties, e.g. Amos Black, Davidson's Eight, Edina, Goliath, Monarch, Victoria, Wellington XXX, Westwick Triumph. With endosulfan and endrin do not exceed the stipulated rate of application or spray later than the stipulated times. Fruit from nursery stock sprayed with endrin must not be used for human or animal consumption or processed for such uses. The interval between last application and harvest for crops treated with endosulfan must not be less than 6 weeks.

**Blackcurrant leaf midge** (*Dasyneura tetensi*). There are three, sometimes four, generations a year; midges of the first appear in April to June, the second late June and July, and the third late July and August. Eggs are laid between folds of young leaves, and feeding by the white or orange maggots causes the leaves to become tightly twisted and folded. Shoot growth may be checked and lateral branches develop, but more important is the masking of symptoms of reversion in nursery stock. Pupation occurs in the soil. Some varieties are more susceptible, e.g. Goliath, Seabrook's Black, Baldwin and Wellington XXX.

Good control is possible by sprays against emerging midges in April but timing is difficult owing to variability of time of emergence. Use DDT at 10–16 oz a.i./100 gal (H.V.) (1 lb with severe infestations) at about first open flower stage, and repeat 3 weeks later. Bushes must be thoroughly drenched. Spray at corresponding times on nursery stock.

A good control has been obtained in trials by a H.V. spray just after flowering, when leaves infested with larvae were present, of demeton-S-methyl or dimethoate at doses as for aphid control, or endosulfan or endrin (non-fruiting bushes only) at doses for gall mite control (see above).

The use of DDT may in some seasons encourage an increase in red spider mite populations and an acaricide may have to be used (see p. 207).

Observe the minimum intervals between last application and harvest.

****Blackcurrant sawfly** (*Nematus olfaciens*) (A.L. 30). There are two or more overlapping generations a year. Adults emerge from overwintering cocoons in the soil in May and June, and those of the later broods from mid June to mid September. Eggs are laid on the undersides of the leaves, especially near the middle of the bush. The green black-spotted caterpillars at first feed gregariously but later spread through the bush and may cause defoliation.

Bushes should be examined in May and early June for caterpillars; if present, spray in early June while the caterpillars are still small, with azinphos-methyl, H.V. at 5·3 oz a.i. (24 fl. oz 22% e.c.)/100 gal, L.V. 10·6 oz a.i./acre; *or* DDT, $\frac{1}{2}$–1 lb a.i./100 gal (H.V.). It is important to see that the middles of the bushes are well sprayed. A second spray can be given 2 weeks later if necessary. The organophosphorus compound should be used in preference to DDT; the latter may also encourage a build up of red spider in some seasons.

****Brown or Peach scale** (*Parthenolecanium corni*) (A.L. 88). The full grown female scales are about $\frac{1}{8}$–$\frac{1}{4}$ in long, tortoise-shaped, and chestnut brown in colour. Eggs are laid beneath the scales in the summer and in the autumn the young escape. They overwinter on the branches, often under loose bark, and after a short period of activity in the spring settle down, becoming adult in June. After laying eggs the female dies.

Through winter washing can give adequate control: DNOC-petroleum oil, 4½% (H.V.), when the buds are breaking, *or* tar oil 5% (miscible) or 5–6% (stock emulsion) H.V.), when the buds are dormant, in December to January, not later.

If control is required out of the dormant season, spray with diazinon, 5 oz a.i. (25 fl. oz 20% e.c.)/100 gal H.V., *or* malathion, 30 oz a.i. (2½ pints 60% e.c.)/100 gal H.V., either in late summer or early spring. This spray is directed against the young 'crawlers' before they have settled down and formed protective scales; good cover of the wood and undersides of the leaves is essential. With heavy infestations a second spray 14 days later may be necessary.

Do not winter wash in windy or frosty weather or when the bushes are wet. The red currant variety Raby Castle is liable to oil injury and should not be winter washed.

**Clay-coloured weevil** (*Otiorhynchus singularis*) (A.L. 57). Adult weevils appear in late spring or early summer, sheltering under clods of soil or in dense vegetation by day and ascending plants at night to feed. Holes are eaten in leaves, and the bark gnawed from stems of currant, gooseberry, raspberry, fruit trees, hops and other plants. Eggs are laid in the soil in summer and the white grubs feed on the roots of various plants until fully grown early in the following summer. Adults also hibernate, and are often common in woodland so that adjoining fruit crops may be invaded.

The adults are not easy to kill and high concentrations of DDT are required. Use DDT at 4 lb a.i. in at least 300 gal/acre, spraying the soil as well as the bushes. Also effective is 20 or 25% DDT w.p. as a dust, applied to the soil and lower part of the bushes at about ¼ oz/plant.

Do not use DDT within 14 days of harvest.

***Common green capsid** (*Lygocoris pabulinus*) (A.L. 154) winters as eggs on currant, gooseberry and other shrubs and occasionally apple. Hatching of the eggs appears to be a little later than with apple capsid and begins at the time of early pink bud of Bramley's Seedling apple, ending at about petal fall. The bugs puncture leaves and shoots, and some may find their way on to herbaceous plants including strawberry, causing the fruit to become misshapen. A second generation occurs in July and August and the bugs when mature fly to woody host plants in the autumn to lay winter eggs.

Full bloom of Bramley's Seedling can be taken as a guide to optimum timing of sprays on other crops, though with the more persistent insecticides somewhat earlier sprays appear to be effective.

The following timings are suggested:

Currant and gooseberry—Dimethoate, H.V. 4·8 oz a.i. (12 fl. oz 40% e.c.)/100 gal *or* malathion, H.V. 18 oz a.i. (1½ pints 60% e.c.)/100 gal, immediately before flowering on currant and at the end of flowering on gooseberry are the most effective. Endosulfan used against gall mite on blackcurrant will give moderate control of capsid.

Strawberry—DDT, 1 lb a.i./acre (H.V.), *or* TDE, ¾ lb a.i. (1½ lb 50% w.p.)/100 gal (H.V.), in late April or early May just before the first flowers open. If malathion or dimethoate are used against aphids at this time, they should also control common green capsid.

Raspberry, blackberry, etc.—Insecticides as for strawberry, in late April, or early May. On blackberry, endrin for 'redberry' control would also control capsid. Dimethoate or malathion used against aphids should also control common green capsid.

Observe the minimum intervals between last application and harvest.

***Currant clearwing moth** (*Synanthedon tipuliformis*). A local pest mainly of blackcurrant, though red currant and gooseberry may also be infested. The fly-like moths, which are active in sunny weather, lay eggs in June and the caterpillars tunnel in the pith of the branches until full grown in the following April. Preliminary trials suggest that azinphos-methyl at 5·3 oz a.i. (24 fl. oz 22% e.c.)/100 gal H.V. after picking will give some control.

***Eelworm.** The leaf and bud eelworm (*Aphelenchoides ritzemabosi*) may also attack blackcurrant, especially the varieties Daniel's September Black and Westwick Choice. The eelworms live in the buds which fail to open in the spring, and lengths of bare branch develop. Damage may be worse following a wet season. There is no effective recommendation for control by nematicides.

***Gooseberry red spider mite** (*Bryobia ribis*). This at one time was a serious pest in many gooseberry plantations, but is now uncommon. Although it will feed on other plants such as apple it is only known as a pest on gooseberry. It overwinters as eggs under loose bark; hatching begins near the beginning of March and continues well into April. There is only one generation, the laying of winter eggs beginning in May. The mites feed on the leaves in warm sunny conditions, retiring to the wood and under bud scales in cold weather and also for moulting. On severely infested bushes the foliage becomes yellowed, and leaves may wither and drop.

Organophosphorus insecticides applied in warm weather in April when the mites are active on the leaves are effective. This treatment could be combined with aphid control by spraying at first open flower (see p. 201), and the use of these insecticides against aphids may account for the decline of *Bryobia* as a pest. Should heavy infestations be allowed to develop, an additional spray may be needed earlier, as soon as mites are seen on the opening buds.

Suitable insecticides are (doses as a.i./100 gal, H.V.): demeton-S-methyl, 3·5 oz (6 fl. oz 58% e.c.); *or* dimethoate, 3·6 oz (9 fl. oz 40% e.c.);

*or* formothion, 7 oz (16 fl. oz 43% e.c.); *or* malathion, 18 oz (1½ pints 60% e.c.); *or* oxydemeton-methyl, 3·4 oz (6 fl. oz 56% e.c.). Quinomethionate, used against American gooseberry mildew at 8 oz 25% w.p./100 gal H.V., will also control red spider mites.

A winter wash of 4½% DNOC-petroleum oil, thoroughly applied when the buds are breaking, is effective against the winter eggs.

Observe the minimum intervals between last application and harvest.

***Gooseberry sawfly** (*Pteronidea ribesii*) (A.L. 30). The adult sawflies first appear in April and May. Eggs are laid on the undersides of the leaves, especially near the centre of the bush. The black-spotted green caterpillars feed together for a few days, later spreading through the bush which may be defoliated. When fully grown they are green with an orange patch behind the head and another near the tail and lack the black spots. There are three overlapping generations in the year. *P. leucotrocha* also occurs, and both species may attack red and white currant but not black.

Examine the bushes in May and early June so that spraying can be started as soon as appreciable numbers of caterpillars are seen. The usual time for treatment is shortly after fruit set. Use one of the following insecticides: azinphos-methyl, H.V. 5·3 oz a.i. (24 fl. oz 22% e.c.)/100 gal, L.V. 10·6 oz a.i./acre; *or* derris, ⅔ oz rotenone/100 gal (H.V.); *or* malathion 18 oz a.i. (1½ pints 60% e.c.)/100 gal (H.V.); *or* nicotine, 8 oz (95–98%)/100 gal, with added wetter (H.V.).

It is important to spray thoroughly the centres of the bushes. Since the eggs hatch over 2–3 weeks the sprays may have to be repeated. Recent evidence has suggested that DDT may not be sufficiently effective in certain circumstances. Some DDT emulsions may damage some gooseberry varieties.

Observe the minimum intervals between last application and harvest.

***Magpie moth** (*Abraxas grossulariata*) (A.L. 65). The moth flies in July and August and lays its eggs on the leaves of currant and gooseberry. The caterpillar feeds for a while then finds suitable shelter for hibernation, reappearing at bud burst to feed on the leaves, pupating in June. A small second brood may occur. It is now an uncommon pest.

The caterpillars will be controlled by DDT ½–1 lb a.i./100 gal (H.V.) at late grape stage or first open flower for capsid and leaf midge control. With heavy infestations an earlier spray may be necessary, using DDT at 1 lb.

****Red spider mite** (*Tetranychus urticae*) (A.L. 226). This is the red spider mite common on many glasshouse and outdoor plants, occasionally infesting top and soft fruit crops. It overwinters as adult females in the soil and other shelter, and these emerge over a fairly long period in April to feed and breed on the foliage. In some seasons the use of insecticides such As DDT on bush fruit encourages a build-up of red spider. Straw mulches also favour hibernating mites.

Where necessary spray just after flowering, or just after picking, with one of the following (doses as a.i./100 gal, H.V.): demeton-S-methyl, 3·5 oz (6 fl. oz 58% e.c.); *or* derris, $\frac{1}{3}$ oz rotenone; *or* dimethoate, 3·6 oz (9 fl. oz 40% e.c.); *or* formothion, 7 oz (16 fl. oz 43% e.c.); *or* oxydemeton-methyl, 3·4 oz (6 fl. oz 56% e.c.); *or* tetradifon, 2 oz (1$\frac{1}{4}$ pints 8% e.c.). Derris is less effective than the other compounds.

Quinomethionate used against American gooseberry mildew (1 lb 25% w.p./100 gal H.V. on blackcurrant, $\frac{1}{2}$ lb on gooseberry) will also control red spider mites. It can be used on most of the important varieties of blackcurrant, but some varieties, and also some gooseberries, are intolerant, see below.

Observe the minimum intervals between last application and harvest.

***Winter moths** (A.L. 11). Caterpillars of the winter moth group (see under Apple, p. 176) may attack currant.

DDT, $\frac{1}{2}$ lb a.i./100 gal (H.V.) at late grape stage of just before the first open flower for control of capsid or leaf midge, and repeated 3 weeks later, will also control these caterpillars. Information on alternatives is lacking but a zinphos-methyl should also be effective.

## 9.8B DISEASES

*****American gooseberry mildew of gooseberry** (*Sphaerotheca mors-uvae*) (A.L. 273). The white powdery (sporing) fungal growth is seen on young leaves, fruits and shoots. Under moist stagnant air conditions, the disease becomes epidemic. In late summer and autumn, the fungal growth becomes a brown felt-like layer containing black spore cases (perithecia). Some fall to the ground and spores are ejected in the following spring and so start a new cycle.

Lime sulphur, sulphur as a wettable powder or colloidal formulation, *or* dinocap, *or* quinomethionate should be sprayed on the foliage (1) just before the flowers open, (2) at fruit set and (3) fruit swelling, 14–21 days later. Lime sulphur should be applied at 2 gal/100 gal (H.V.) at pre-blossom and at 1 gal/100 gal (H.V.) at fruit set. The sulphur sprays may be prepared from w.p. as colloidal sulphur at 1·5–2 lb sulphur/100 gal (H.V.). The dinocap spray should contain 0·25 lb a.i. (1 lb 25% w.p.)/100 gal (H.V.) or 0·5 lb a.i./acre (L.V.); the quinomethionate spray should contain 2 oz a.i. (8 oz 25% w.p.)/100 gal (H.V.).

Do not use lime sulphur or sulphur on the varieties Careless, Leveller, Lord Derby, Early Sulphur, Golden Drop, Roaring Lion and Yellow Rough which are sulphur-sensitive. Further, sulphur must not be used when fruit is for canning.

Infected shoot tips should be removed as soon as the wood is ripe in late August or September, and excessive nitrogenous manuring should be avoided.

***Blackcurrant leaf spot** (*Pseudopeziza ribis*). From about May onwards, brown spots or patches appear on the leaves. on these patches spores (conidia) are formed which spread infection through the plantation. When the disease is unchecked, premature defoliation occurs, weakening subsequent growth and reducing the yield. The fungus overwinters in dead leaves and spores (ascospores) from these restart the cycle the following spring.

A spray containing dodine, *or* mancozeb/zineb, *or* quinomethionate, *or* zineb should be applied immediately after flowering and repeated twice at 10–14 day intervals to within a month of picking. A final spray of copper, *or* dodine, *or* mancozeb/zineb, *or* quinomethionate, *or* zineb should be applied immediately after the fruit has been picked. Use dodine at 5·2 oz a.i. ($\frac{1}{2}$ lb 65% w.p.)/100 gal (H.V.); mancozeb/zineb according to the maker's instructions; quinomethionate at 4 oz a.i. (1 lb 25% w.p.)/100 gal; zineb at 21 oz a.i. (2 lb 65% w.p.)/100 gal (H.V.). The copper sprays should be used only at post harvest and may be prepared from w.ps containing copper oxychloride, copper oxide or Bordeaux mixture to contain 1·5–2·5 lb metallic copper/100 gal (H.V.). Application should be at a rate of 200 gal/acre and, with mancozeb/zineb, the last treatment should be given at least 6 weeks before harvest.

****Blackcurrant rust** (*Cronartium ribicola*). This rust spends part of its life cycle on blackcurrants and part on five-needled pines, particularly the Weymouth pine (*Pinus strobus*). Spores (aecidiospores) from pine trees infect nearby currant bushes, the fungus appearing as yellow outgrowths on the underside of the leaf in early summer. From these outgrowths, the spores (uredospores) are produced to spread the disease within the currant plantation. Later, yet other kinds of spores (teleutospores and basidiospores) are produced on the leaves and these cause reinfection of pines.

Where rust is troublesome, the bushes should be sprayed post harvest with copper, *or* mancozeb/zineb, *or* zineb, following the dosage rates and programme recommended for blackcurrant leaf spot (see above).

***European gooseberry mildew** (*Microsphaera grossulariae*) is much less serious than American gooseberry mildew (see above). It is seen as a delicate sporing mould, mainly on upperside of leaf; rarely on berries. Overwintering spore cases fall to ground with leaves and restart cycle by ejecting spores (ascospores) the following spring. Spray with one of the sulphur preparations *or* with dinocap *or* quinomethionate as listed for American gooseberry mildew.

****Gooseberry cluster cup rust** (*Puccinia pringsheimiana*) (A.L. 198).

P

This is seen in early summer as dark red or orange coloured blister-like swellings on the leaves and fruits. Later the blisters become covered with minute pits with raised, reflexed, yellow margins (cluster cups). The spores which develop within the pits can infect sedges (*Carex* spp.) but not gooseberries. On sedges, the disease appears in a different form but in the following summer spores are developed which infect gooseberries and so start the cycle again.

A single application of a copper-containing spray should be applied about 14 days before flowering, using Bordeaux mixture w.p., 1·5–2 lb metallic copper/100 gal (H.V.).

Copper-containing fungicides should not be used on the varieties Careless, Early Sulphur, Freedom, Golden Drop, Leveller, Lord Derby, Roaring Lion and Yellow Rough. With canning and other forms of processing the manufacturers should be consulted before any sprays are used.

Because the disease cannot occur without the presence of infected sedges, those in the neighbourhood should be destroyed.

****Powdery mildew of blackcurrant** (American gooseberry mildew) (*Sphaerotheca mors-uvae*), is usually first seen in May or June as a white powdery (sporing) fungal growth on both surfaces of the leaves and on the fruit. In late summer and autumn, the fungal growth becomes a felt-like layer containing black spore cases (perithecia) which overwinters on shoots and fallen leaves; it is possible that these can start infection the following year.

Apply three sprays of dinocap, *or* quinomethionate, at intervals of 10–14 days commencing in mid May to late May or as soon as the mildew appears. Where mechanical harvesting is practised, 10% lime sulphur with appropriate wetter should be applied immediately after cutting down, followed by a single application of quinomethionate in mid August.

Use dinocap at 4 oz a.i. (1 lb 25% w.p.)/100 gal *or* quinomethionate at 4 oz a.i. (1 lb 25% w.p./100 gal, both at 200 gal/acre for uncut bushes but where post harvest lime sulphur sprays have been applied to cut-down bushes, quinomethionate should be applied at half the concentration shown above.


## 9.9 BLACKBERRY, LOGANBERRY AND RASPBERRY

### 9.9A PESTS

Annual sprays are required for control of raspberry beetle and aphids, and on susceptible varieties for cane midge.

*****Aphids.** There are several species, overwintering as eggs on the canes. The rubus aphid (*Amphorophora rubi*), which begins hatching in March, is a large pale green aphid and causes slight leaf curling, dispersing between *Rubus* spp. in the summer. The raspberry aphid (*Aphis idaei*) is smaller and grey green. It causes pronounced curling of young leaves, and infests fruiting laterals, dispersing to raspberry and raspberry hybrids in June and July. Both are important as virus vectors.

The blackberry aphid (*Macrosiphum fragariae*) hatches in February and March, feeding on the tips of the buds, and heavy infestations cause severe leaf curling. Dispersal to grasses occurs in May and June. Only blackberry is severely attacked.

For control either winter washes or summer sprays may be used.

The winter washes are: tar oil, 5% (miscible) or 5–6% (stock emulsion) (H.V.), when the buds are dormant *or* DNOC-petroleum oil $4\frac{1}{2}$% (H.V.), up to bud break.

For spring sprays, use one of the following when infestations seen in April (doses as a.i./100 gal, H.V. or a.i./acre, L.V.): demeton-S-methyl, H.V. 3·5 oz (6 fl. oz 58% e.c.), L.V. 7·0 oz; *or* dimethoate, H.V. 4·8 oz (12 fl. oz 40% e.c.); *or* formothion, H.V. 7 oz (16 fl. oz 43% e.c.); *or* malathion, H.V. 18 oz ($1\frac{1}{2}$ pints 60% e.c.); *or* oxydemeton-methyl, H.V. 3·4 oz (6 fl. oz 56% e.c.), L.V. 6·8 oz. Dimethoate, formothion or malathion are probably also effective L.V.

Follow the precautions against poisoning bees given under Apple aphids on p. 162. See pp. 28–61 for minimum interval between last application of insecticides and harvest.

***Bramble Shoot Moth** (*Notocelia uddmanniana*). Moths appear from late June into August, laying eggs on the leaves and shoots of blackberry, loganberry and phenomenal berry. The caterpillars which hatch hibernate in the second instar, reappearing in April to feed on the blossom and web together young leaves on the shoots. Pupation occurs in the webbed leaves in June.

The caterpillars can be controlled by spraying at bud burst with azinphos-methyl, 2·6 oz a.i. (12 fl. oz 22% conc.)/100 gal (H.V.), *or* DDT 1 lb a.i./100 gal (H.V.).

***Clay-coloured weevil** (*Otiorhynchus singularis*) (A.L. 57). See under Currant, p. 205.

***Common green capsid** (*Lygocoris pabulinus*) (A.L. 154). See under Currant, p. 205.

*****Raspberry beetle** (*Byturus tomentosus*) (A.L. 164). The most important pest of cane fruits. The beetles hibernate in the soil, emerging April and May, and are active in sunny weather frequenting flowers of apple, hawthorn, raspberry, etc. Eggs are laid in the flowers of raspberry and other *Rubus* spp., hatching in 10–12 days. The larvae feed on the surface of the fruit, and as it begins to ripen tunnel into the plug. When full grown they pupate in the soil.

To control this pest, use one of the following insecticides to kill the young larvae, at the times given below: DDT, $\frac{1}{2}$ lb a.i./100 gal (H.V.) or $1\frac{1}{4}$ lb a.i./acre (L.V.), *or* derris, $\frac{1}{3}$ oz rotenone/100 gal (H.V.), *or* malathion, 18 oz a.i. ($1\frac{1}{2}$ pint 60% e.c.)/100 gal (H.V.) or 36 oz a.i./acre (L.V.).

On raspberry and loganberry spray with derris or malathion when most of the blossom is over (mid June) and about 2 weeks later at first pink fruit. On raspberry a single spray at first pink fruit often gives adequate control. Where because of risk of heavy infestations it is necessary to prevent damage to buds and flowers, spray at white bud, with a second spray at first pink fruit as above.

On blackberry, spray as soon as the first blossom opens with any of the three insecticides. A second spray may be needed about a fortnight later, using derris or malathion.

Malathion will also control the leafhopper transmitting rubus stunt virus.

Preference should be given to the use of derris or malathion, and in any case the use of DDT may in some seasons encourage an increase in red spider infestations. Do not spray during full bloom because of danger to bees. Observe the minimum intervals between last application and harvest.

****Raspberry cane midge** (*Thomasiniana theobaldi*). Midges emerge in late April and early May and lay eggs on the young spawn, the females seeking breaks in the rind, such as growth splits, for the purpose. The pink larvae feed under the rind, and the damaged tissues are susceptible to fungal attack which may lead eventually to the death of the cane ('midge blight'). Two further generations of midges appear in July and again in August, but these overlap considerably, and a partial fourth brood may occur. The later generations are larger and considerable damage may result. The winter is passed as larvae in cocoons in the soil, pupating in the spring. Varieties, the rind of which splits freely, such as Bath's Perfection, Malling Enterprise and Malling Promise, are most susceptible. Norfolk Giant, Malling Landmark, Jewel and Exploit are lightly attacked, while Lloyd George is intermediate.

The object of control is to protect future fruiting canes from damage by spraying the young spawn when the first generation midges are emerging to lay eggs. Use gamma-BHC 2 oz a.i. (10 fl. oz 20% e.c.)/100 gal (H.V.), with added wetter, when most spawn on Malling Promise or Exploit are 9–12 in high (usually first week of May) and again 2 weeks later. Only the base of the canes need be sprayed and the spawn should be well wetted.

Information is lacking on the value of alternative insecticides. Routine spraying of the lightly attacked varieties is probably unnecessary.

Do not spray during or after flowering with gamma-BHC because of risk of tainting fruit.

***Raspberry moth** (*Lampronia rubiella*) (A.L. 66). The moths fly in May and June and eggs are laid in the flowers. The young caterpillars feed in the plug, causing little damage, and leave as the fruit ripens to spin cocoons for the winter in the soil, in crevices in canes, stakes, etc. In April they emerge and tunnel in the shoots which shrivel and die.

To control, spray with 8% tar oil in late February or March at least 150 gal/acre, directed to wet thoroughly the soil, and bases of canes and stakes, *or* with DDT w.p. at 4 lb a.i./100 gal (H.V.) in early April to kill emerging caterpillars.

****'Redberry'.** In this condition of blackberries, which appears to be associated with the gall mite (*Aceria essigi*), some of the berries fail to ripen normally, only some of the drupelets turning black while the remainder stay red. The mites overwinter as eggs, mature and young stages under the shoot scales near the buds, and in spring and summer probably move on to the expanding shoots, and later into the flowers and fruit.

A useful reduction of the incidence of redberry can be obtained by spraying with endrin 6 oz a.i. ($1\frac{1}{2}$ pints 20% e.c.)/100 gal (H.V.) with added wetter, in late April or early May and again 3–4 weeks later. Annual sprays should not be necessary.

Endrin may only be used on blackberry before flowering.

***Red-legged weevil** (*Otiorhynchus clavipes*) (A.L. 57). See under plum, p. 200, and currant, p. 205.

****Red spider mite** (*Tetranychus urticae*) (A.L. 226). See under currant, p. 205.

****Strawberry rhynchites** (*Caenorhinus germanicus*). This pest may also attack raspberry and blackberry, the weevils girdling the tips of new growth and the stems of blossom trusses. It is controlled by spraying with DDT, 8–10 oz a.i./100 gal (H.V.) *or* $1$–$1\frac{1}{2}$ lb a.i./acre (L.V.), *or* 5% DDT dust, when damage is first seen. Also effective is TDE, $\frac{3}{4}$ lb a.i. ($1\frac{1}{2}$ lb 50% w.p.)/100 gal (H.V.). Information is lacking on the value of alternative insecticides.

Observe the minimum intervals between last application and harvest.

## 9.9B DISEASES

****Loganberry cane spot** (*Elsinoe veneta*). Spotting of stems, leaves, flower stalks and fruits occurs showing symptoms like those caused on raspberry by the same fungus (see above).

Control measures are similar to those recommended for raspberry cane spot except that lime sulphur should not be used on loganberries, and the times of application are later, viz. (1) pre-blossom, (2) as soon as the fruit has set.

***Raspberry cane spot** (*Elsinoe veneta*). New infections are seen as small purple spots on the young canes from early June onwards. The spots enlarge later becoming elliptical up to ¼ in long having a light grey centre with a purple border; the centres of the spots split leaving cavities which give the fruiting canes a rough and cracked appearance. Where spots have coalesced, the tips of canes may be killed. Leaves and fruits are sometimes attacked; the latter become distorted.

Spores (conidia) of the cane spot fungus are released from the spots, so spreading the disease during the growing season. The fungus over-winters in the canes and produces a second type of spore (ascospore) in spring to restart the cycle.

An application of copper, lime sulphur (with wetter) *or* thiram should be made when the buds are not more than ½ in long and again just before blossom to protect the young canes.

Use lime sulphur at 5 gal lime sulphur/100 gal (H.V.) in bud burst applications and at 2½ gal lime sulphur/100 gal (H.V.) at white bud.

Prepare the copper sprays from either copper oxychloride w.p. *or* liquid *or* Bordeaux w.p. at 2 lb metallic copper/100 gal (H.V.) at bud burst and at 1·5 lb metallic copper/100 gal (H.V.) at white bud. Add cotton seed oil in the white bud spray as given under cherry bacterial canker (see p. 196). The thiram spray should be used at 1·6 lb a.i. (2 lb 80% w.p.)/100 gal (H.V.), but not when the fruit is to be canned or quick frozen.

Badly cankered and spotted canes should be cut out and burnt.


## 9.10 STRAWBERRY

### 9.10A PESTS

The pests of strawberries are discussed in M.A.F.F. Bull. 95.

***Aphids.** Several species are found on strawberry, but the strawberry aphid (*Chaetosiphon fragaefolii*) is the most important as it is the main vector of virus diseases. It is a creamy white aphid with knobbed hairs on its back. It occurs on the plants all the year, with peak numbers in early summer on established fruiting plants, and in September on first year plants. Winged forms appear in May and June, dispersing to other strawberry plants, and small numbers also occur in October to December.

The shallot aphid (*Myzus ascalonicus*), a light greenish brown species, sometimes colonizes the plants in autumn and may cause severe damage in the following spring, especially after a mild winter, distorting leaves and blossom and destroying the crop.

Organophosphorus insecticides give the best control, applied at the times given below (doses as a.i./100 gal, H.V., or oz/acre, L.V.): deme-phion, H.V. 3·6 oz (12 fl. oz 30% e.c.), K.V. 4·8 oz; *or* demeton-S-methyl,

H.V. 3·5 oz (6 fl. oz 58% e.c.), L.V. 7·0 oz; *or* dimethoate, H.V. 4·8 oz (12 fl. oz 40% e.c.); *or* disulfoton, 16·0 oz (14 lb 7·5% granules)/acre; *or* ethoate-methyl, H.V. 4·8 oz (24 fl. oz 20% e.c.); *or* formothion, H.V. 7 oz (16 fl. oz 43% e.c.); *or* malathion, H.V. 18 oz (1½ pints 60% e.c.); *or* mevinphos, H.V. 2 oz (2 oz 99%)/200 gal/acre; *or* oxydemeton-methyl, H.V. 3·4 oz (6 fl. oz 56% e.c.), L.V. 4·6 oz (8 fl. oz 56% e.c.); *or* phorate, 24 oz (15 lb 10% granules)/acre; *or* phosphamidon, H.V. 4·8 oz (16 fl. oz 30% s.c.), L.V. 9·6 oz; *or* schradan, H.V. 6·6 oz (12 fl. oz 55% s.c.), L.V. 7·7 oz (14 fl. oz 55% s.c.). If aphid control is required a few days before picking, use dichlorvos, H.V. 8 oz a.i. (16 fl. oz 50% e.c.)/100 gal (still weather conditions with temperatures 55–60°F or over for best results), *or* mevinphos.

Treat fruiting plants (except maidens) shortly before flowering, usually mid or late April, or early April for plants under cloches; spray maidens as soon as picking is over; H.V. sprays are advised if red spider mites are also present.

Treat runner beds in late May and again in early July. Runners should be dipped before planting in demeton-S-methyl, dimethoate, formothion, malathion, nicotine (8 oz 95–98%/100 gal), *or* oxydemeton-methyl at concentrations given above.

Do not spray open blossoms; observe the minimum intervals between last application and harvest.

***Chafer grubs** (A.L. 235). The large white grubs attack the roots, causing wilting and death of plants. Damage usually only occurs where crops are planted after pasture. (See also p. 113.) If chafer grubs are seen during cultivations, work gamma-BHC into the soil before planting, as for wireworm control (see p. 219).

***Common green capsid** (*Lygocoris pabulinus*) (A.L. 154). See under Currant, p. 205.

****Cutworms** (*Agrotis* spp. etc.) (A.L. 225). In some seasons these plump greenish brown caterpillars feed on the roots and crowns, and may eat away the growing point. They feed at night and are most likely to be troublesome from late July to September. (See also p. 127.)

To control, drench with DDT ½ lb a.i./100 gal (as emulsion) at ¼ pint/ plant. Observe the minimum intervals between last application and harvest. Information is lacking on the value of alternative insecticides.

*****Eelworms** (A.L. 414). The leaf and bud eelworms (*Aphelenchoides fragariae* and *A. ritzemabosi*) feed in the crowns and in the folds of young unopened leaflets. Damaged leaves may be puckered, some showing on expansion a pale grey or silver patch near the base of the midrib. The main crown may become blind, secondary crowns developing. The stem eelworm (*Ditylenchus dipsaci*) causes a marked corrugation of leaves, with shortening and thickening of the stalks of leaves and blossom trusses. Strawberries are attacked by the stem eelworm races affecting onions, oats, red clover, narcissus, parsnips and other vegetables.

Runners should only be taken from healthy crops. If need be, control may be obtained by hot water treatment (115°F for 10 minutes for leaf and bud eelworms, 115°F for 7 minutes for stem eelworm). Hot water treatment needs to be undertaken with great care to avoid killing the plants. In experiments, drenching established plants already infested with stem eelworm with nematicides (parathion or thionazin) has given disappointing results.

For control of *Xiphinema*, vector of arabis mosaic, see p. 219.

****Leafhoppers** (*Aphrodes* and *Euscelis* spp.). These insects are vectors of the green petal virus disease, which may be carried from clover and certain weeds.

Their control has not been studied and there is no standard recommendation but spraying of valuable runner beds in areas where the disease occurs with DDT 1 lb a.i./100 gal, *or* dimethoate, 4·8 oz (12 fl. oz 40% e.c.)/100 gal. H.V., *or* malathion 18 oz a.i. (1½ pints 60% e.c.)/100 gal (H.V.) in July and twice more at 14 day intervals is suggested.

*****Red spider mite** (*Tetranychus urticae*) (A.L. 226). This is the red spider mite which is a glasshouse pest and also occurs on a variety of soft and top fruit crops. It overwinters as adult females in the soil and other shelter; these emerge in April to feed on the foliage, on which eggs are laid and up to 7 generations may follow during the summer. Some varieties under cloches, e.g. Cambridge Favourite, are particularly susceptible to severe damage.

The systemic organophosphate insecticides used for aphid control will also control red spider. Other acaricides, not effective against aphids, may also be used where required, e.g. chlorbenside 4 oz a.i./100 gal (H.V.) *or* dicofol, 6·4 oz (32 fl. oz 20% e.c.)/100 gal (H.V.); *or* summer petroleum oil 1%; *or* tetradifon 2 oz (1¼ pint 8% e.c.)/100 gal. (H.V.); *or* tetrasul, 6·4 oz (2 lb 20% w.p.)/100 gal (H.V.). These are useful against red spider mites resistant to organophosphorus insecticides, which have occurred several times in recent years.

Do not spray open blossoms. Observe the minimum intervals between last application and harvest.

****Strawberry blossom weevil.** (*Anthonomus rubi*). Adult weevils emerge from hibernation in April and May. The female lays eggs in the unopened flower buds, then partially severs the flower stalk below. The larva develops inside the flower bud, adult weevils emerging in July.

For control apply DDT 1 lb a.i./100 gal (H.V.), *or* 5% DDT dust, *or* TDE ¾ lb a.i. (1½ lb 50% w.p.)/100 gal, as soon as the first damage is seen, usually just before the flowers open and again about 10 days later if necessary. Recent trials have indicated that azinphos-methyl, 2·64 oz

(12 fl. oz 22% e.c.)/100 gal H.V., 5·3 oz/acre L.V.; *or* malathion, 18 oz (1½ pints 60% e.c.)/100 gal H.V. may give control equal to or better than that by DDT.

***Strawberry mite** (*Steneotarsonemus pallidus*). These microscopic colourless mites feed amongst the young folded leaflets, which may remain undersize, wrinkled and turn brown, and with heavy infestations plants are stunted. The mites breed and overwinter on the plants. The pest occurs sporadically, and is usually more in evidence in hot dry summers. Except sometimes with crops under cloches attacks do not usually become severe until after cropping.

Where the pest occurs, spray thoroughly after picking, when new growth appears after mowing or burning off, with endosulfan, 8 oz a.i. (2 pints 20% e.c.)/100 gal H.V.; *or* endrin, 4 oz a.i. (1 pint 20% e.c.)/ 100 gal H.V. Dicofol, 6·4 oz a.i. (32 fl. oz 20% e.c.)/100 gal H.V. will control light attacks. Good cover is important and drenching sprays should be used with any of these insecticides. Endosulfan or endrin may be used only after the crop has been picked and before the next season's flowers open.

****Strawberry rhynchites** (*Caenorhinus germanicus*). The weevils bite through the stems of leaves and fruit trusses in May and June, causing them to wither. For control, use DDT, 8–10 oz a.i./100 gal (H.V.) or 1¼ lb/acre (L.V.), or 5% DDT dust 40–60 lb/acre, *or* TDE, ¾ lb a.i./ 100 gal (H.V.), in mid to late April. Repeat after 10 days if required. Information is lacking on alternatives to these insecticides.

Observe the minimum intervals between last application and harvest.

****Strawberry seed beetle** (*Harpalus rufipes*). The adult beetles overwinter under rough vegetation, and may enter strawberry fields when the fruit is forming. They bite the seeds from the fruit, spoiling its appearance and market value. Eggs are laid in late summer in the soil in rough weedy places, and the larval stages feed upon seeds, with a partiality for fat-hen, pupating the following spring and giving rise to further adults.

Other ground beetles, e.g. *Pterostichus* (*Feronia*) spp., eat holes in the fruit which resemble slug damage, but are less important.

Surveys have shown that strawberry seed beetle is a local and sporadic pest and that much of the damage attributed to it is in fact caused by linnets. Linnets pick the seeds out cleanly, usually only from the upper surfaces of exposed fruits, and cause little damage to the flesh. The beetles also remove the seed or bite it open on the spot and in doing so cause severe injury to the adjoining flesh of the fruit; they may damage any part of the fruit but more often the lower surface next to the ground.

It is not possible to predict in what crops damage may occur though it is more likely in crops which were allowed to become weedy in the previous season or are surrounded by rough vegetation. Maiden crops also seem more likely to be attacked. The standard control measure is the application of aldrin to the soil just before strawing, at 5 lb a.i./acre ($3\frac{1}{2}$ cwt $1\frac{1}{4}$% dust, $2\frac{1}{4}$ cwt 2% dust, or $13\frac{1}{2}$ pints 30% e.c. in not less than 300 gal/acre) cultivated in between the rows. The headlands should also be treated. This treatment will, however, be reviewed at the end of the 1965 season and possibly modified. There is no experimental evidence to show whether or not restricting the application of aldrin to the headlands can give adequate protection. A minimum of 3 weeks must elapse between last application of aldrin and picking.

Aldrin treatment should only be used as a preventive measure where previous experience suggests that there is a reasonable certainty that seed beetle attacks are likely to occur.

Promising results have been obtained with a poison bait made with 1 pint 60% malathion e.c./gal water/40 lb crushed oats/acre, which can be used if an attack is discovered during fruiting. The diluted malathion is applied by sprayer or watering can with a fine rose while the heap of oats is turned until the mass is evenly moistened. The bait is placed in small handfuls between the plants or between the rows at about 1 yd intervals, scuffing the straw aside if necessary. The bait kills a high proportion of the beetles, giving a reduction of fruit damage, and remains effective for 5–6 days under moist conditions or about 10–12 days in dry weather. It is doubtful if it would be effective as a preventive treatment applied before strawing, and it is not known if baiting headlands at the start of fruit formation would give adequate control.

**Strawberry tortrix moth** (*Acleris comariana*). The moth has two broods, in June to July and August to September. It overwinters as eggs on the plants, these hatching in April or May and the caterpillars feed on leaves and flowers. It is controlled by azinphos-methyl, 5·3 oz (24 fl. oz 22% e.c.)/acre, in not less than 50 gal water; *or* DDT, 1 lb a.i./100 gal H.V., 2 lb/acre L.V.; *or* malathion, 18 oz ($1\frac{1}{2}$ pints 60% e.c.)/100 gal H.V.; *or* mevinphos, 4 fl. oz 99%/200 gal/acre; *or* TDE, 12 oz ($1\frac{1}{2}$ lb 50% w.p.)/100 gal H.V.; *or* trichlorphon, 1 lb a.i. ($1\frac{1}{4}$ lb 80% s.c.)/100 gal H.V. Azinphos-methyl, DDT and mevinphos appear to be the most effective. Spray if necessary as soon as caterpillar damage is seen; a second spray in late summer after the fruit is picked may be needed against the second generation. Preference should be given to the organophosphorus insecticides.

**Swift moths** (*Hepialus* spp.) (A.L. 160). The whitish caterpillars dotted with black are occasionally injurious, biting through roots and excavating cavities in the crowns. If the caterpillars are seen during breaking of grassland for strawberries, work gamma-BHC into the soil before planting as for wireworm control (see p. 219). With attacks on established crops drenching with DDT $\frac{1}{2}$ lb a.i./100 gal (as emulsion) at $\frac{1}{4}$ pint/plant, is effective.

***Vine weevil** (*Otiorhynchus sulcatus*). See Wingless weevils, below.

****Wingless weevils** (*Otiorhynchus* spp.) (A.L. 57). Several species occur as pests, the larvae killing or weakening plants by feeding on the roots. Two common species are the vine weevil (*O. sulcatus*) and strawberry root weevil (*O. rugosostriatus*). Eggs are found mostly in late July, August and September, in the surface soil beneath the canopy of leaves. The larvae hatch in the late summer and autumn and feed on the roots during the winter and spring, then pupate in the soil. Vine weevil adults emerge in May to June and strawberry root weevil in June to July. Plants may collapse during the fruiting period. The adults also feed on the foliage at night but damage is not serious. A few adults overwinter in the soil under the plants but most disappear in the late summer.

In parts of the south-west the red-legged weevil (*O. clavipes*) is a serious pest. Adults appear in two waves, emerging *en masse* in the spring from pupae formed in the autumn, and in a succession from mid June to end August from pupae formed in late spring and summer. Eggs are laid from late May to end August, and the larvae feed on the roots.

It is not worth trying to treat plants when an attack is discovered during flowering or fruiting. The object is to eradicate an infestation and prevent it spreading to adjoining strawberry crops. After fruiting, apply DDT dust at 5 lb DDT/acre to the crowns and surrounding collar of soil to kill grubs hatching from the eggs and before they can penetrate deeply in the soil. This should be done about the second week of August for vine weevil and strawberry root weevil, and the first week of September for red-legged weevil. Drenching the crowns and surrounding collar of soil with DDT 1 lb/100 gal (as emulsion), $\frac{1}{2}$–1 pint/plant, is also effective.

Before replanting in infested land work gamma-BHC into the soil as for wireworm control (see below).

Information is lacking on the value of insecticides alternative to these organochlorine compounds.

****Wireworms** (*Agriotes* spp.) (A.L. 199). Strawberries are susceptible to damage by these pests which may occur where crops are grown in broken up grassland. Roots are bitten through and holes drilled into the crowns. (See also Chapter 5.) When breaking up grassland for strawberries work gamma-BHC at $\frac{3}{4}$–$1\frac{1}{4}$ lb a.i./acre into the soil at least 2–3 weeks before planting. Do not plant potatoes, carrots or onions for at least 18 months because of risk of taint.


## 9.10B DISEASES

***Arabis Mosaic** (virus) (A.L. 530). Yellow spots or blotches usually appear on the foliage and in some varieties the spots become bright red. Leaf crinkling and stunting may occur. The leaf symptoms are most pronounced in late spring and autumn, tending to fade during summer. The causative virus is transmitted by the eelworm *Xiphinema diversicaudatum* which lives in the soil and attacks the roots.

Unless plants were infected at the time of planting, the disease is usually seen in well defined patches corresponding to pockets of soil infested with *X. diversicaudatum*. In the latter instance, control of the disease has been obtained by fumigating the affected soil with D-D mixture. This should be applied at the rate of 400 lb/acre (1 lb/100 ft²) by injecting 3·5 ml ($\frac{1}{8}$ fl. oz) at a depth of 6 in at 12 in spacings. Best results are obtained if application of the nematicide is made in autumn.

***Grey mould** (*Botrytis cinerea*) (M.A.F.F. Bull. 95). The fungus invades the floral parts and is also common on decaying plant remains and, under moist conditions, infects strawberry fruits on which it is seen as a grey furry mould. Loss of crop can be very high. Spray from early flowering (about 20%) with captan 2 lb a.i. (8 lb 50% w.p.)/100 gal; *or* dichlofluanid 12 oz a.i. (1$\frac{1}{2}$ lb 50% w.p.)/100 gal; *or* thiram 1·6 lb a.i. (2 lb 80% w.p.)/ 100 gal and repeat at 10-day intervals until harvest. Alternatively spray twice only, at early flowering and early petal fall with captan 4 lb a.i./100 gal *or* dichlofluanid 1$\frac{1}{2}$ lb a.i./100 gal *or* thiram 3·2 lb a.i./100 gal and omit the later normal rate sprays unless *Botrytis* is seen building up. The flower parts must be well covered, applying at least 200 gal/acre; corresponding rates for spraying three times are given on the labels of some products. Lance spraying should be considered if good cover from boom spraying is doubtful.

Thiram, captan, lime sulphur and copper are liable to make fruit unsuitable for processing, so that the grower should consult processors before adopting a particular spray programme.

**Mildew** (*Sphaerotheca macularis*) (M.A.F.F. Bull. 95). The disease is seen during spring as dark patches on the upper side of the leaf. These patches correspond to a whitish-grey sporing growth on the underside. Affected leaves may curl upwards as if with drought, and the mildew spreads to other leaves, to the blossoms and the berries. The latter may become shrivelled or otherwise unmarketable. The fungus overwinters on old green leaves.

Royal Sovereign and Cambridge Vigour are very susceptible to mildew; Cambridge Favourite is resistant.

Where mildew is likely to occur, an application of lime sulphur, or one of the other forms of sulphur, *or* dinocap, should be made just before flowering and at 10–14 day intervals. A wetter should be added to lime sulphur, which should be used at 1 gal lime sulphur/100 gal (H.V.). The sulphur sprays should contain 1·5–3 lb sulphur/100 gal (H.V.) but neither lime sulphur nor sulphur should be applied to the ripening fruit if it is to be used for processing. The dinocap spray should contain 0·25 lb a.i. (1 lb 25% w.p.)/100 gal, applied H.V.

Foliage should be burnt off after harvesting. If this is not done, post-harvest applications of fungicide may be needed.

## 9.11 GRAPE VINE

### DISEASES

*Downy mildew* (*Plasmopara viticola*), mainly on outdoor grapes. Light green patches occur on the upper surfaces of leaves and these correspond to growths of the sporulating fungus on the undersides; diseased areas later become dry and brittle. Berries can also be affected and, when severe, may shrivel. Overwintering spores are produced in affected leaves and these can renew the disease the following spring. Where infection is expected to occur a protective spray of zineb at 21 oz a.i. (2 lb 65% w.p.)/100 gal H.V. should be applied at 10–14 day intervals. Diseased leaves and tendrils should be burnt.

***Powdery mildew** (*Uncinula necator*) (A.L. 207). Powdery mildew is common on vines growing both outdoors and in the glasshouse. The mildew forms white sporulating patches on young leaves and shoots but its development may be so sparse that the grey or purplish discoloration of the diseased parts is the most obvious symptom. Flowers and berries may also become infected causing them to drop. At a later stage berries become distorted and cracked.

With vines, both outdoors and indoors, sulphur dusts or sprays *or* dinocap should be applied to the foliage as soon as the mildew appears. Where it has been troublesome in previous years the first application should be made 10–14 days before it is expected. Outdoors a total of up to four applications may be needed to prevent a build-up. In glass-houses the first application is commonly made when the laterals are 12–15 in long with a second just before the flowers open, and a third after the berries have set.

The following rates are recommended: dinocap, 2 oz a.i. (8 oz 25% w.p.)/100 gal H.V. (outdoors) or 1 oz a.i. (4 oz 25% w.p.) 100 gal H.V. (indoors); *or* sulphur, 3 lb a.i. (4 lb 80% w.p.)/100 gal H.V. or 2 lb a.i. (2 pints 80% colloidal)/100 gal H.V.

## 9.12 HOP

### 9.12A PESTS

Annual routine control measures are necessary against damson-hop aphid.

*Clay-coloured weevil** (*Otiorhynchus singularis*). See under Currant (p. 205).

***Damson-hop aphid** (*Phorodon humuli*) (M.A.F.F. Bull. 164). This aphid winters as eggs on *Prunus*, especially *P. spinosa* and *P. domestica*. These hatch in April and dispersal of winged aphids to hops begins in earnest in late May or early June, rising to a maximum in late June and early July. It then declines, normally ending by mid or late July, but in exceptional seasons, such as those occurring from 1962–64, it may persist well into August. A return flight to *Prunus* occurs in September and October; no dispersal between gardens in the summer is known. If adequate control is not obtained by the burr stage in late July, the cones may become infested.

Although a low degree of resistance to organophosphorus has been shown to occur, these are still the most effective compounds, but thorough spraying at the full recommended rates is essential. It is most important to obtain good control by the burr stage, as once the cones become infested it is very difficult to get satisfactory results. Not only does the dense canopy of foliage after late July present a difficult target, but the movement of foliar-applied systemic insecticides once growth of the bine has ceased appears to be restricted.

Insecticides may be applied either as foliar sprays or soil drenches, or a combination of both. Soil drenches have the advantage that they can be applied when the weather is unsuitable for spraying; possible disadvantages are that under dry conditions uptake by the bine may be slow, and also with large diffuse hills and those by poles some bines may escape treatment and become infested. In general, soil drenches probably result in a better uptake of systemic insecticide and greater persistence than with sprays, so that although a programme based on the latter can give good results there is much to commend the following combination of sprays and drenches which is often adopted—spray in late May or early June when the aphids are sufficiently numerous, repeating 2–3 times later if required, and apply a soil drench in mid July. The use of a soil drench in July is particularly advantageous in seasons when aphid dispersal to hops is late and continues well into August. Prediction of the end of dispersal may prove possible so that the July soil drench can be timed to the best advantage.

**Sprays:** (doses as a.i./100 gal H.V., or a.i./acre L.V.): demephicen, L.V. 4·5 oz (16 fl. oz 30% e.c.) for the first spray, 7·2 oz (24 fl. oz 30% e.c.) for subsequent sprays; demeton-S-methyl, H.V. 3·5 oz (6 fl. oz 58% e.c.), L.V. 4·5 oz (8 fl. oz 58% e.c.) for the first spray (bines up to breast wire) and 7·0 oz (12 fl. oz 58% e.c.) for subsequent sprays; *or* menazon, H.V. 2 oz (¼ pint 40% e.c.), L.V. 4 oz (½ pint 40% e.c.) for the first spray, 8 oz for subsequent sprays; *or* morphothion, H.V. 4 oz (1 pint 20% e.c.) L.V. 4 oz for the first spray, 6 oz (1½ pints 20% e.c.) for later ones, repeat sprays at 10–14 day intervals; *or* oxydemeton-methyl, H.V. 3·4 oz (6 fl. oz 56% e.c.), L.V. 4·6 oz (8 fl. oz 56% e.c.) for the first spray, 6·8 oz

(12 fl. oz 56% e.c.) for subsequent ones; *or* phosphamidon, H.V. 4·8 oz (16 fl. oz 30% s.c.), L.V. 4·8 oz for the first spray, 9·6 oz for subsequent sprays; *or* schradan, H.V. 6·6 oz (12 fl. oz 55% s.c.) for first spray, repeat late July with intermediate spray if needed, L.V. 7·7 oz (14 fl. oz 55% s.c.) for first spray, repeat end June and late July; *or* vamidothion, H.V. 4·8 oz (12 fl. oz 40% e.c.) for first spray, 6·4 oz (16 fl. oz 40% e.c.) for subsequent sprays, L.V. (do not use less than 50 gal/acre): 9·6 oz (24 fl. oz 40% conc.) for first spray, 12·8 oz (30 fl. oz 40% e.c.) for subsequent sprays. See pp. 28–61 for minimum periods between last application and harvest.

Dimethoate is also effective but phytotoxic to some varieties. It can, however, be used on Fuggle, at 4·8 oz (12 fl. oz 40% e.c.)/100 gal/acre, in late May or early June as with the other insecticide sprays given above, and repeated at 2–3 week intervals until late July or early August as required.

Mevinphos at 2–3 fl. oz 99% conc./100–200 gal/acre (the higher rate in cool weather) is also effective but has short persistence. Malathion, 18 oz a.i. (1½ pints 60% e.c.)/100 gal H.V.; *or* nicotine, 8 fl. oz (95–98%)/100 gal H.V.; *or* TEPP 4 fl. oz 40% e.c./100 gal plus spreader H.V. may also be used where sprays near picking are required, but as already mentioned control is very difficult once aphids are in the cones.

All the above insecticides, except malathion and nicotine, are systemic, and demeton-S-methyl, dimethoate, morphothion and oxydemeton-methyl also have contact action. Under slow growing conditions, such as with dry soil or cold weather, insecticides with contact effect sometimes give better control than those which are almost entirely systemic in action. With schradan 24 hours of dry weather should follow spraying to ensure adequate uptake of insecticide.

High volume sprays are usually 100–150 gal/acre for the first two applications and 200–250 gal/acre for later ones. Low volume sprays are usually 30–75 gal/acre, and should be at least 50 gal/acre for the second and subsequent applications. In either case the first spray is often applied up alternate alleys, but with the following applications every row should be sprayed. If late sprays in August are required at least 250 gal/acre should be applied and an insecticide with good contact action used. Certain varieties, such as Northern Brewer, John Ford, Bullion and College Cluster do not appear to absorb insecticides readily and may need more frequent sprays or high volume applications.

**Soil drench:** For a drench in mid July following foliar sprays in June, use dimefox, 20 oz a.i. (2 pints 50% e.c.)/25 gal, pouring ¼ pint of the diluted insecticide on to the crown of each hill. An alternative is to rely entirely on dimefox, applying it at the same rate, 2 pints 50% e.c./25 gal at ¼ pint/hill, in late May or early June and again in mid July.

Dimefox is best applied by special lances fed from a tractor mounted tank, giving the correct quantity to each hill when the trigger is released. On first and second year hops it should always be used at a lower rate, i.e. 20 oz (2 pints 50% e.c.)/50 gal, and it should not be applied to newly planted hops.

Also effective as drenches but relatively more expensive are demeton-S-methyl, 4·5 oz a.i. (8 fl. oz 58% e.c.)/10 gal, *or* oxydemeton-methyl, 4·6 oz (8 fl. oz 56% e.c.)/10 gal, each at ¼ pint/hill.

Do not handle hops for 4 days after treatment with dimefox or schradan unless rubber gloves are worn. Observe the minimum intervals between last application and picking.

****Hop capsid** (*Calocoris fulvomaculatus*) winters as eggs laid in softer parts of hop poles; these hatch late May and the bugs puncture the leaves which become perforated and split open; the bine may also be scarred and become twisted in growth. This is no longer a serious pest, and appears to be well controlled by sprays used against damson-hop aphid. It may also be controlled with DDT, 1–1½ lb a.i./acre in not less than 30 gal, in late May or early June.

***Hop flea beetle** (*Psylliodes attenuata*). Adults hibernate under rough vegetation, emerging in the latter part of April and in May; they eat holes in the leaves and damage young shoots. The second brood in late summer feeds on the leaves and cones. This has become a relatively uncommon pest. If required spray when beetles first seen with DDT, 1 lb a.i./acre, in not less than 30 gal, or apply DDT dust at rates equivalent to 2–3 lb a.i./acre. Repeat against second brood if necessary.

****Hop root weevil** (*Epipolaeus caliginosus*). Adult weevils are found throughout the year, being most numerous in late summer and autumn. Eggs are laid in punctures made in the rootstock from mid August to May, and the grubs tunnel in the rootstock and base of the bine for 9–18 months. The bine is weakened or killed, and feeding by adults in spring may also weaken growth.

Dust the hills with gamma-BHC dusts (0·2–0·65% a.i.) at about 1 lb/10 hills in late May or early June immediately before earthing up.

***Hop strig midge** (*Contarinia humuli*). Midges emerge from pupae in the soil in the latter part of July and in August, and lay eggs in the cones from early August. The maggots feed at the bases of the bracts and along the strig, and the cones become brown and undersized. The maggots are fully fed by the end of August and drop to the soil. This pest is now rarely seen, but if control is necessary, spray with DDT, 1 lb a.i./100 gal (H.V.), in late July.

****Red spider mite** (*Tetranychus urticae*). The mite winters as adult females in the soil, crevices in poles and wirework. It emerges in late April and feeds on the leaves, where eggs are laid and up to 7 generations may follow during the summer.

This pest appears to have been virtually eliminated from hop gardens by the organophosphorus insecticides used against damson-hop aphid. All those mentioned for aphid control are effective, except menazon and morphothion. Tetradifon at 2 oz a.i. ($1\frac{1}{4}$ pints 8% e.c.)/100 gal H.V. or 4 oz a.i./acre L.V. is also effective.

****Rosy rustic moth** (*Hydraecia micacea*) is usually a local and minor pest although several heavy infestations have occurred in recent years. The moths are on the wing from late July to late October, being most numerous in September. Eggs are laid in the autumn and there is some doubt whether they hatch in the autumn or the following spring. The young caterpillars begin to ascend the bine in late April or early May; they bore into the stem and tunnel in the pith. During June they leave the aerial part of the bine to enter the soil and complete their growth by feeding on the crown or the base of the bine. All appear to have descended by the end of June and the first pupae occur at this time.

No experiments on control have been carried out, but a DDT spray at the beginning of May with a second 14 days later should protect the bine from attack.

***Wireworms** (A.L. 199). Hop sets planted in land ploughed from old grass may be attacked by wireworms, and these are sometimes troublesome when replanting in old gardens.

When breaking up grassland for hops work gamma-BHC into the soil at $\frac{3}{4}$–$1\frac{1}{2}$ lb a.i./acre at least 2–3 weeks before planting. An alternative measure is the application of gamma-BHC wireworm dust at $\frac{1}{4}$ oz of dust to the planting hole before planting. Where established hops are attacked sprinkle this amount of dust over the soil surrounding the hill.

## 9.12B DISEASES

*****Downy mildew** (*Pseudoperonospora humuli*). The disease begins each spring from systemically infected shoots (basal spikes) arising from the crown of the rootstock. Spores from these result in more basal spikes. In suitable weather, the disease progresses upwards throughout the growing season. Terminal and lateral shoots become 'spikes', young leaves are killed, old leaves are spotted, and cones may become infected at any stage of development. If routine preventive methods are not adopted, the whole crop may become worthless. Downy mildew too is probably the most common cause of death of rootstocks, being introduced through short shoots (secondary basal spikes) and through the base of the bine; some varieties, particularly W.G.V., are very prone to this form of the disease.

After hand removal of spikes, Bordeaux powder, copper oxychloride or other specially formulated dust, based on 6% metallic copper, *or* zineb dust containing 14% a.i., should be applied with a muslin bag or hand

R

machine at about 15 lb/acre as soon as the first shoots appear, the object being to hit the crown of the hill and to give the shoots a good coverage. Two or three further dustings should be given at intervals of 10–14 days, making sure that all new growth is covered. Alternatively, after the first application of dust, streptomycin w.p. as a spray may be applied to the young shoots so that their removal from the hill may be safely delayed for 2 weeks or more. It is especially important to use streptomycin in accordance with the maker's instructions.

Copper, metiram, propineb, zineb or other dithiocarbamates formulated for hop as wettable powders or liquids, should be applied to the foliage 10 days after the last hill dusting and then about every 14 days until immediately after burr. More frequent and later applications may be given on susceptible varieties if the weather is warm and humid. The following rates are recommended: copper oxychloride. $1–1\frac{1}{2}$ lb metallic copper/100 gal H.V. or 2–3 lb/acre L.V., *or* cuprous oxide, $\frac{3}{4}–1\frac{1}{2}$ lb metallic copper/100 gal H.V. or $1\frac{1}{2}–2$ lb/acre L.V.; *or* Bordeaux mixture, $1–1\frac{1}{2}$ lb metallic copper/100 gal H.V. or 2–3 lb/acre L.V.; *or* metiram, 22 oz a.i. (2 lb 70% w.p.)/100 gal H.V., or 22 oz a.i./acre L.V. up to burr and 22 oz a.i./100 gal H.V. or 45 oz a.i./acre L.V.; *or* propineb, 22·4 oz a.i. (2 lb 70% w.p.)/100 gal H.V. or 22·4–56 oz a.i. (2–5 lb 70% w.p.)/acre L.V., in not less than 30 gal, the lower rate being used at the beginning of the season; *or* zineb 21 oz a.i. (2 lb 65% w.p.)/100 gal H.V. or 42–52 oz a.i./acre L.V. The dithiocarbamate products should be used according to the maker's instructions.

Copper-containing dusts, as for hill treatment, or a dust containing 8% zineb may be used for foliar application instead of a spray where the latter is inappropriate. Dusts should be applied at 30–50 lb/acre, and the intervals shortened to 7 days in rainy weather.

To control downy mildew, a protective coating of fungicide is needed throughout growth so that a full programme is necessary in most years. Coverage of all growth, particularly the highest, is of utmost importance.

Cones with fungicidal deposit are not always acceptable and before late applications are made, the grower is recommended to consult his hop factor.

Additional measures include the removal of all spikes (basal, terminal and lateral) as soon as identifiable. Among commercial varieties, the most susceptible include Goldings, Golding varieties, the new Wye varieties and W.G.V.; Fuggles and Bramling Cross are less susceptible.

***Hop mould** or **Powdery mildew** (*Sphaerotheca macularis*). Mould, a single fungus, appears in two forms. From May onwards the white powdery stage develops on the leaves and sometimes on the young shoots. The burr and cones may likewise be attacked and become distorted and useless. If effective treatment has not been given, and if the weather is humid and warm, the late summer stage develops. This is seen as foxy-red spots or patches on leaves and cones. On these red patches, minute black spore cases (perithecia) form. When cones shatter, spore cases fall to the ground, where they overwinter. In spring, usually May, the spores within the cases mature and are ejected to the shoots or lower leaves of the plant where they germinate and so begin the cycle again.

Sulphur, as a fine dust, a wettable powder or as a colloidal formulation, *or* dinocap as a wettable powder, should be applied in mid-May and then at intervals of 10–14 days according to the proneness of the garden to mould. If the full programme has been carried out and if no mould can be detected, applications may cease at burr. Otherwise they should be continued for another few weeks. The following rates are recommended: sulphur dust, 25–50 lb/acre, *or* colloidal sulphur *or* sulphur w.p., 3–4 lb sulphur/100 gal H.V., *or* 7·5 lb sulphur/acre L.V.; *or* dinocap 4 oz a.i. (1 lb 25% w.p.) 100 gal H.V. or 8–10 oz a.i./acre L.V.

A wetting agent may be needed when using some proprietary preparations at high volume.

When a garden has not been picked because of excessive mould, the bines should be cut and burnt before the cones shatter. The variety Northern Brewer is particularly susceptible.

# CHAPTER 10

# PEST AND DISEASE CONTROL IN GLASSHOUSE CROPS

## 10.1 GENERAL

### SOIL PESTS AND DISEASES

Soil borne pests and diseases are peculiarly apt to become limiting factors in the economic production of glasshouse crops, for these crops are usually grown without rotation. Moreover, pathogenic fungi such as *Phytophthora* spp., *Rhizoctonia solani* and *Pythium* spp., and pests such as nematodes, leatherjackets and wireworms may occur in maiden loams used in propagating and potting composts causing serious losses of seedlings and young plants.

Soil borne pathogens and pests can be eliminated from composts by partial disinfection with heat, most commonly by 'steaming', but their complete eradication in glasshouse border soils by steaming is seldom accomplished as the heat does not penetrate to a sufficient depth.

Chemical soil sterilants can be used as substitutes for steaming but are more specific in their fungicidal, nematicidal and insecticidal action than heat, and as they, or their decomposition products, are highly phytotoxic, a lengthy time period between treatment and planting may be necessary to ensure their release from treated soils. Methyl bromide however, disappears rapidly. Their vapours may diffuse through partitions, or be carried via heating ducts or ventilators into adjoining glasshouses in sufficient amounts to injure plants.

For the successful use of chemical soil sterilants the soil must be thoroughly worked before treatment and should be evenly moist but not wet. Soils with a high organic content absorb chemicals more strongly than light sandy soils and retain the chemical longer, and longer intervals are required between treatment and planting. Adequate distribution of chemical sterilants in the soil is essential and exceptionally heavy soils which cannot be broken up to a reasonable tilth are unsuitable for treatment by chemicals as clods are not penetrated. Where the chemicals are mixed into the soil by rotary cultivation, L-shaped tines should be use with a rotor speed set between 120–150 r.p.m. and a forward speed not exceeding 15 ft/min. Most chemical sterilants are volatile or decompose to produce volatile compounds, and with these it is necessary to 'seal' the surface for a few days by compacting and wetting it to a depth of $\frac{1}{2}$–1 in. Methyl bromide, which is gaseous above 40°F, diffuses into the soil from the surface when released from canisters or cylinders under suitable plastic sheets which have their edges buried in the soil to a depth of about 5 in to prevent the escape of the gas into the atmosphere. The rate at which sterilization is accomplished and the speed of elimination of the chemical from treated soil are greater at higher temperatures, consequently it is inadvisable to treat with chemicals during the late autumn and winter. Their release from treated soils is hastened by forking or rotary cultivating the soil to a depth of 10–12 in 3 weeks after application in the autumn or 1–2 weeks in the summer, and repeating at 14 day intervals once for summer and twice for autumn applications. Before using or planting up soil treated with metham-sodium, dazomet or methyl isothiocyanate, freedom from phytotoxic residues should be established by seeing that cress seed germinates normally when sown on the moistened surface of a representative sample of about 2 lb of the soil in a closed glass jar.

Where autumn applications are made to normal loam soils before November the following minimum intervals should be allowed before planting: chloropicrin 10 weeks, dazomet 10 weeks, 'D-D mixture' 6 weeks, formaldehyde 6 weeks, metham-sodium 10 weeks, methyl isothiocyanate 8 weeks. For November or March applications the intervals with chloropicrin, dazomet, metham-sodium and methyl isothiocyanate should be increased to at least 12 weeks and formaldehyde to 7 weeks.

For summer applications, the intervals for chloropicrin, dazomet, metham-sodium and methyl isothiocyanate can be reduced to 5 weeks and with formaldehyde 4 weeks. The interval required with methyl bromide is about 5 days in summer and 8 days in winter. The absence of hazardous contamination of the glasshouse atmosphere should be decided by the contractors. Methyl bromide treated soils must not be forked over to hasten the release of the gas.

The following chemicals are used for general soil disinfection, and the rates and methods of application are those which have been found satisfactory under most conditions.

## 10.1A FOR THE TREATMENT OF BORDER SOILS

**Chloropicrin:** Applied by injecting 3·5 ml ($\frac{1}{8}$ fl. oz) at a depth of 6 in at 12 in spacings (560 lb/acre), followed by surface sealing. A respirator and rubber boots must be worn when injecting chloropicrin, and rubber gloves when filling the injector. The injector should preferably be filled outside the glasshouse.

**Dazomet** applied by sprinkling 1 lb 85% dust/100 ft$^2$ evenly on the surface and immediately admixing with the soil by rotary cultivation, or by forking twice to a depth of 10–12 in; follow by surface sealing.

**'D-D mixture'** (dichloropropane-dichloropropene): normally used at 300–400 lb/acre, and applied by injecting 3·5 ml ($\frac{1}{8}$ fl. oz) at a depth of 6 in and at 12 in spacings in normal soils or 5·5 ml ($\frac{1}{5}$ fl. oz) at 15 in spacings in light, open soils, and followed by surface sealing. Its regular use on heavy soils is inadvisable, in view of the possibility of tainting.

**Formaldehyde** applied normally as a 1 in 50 dilution of commercial (38–40%) formalin at 50 gal diluted liquid/10 yd². Better results are obtained on light open soils by applying an additional 40–50 gal of the liquid/10 yd² to the second spit by double digging.

**Metham-sodium:** the normal dose is 13 oz a.i./100 ft², applied as 2 pints 33% sol/25 gal/100 ft² or applied as a 2% solution at 4 gal/100 ft² and immediately admixed with the soil by rotary cultivation or thorough digging, and followed by surface sealing or by injecting 11 ml (1 pint/50 injections) of concentrate at a depth of 8 in at 12-in spacings and incorporating in the soil by cultivation as with dazomet.

**Methyl bromide** is widely used in America, alone or in combination with other chemicals, for the control of nematodes in glasshouse soils. It has the advantage of rapid diffusion into and from the soil, consequently the interval between treatment and planting is short. Recent trials in Britain have confirmed its efficacy at $\frac{1}{2}$ lb/50 ft². Methyl bromide has a high mammalian toxicity and special equipment, including impermeable sheeting to retain it in the soil, is required. Chloropicrin (2%) may be added as a warning agent to decrease risks. At present methyl bromide should only be applied by contractors or trained operators.

**Methyl isothiocyanate:** Experimental results overseas and in this country indicate that methyl isothiocyanate injected at 160 lb/acre gives results comparable with those given by metham-sodium. 'D-D mixture' may be mixed with methyl isothiocyanate to improve the control of root knot eelworm. Such mixtures are available.


## 10.1B FOR THE TREATMENT OF SMALL QUANTITIES OF SOIL FOR PROPAGATING AND POTTING

### (M.A.F.F. Bull. 22)

**Dazomet:** Add to warm moist soil 6 oz dazomet (i.e. 7 oz 85% dust)/yd³, thoroughly mix, cover for 3 weeks, then turn three times at fortnightly intervals. Apply cress germination test before use, and beware of phytotoxic fumes affecting nearby plants.

**Formaldehyde:** 25–50 gal of a 1 in 50 dilution of commercial (38–40%) formalin per yd³ of soil applied to saturate the soil in 6 in layers as the heap is built. Cover with plastic sheeting or similar material for at least 2 days, uncover and turn. Allow 4–6 weeks before use depending on temperature, lightness of soil, and number of turnings.

**Metham-sodium:** apply 2 pints of a 33% solution of metham-sodium in 25 gal at 5 gal/yd³ to moist soil in 6 in layers as heap is built. After 14 days turn at least 3 times at 7–14 day intervals. Apply the cress germination test (see p. 229) before use, and beware of fumes affecting nearby plants.

## 10.1C SUSCEPTIBILITY OF PLANTS TO CHEMICALS

Plants grown in glasshouses during the winter under adverse light conditions have a 'soft' growth which is abnormally susceptible to chemicals applied at concentrations which are readily tolerated by older plants. High volume sprays, dusts and smokes are usually less phytotoxic than atomized solutions (aerosols) but considerable caution should be exercised in applying pesticides in any form to seedlings and recently potted or planted young plants or cuttings.

Varieties (cultivars) of certain plants, in particular chrysanthemums, differ considerably in their tolerance to chemicals and preliminary tests should be made on new varieties.

## 10.1D ATOMIZED FLUIDS (AEROSOLS) AND SMOKES

The quantities of pesticides required, when applied by atomization of solutions or as smokes, are expressed in g/100 ft³ of active material in the atmosphere. (1 g/1,000 ft³ is equivalent to 1 oz/28,000 ft³.) To obtain adequate distribution of smokes, four or more generators should be used to each 100 ft of glasshouse length in narrow, low houses and three per 100 ft in wide, higher houses.

Atomized solutions and smokes are convenient to use but are generally less effective than high volume sprays of the same pesticide. As atomizing solutions vary in content of a.i. between different manufacturers, no general amount of formulation to be used can be given; label instructions should be followed.

## 10.1E GLASSHOUSE HYGIENE

Strict attention to hygiene is important in reducing the incidence of pests and diseases in glasshouses. When diseases have occurred, the internal structure of glasshouses and mushroom houses (including any staging) and the soil surface should be disinfected, after the removal of any debris, by rapidly spraying with a 1 in 50 dilution of 40% formaldehyde and closing the house for at least 24 hours. Protective clothing and a respirator are necessary. Alternatively they may be fumigated by pouring 15 fl. oz

40% formaldehyde onto 8 oz potassium permanganate (in glass or earthen-ware containers)/15,000 ft³. The containers should be distributed in the house, and the above amounts not much exceeded in each container. Boxes, pots, etc., may be disinfected by steam, boiling water or formalde-hyde. Glass partitions cannot be relied on to prevent ingress of the highly phytotoxic formaldehyde vapour into adjoining houses.

## 10.2 SPECIFIC CROPS

### ARUM (CALLA) LILY

PESTS

****Aphids** (*Myzus persicae, Macrosiphum euphorbiae, Aphis gossypii*). These, and occasionally other spp., cause stunting and disfiguration of the young growth and severe spotting of the flower spathes. Control before the flowers open by adopting methods recommended for Carnation—Aphid, p. 236.

****Red spider mite** (*Tetranychus urticae*). For control measures see Carnation—Red spider mite, p. 237. Petroleum emulsions (glasshouse grade) may be used on arum at 1 gal/100 gal.

***Thrips** (*Heliothrips femoralis, H. haemorrhoidalis, Thrips tabaci*) breed on the foliage and flowers of arum causing pale, whitish spots on the leaves and brown spots on the flowers. For control see Carnation—Thrips, p. 237.

DISEASES

***Corm rot** (*Pectobacterium carotovorum*). Infected plants are attacked at the collar and the corms become invaded causing yellowing and death of the foliage. Badly diseased corms should be destroyed. Slightly affected corms may be saved by washing away all the soil when they are dormant, scraping away all diseased tissue and steeping for 2 hours in formaldehyde, 1 pint 40% formaldehyde/6 gal. Ensure that the corms are not recontami-nated and pot without delay into soil sterilized by heat or formaldehyde. Disinfect the glasshouse if vacant, see p. 231.

***Root rot** (*Phytophthora richardiae*). Leaves of attacked plants first become yellowish and streaky, and gradually turn brown and die. Flowers become brown at their tips and may be deformed. Treat as for corm rot above.

### ASPARAGUS FERN (*Asparagus plumosus, A. sprengeri*)

PESTS

***Caterpillars.** Foliage-eating caterpillars may be controlled with DDT sprays, dusts, atomized solutions and smokes *or* TDE sprays, as for Carna-tion—Tortrix, p. 237.

***Red spider mite** (*Tetranychus urticae*) causes browning and abscission of leaflets. Control with demeton-S-methyl, diazinon, dicofol, dimethoate, naled, oxydemeton-methyl *or* tetradifon as for Carnation—Red spider mite, p. 237. Azobenzene smokes, 3 g a.i./1,000 ft³ may be used.

*****Thrips** (*Thrips tabaci*). Severe attacks cause a silvering of the leaflets. To control, use DDT *or* diazinon as for Carnation—Thrips, p. 237, *or* DDT smokes 3·3 g/1,000 ft³ *or* naled fumigation 1 fl. oz 50% e.c./1,000 ft³.

*****Whitefly** (*Trialeurodes vaporariorum*) (A.L. 86). For description see Cucumber—Whitefly, p. 248. Control as for thrips above.

### DISEASES

*****Basal rot** (*Pythium* spp.). The fungus attacks the stems near soil level. Control as for root rot (below). It is possible that watering with captan 1 lb a.i. (2 lb 50% w.p.)/100 gal will check the spread of the disease.

*****Root rot** (*Fusarium* spp.). Branches become yellow and finally desiccated, basal shoots wither and roots decay. To prevent re-occurrence, sterilize the soil, preferably by steaming.

### AZALEA

#### PESTS

*****Red spider mite** (*Tetranychus cinnabarinus, T. urticae*). Mites may be a pest on azaleas if plants to be forced are placed in houses containing infected plants or hibernating mites of the second species. To control use demeton-S-methyl, *or* diazinon, *or* dicofol, *or* oxydemeton-methyl, as for Carnation—Red spider mite, p. 237.

*****Thrips** (*Heliothrips haemorrhoidalis*). This thrip causes silvery or whitish patches with reddish-brown spotting on the undersides of the leaves and on the petals of flowers. Attacks almost invariably result from existing infestations within the glasshouse and are rare in commercial houses. Control as for Asparagus—Thrips, above.

#### DISEASES

*****Grey mould** (*Botrytis cinerea*). Rotting of flowers by Botrytis only occurs under excessively damp conditions and is not frequent on plants grown commercially. Ample ventilation and prompt removal of all diseased flowers will usually prevent its spread but it is advisable, especially in unheated houses, to spray with captan at 1 lb a.i. (2 lb 50% w.p.)/100 gal, repeating this if necessary after 10 days. The tolerance of azaleas to dichlofluanid (see Chrysanthemum—Grey mould, p. 243) is not yet known.

## BEANS—(FRENCH AND CLIMBING FRENCH)

***Aphids:** Atomize solutions of dichlorvos *or* gamma-BHC *or* malathion *or* diazinon; or smoke with gamma-BHC *or* nicotine *or* parathion; or spray with gamma-BHC *or* malathion *or* nicotine as for Tomato—Aphids, and allow the same intervals between application and picking.

****'French-fly'** (*Tyrophagus dimidiatus*). For description and control see Cucumber—French-fly, p. 246.

****Red spider** (*Tetranychus urticae, T. cinnabarinus*). On young plants use parathion smokes *or* H.V. sprays of dicofol as for young tomatoes. On established plants use H.V. sprays of dicofol *or* tetradifon *or* atomized solutions of chlorfenson *or* dicofol *or* smokes of tetradifon as for Tomato—Red Spider mite, p. 268.

***Symphylids.** For description and control see Tomato—Symphylids, p. 269.

***Whitefly**—see Tomato—Whitefly, p. 269.

***Wireworms**—see Tomato—Wireworms, p. 270.

## BEGONIA

Begonia flowers are susceptible to injury by many pesticides, particularly those applied as sprays. Do not spray plants in sunshine or under slow drying conditions.

PESTS

****Aphids.** A number of species including *Myzus persicae, Aulacorthum solani* and *Aphis gossypii* breed on begonias under glass and may cause serious damage to foliage and flowers. Control before flowering with gamma-BHC, *or* malathion, *or* naled, *or* nicotine as for Carnation—Aphids, p. 236, or by dipping the plants in gamma-BHC, *or* malathion, *or* nicotine at concentrations as for H.V. sprays.

***Mealy bug** (*Planococcus citri*). This species, and occasionally *Pseudococcus adonidum*, attack plants kept in private glass houses but are not common in commercial houses. Several treatments, at 10–14 day intervals, with malathion at 18 oz a.i. (30 fl. oz 60% e.c.)/100 gal by spraying or dipping will control this pest.

***Mites.** The Tarsonemid (cyclamen) mite (*Steneotarsonemus pallidus*) and the broad mite (*Hemitarsonemus latus*) cause stunting, crinkling and curling of the leaves and malformed flowers. The control of both is as for Cyclamen—Tarsonemid mite, p. 250.

***Thrips.** Several species, e.g. greenhouse thrips (*Heliothrips haemorrhoidalis*), banded greenhouse thrips (*Hercinothrips femoralis*) and onion thrips (*Thrips tabaci*) infest begonias, the leaves of which may show irregular brownish-red streaks on the upper surface and roughened brownish spots on the underside. Control, before flowering, with DDT, *or* gamma-BHC, *or* malathion, *or* naled, *or* nicotine as for Carnation—Thrips, p. 237.

***Whitefly** (*Trialeurodes vaporariorum*). For description see Cucumber—Whitefly, p. 248. Control with DDT, H.V. sprays, dips or smokes, *or* malathion, *or* naled as for aphids (above).

DISEASES

***Grey mould** or **Blotch** (*Botrytis cinerea*). Both foliage, stems and flowers may be attacked if begonias are grown under excessively damp and cool conditions. For control see Azalea—Grey mould, p. 233.

***Powdery mildew** (*Oidium begoniae*). A whitish, powdery mould on the young leaves, stems and flower buds causes severely attacked tissues to desiccate and is usually associated with large fluctuations in humidity and temperature. Control with chemicals is difficult to obtain without phytotoxic effects. Sulphur vaporization at 0·5 g/1,000 ft$^3$ nightly over a 5–10 day period, *or* spraying with dinocap at 2 oz a.i. (8 oz 25% w.p.)/100 gal are effective but some cultivars may be slightly injured. The tolerance of begonias to quinomethionate is unknown.

***Root** and **Stem rots.** Several fungi including *Corticium* (*Rhizoctonia*) *solani*, *Thielviopsis basicola* and *Pythium* spp., cause rotting of the roots, tubers or stems, particularly if the plants are frequently over-watered. Prevention by using sterilized composts is the only reliable control.


CALCEOLARIA

Sprays must not be applied in bright sunshine or to plants requiring watering.

PESTS

****Aphids** (*Aulacorthum circumflexum*, *A. solani* and *Myzus persicae*). These, and occasionally other species, breed on calceolaria. Sprays of gamma-BHC, *or* malathion, *or* nicotine, *or* smokes of gamma-BHC, *or* nicotine, *or* drenches of demeton-S-methyl, *or* oxydemeton-methyl, as for Carnation—Aphids, p. 236, will control them, except gamma-BHC against *A. circumflexum*.

***Leaf hopper** (*Zygina pallidifrons*). The feeding of these insects on the undersides of the leaves causes mottled or bleached spotting. They are easily controlled with DDT, *or* gamma-BHC, *or* malathion, as for Carnation—Aphids, p. 236, *or* with DDT smokes at 3·3 g a.i./1,000 ft$^3$.

***Whitefly** (*Trialeurodes vaporariorum*). Calceolarias are often badly checked in growth and soiled by whitefly. Control with malathion sprays, *or* atomized solutions of malathion, *or* sprays or smokes of DDT, as for Tomato—Whitefly, p. 269.

DISEASES

***Root** and **Stem rots.** Calceolarias are susceptible to several soil-borne fungal diseases. Seed and potting composts should be made with steam-sterilized soil.

CARNATION (see M.A.F.F. BULL. 151)

PESTS

*****Aphids.** Peach-potato aphid (*Myzus persicae*) and occasionally the glasshouse potato aphid (*Aulacorthum solani*). These aphids are pale to dark yellowish-green when wingless, and mainly confined to the young leaves; the young stages feed within the growing tip, thus protected from contact insecticides. The best control is with systemic insecticides which are marked with an asterisk;

H.V. Sprays. Diazinon 3·2 oz a.i. (16 fl. oz 20% e.c.)/100 gal *or* dimethoate* 4·8 oz a.i. (12 fl. oz 40% e.c.)/100 gal *or* formothion* 7 oz a.i. (16 fl. oz 43% e.c.)/100 gal *or* gamma-BHC 2 oz a.i. (10 fl. oz 20% e.c.)/100 gal (spring and summer only) *or* malathion 18 oz a.i. (30 fl. oz 60% e.c.)/100 gal *or* nicotine 10 fl. oz (95–98%)/100 gal *or* oxydemeton-methyl* 3·4 fl. oz (6 fl. oz 57% e.c.)/100 gal *or* phosphamidon* 3·2 oz a.i. (16 fl. oz 20% s.c.)/100 gal. Sprays of demeton-S-methyl* are effective but impart an unpleasant smell to blooms. Additional wetting agent is usually required in sprays for carnations. Soil drenches of the following are effective, particularly on young plants, when the diluted product is applied at 2 gal/yd² to first year plants, 3 gal/yd² to older plants and 3–4 fl. oz/6 in pot; demeton-S-methyl* 2·3 oz a.i. (4 fl. oz 58% e.c.)/100 gal *or* dimethoate* 4·8 oz a.i. (12 fl. oz 40% e.c.)/100 gal *or* formothion* 7 oz a.i. (16 fl. oz 43% e.c.)/100 gal *or* oxydemeton-methyl* 2·3 oz a.i. (4 fl. oz 57% e.c.)/100 gal.

Control can also be obtained by frequent applications of atomized solutions (see p. 231) of dichlorvos 1 g a.i./1,000 ft³ *or* gamma-BHC 0·7 g a.i./1,000 ft³ (spring and summer only) *or* diazinon 1·8 g a.i./1,000 ft³ *or* malathion 2·4 g a.i./1,000 ft³; by frequent applications of malathion dusts (4%) at 25 lb/acre or smokes giving gamma-BHC 0·7 g/1,000 ft³ *or* nicotine 4 g/1,000 ft³ *or* parathion 0·25 g/1,000 ft³. If carnation tortrix is also present DDT can be atomized together with BHC or malathion. If conditions permit the heating of the closed glasshouse without detriment to flower quality, control may be obtained by applying naled to the cold pipes at the rate of 1·5 g/1,000 ft³ (1 fl. oz 50% e.c./10,000 ft³) and then turning on heat to keep the pipes at a minimum temperature of 140°F overnight or 180°F for 3 hours. If naled is to be applied to hot pipes, a respirator must be worn.

Occasionally slight scorch of the tips of the leaves may follow smoking with nicotine. The 'Ashington' varieties are susceptible to BHC during the autumn and winter.

***Carnation tortrix** (*Cacoecimorpha pronubana*). The yellowish-olive green caterpillars feed on upper leaves, tying them together with silken threads, and later enter flower buds and destroy blooms. For control, use DDT dust at 12 oz a.i. (15 lb 5% dust)/acre *or* DDT sprays 10 oz a.i. ($2\frac{1}{2}$ pint 20% e.c.)/100 gal *or* DDT atomizing solution 3·3 g/1,000 ft$^3$ *or* DDT smokes 3·5 g a.i./1,000 ft$^3$ *or* TDE sprays 12 oz a.i. ($1\frac{1}{2}$ lb 50% w.p.)/100 gal. Apply immediately the pest is seen and repeat after 14 days.

**Earwigs**—see Chrysanthemum—Earwigs, p. 241.

*****Red spider mite** (*Tetranychus cinnabarinus*). A reddish-brown mite which causes initially small whitish spots on the leaves; as the infestation increases, the leaves become pale, stippled and desiccated. The mite breeds on carnations throughout the year. Occasionally *T. urticae* attacks carnations causing similar damage.

They are controlled by H.V. sprays of diazinon *or* dimethoate *or* formothion *or* oxydemeton-methyl, and by soil drenches of demeton-S-methyl *or* dimethoate *or* formothion *or* oxydemeton-methyl (as for Carnation Aphid).

Control can also be obtained with H.V. sprays of chlorbenside 4 oz a.i. (1 pint 20% e.c.)/100 gal *or* dicofol 2 oz a.i. (8 fl. oz 25% e.c.)/100 gal *or* fenson 8 oz a.i./100 gal *or* tetradifon 1·6 oz a.i. (1 pint 8% e.c.)/100 gal, and by atomized solutions giving chlorfenson 6 g a.i./1,000 ft$^3$ *or* dicofol 2 g a.i./1,000 ft$^3$ *or* fenson 6 g a.i./1,000 ft$^3$ *or* tetradifon 1 g a.i./1,000 ft$^3$.

Smokes of tetradifon or azobenzene plus parathion can be used. Azobenzene smokes are available for amateur use.

***Thrips** (*Thrips tabaci*) are slender small insects, brown-brownish black and very active when adult but yellowish and less active when immature. They feed on petals causing silvery streaks and sometimes distortion, and often invade houses from outside in summer. They are controlled by DDT dust (5%) at 15 lb/acre *or* BHC dust (2%) at 15 lb/acre, *or* malathion dust (4%) at 15 lb/acre, repeated at 10 day intervals; *or* by atomized solutions of DDT 3·3 g a.i./1,000 ft$^3$ *or* gamma-BHC 1·0 g a.i./1,000 ft$^3$ *or* malathion 2·4 g a.i./1,000 ft$^3$ *or* diazinon 1·8 g a.i./1,000 ft$^3$. Also effective is spraying with DDT 10 oz a.i. ($2\frac{1}{2}$ pints 20% e.c.)/100 gal H.V. *or* gamma-BHC 2 oz a.i. (10 fl. oz 20% e.c.)/100 gal H.V. *or* diazinon 3·2 oz a.i. (16 fl. oz 20% e.c.)/100 gal H.V. *or* malathion 18 oz a.i. (30 fl. oz 60% e.c.)/100 gal H.V. Preliminary results indicate that phosphamidon 3·2 oz a.i. (16 fl. oz 20% s.c.)/100 gal H.V. gives good control but insufficient information is available for firm recommendations.

***Wireworms**—see Tomato—Wireworms, p. 270.

***Woodlice**—see Tomato—Woodlice, p. 270.

DISEASES

***Fusarium wilt** (*Fusarium oxysporum* f. *dianthi*) invades the vascular tissues causing whole plants or parts of them to wilt. Use steam-sterilized soil for raised beds and borders or pots, and cuttings from disease-free plants.

*****Greasy-blotch** (*Zygophiala jamaicensis*) appears on leaves as green fibrillate patches which may extend and join to give the leaves an oily appearance; encouraged by excessive humidity. When the disease persists in spite of efforts to reduce humidity, overseas experience suggests that captan 1 lb a.i. (2 lb 50% w.p.)/100 gal H.V. at 7–10-day intervals is of value but insufficient U.K. evidence is available for a firm recommendation.

****Leaf spot** or **alternaria blight** (*Alternaria dianthi*) begins as small purple spots which enlarge under moist conditions into typical leaf spots with purple edges and brown-black sporulating centres. The disease may also infect stems.

Zineb 1¼ lb a.i. (2 lb 65% w.p.)/100 gal H.V., *or* mancozeb 1 lb a.i. (1¼ lb 80% w.p.)/100 gal, *or* maneb 1¼ lb a.i. (1½ lb 80% w.p.)/100 gal H.V. *or* captan 1 lb a.i. (2 lb 50% w.p.)/100 gal H.V. have shown promise but additional tests are necessary before firm recommendations can be made. Two sprayings at weekly intervals of stock plants prior to taking cuttings are advised in America.

*****Mildew** (*Oidium* spp.) appears as a whitish powder on the leaves and occasionally on the calyces. It is controlled by dinocap at 1¼ oz a.i. (2½ fl. oz 50% e.c.)/100 gal H.V., two or more applications at 10–14 day intervals (the addition of a wetter is necessary) *or* sulphur dust—at 25 lb a.i./acre at 7 day intervals. Atomized solutions of dinocap *or* H.V. sprays of quinomethionate at 2 oz a.i. (8 oz 25% w.p.)/100 gal should be suitable for use on carnations but experience in their use is necessary before they can be recommended.

****Rust** (*Uromyces dianthi* (*caryophyllinus*)) causes small blisters on the lower leaves and stems, which rupture forming pustules filled with brown, powdery spores. The disease develops when the foliage remains wet or dew forms overnight. For control, reduce the humidity within the glasshouse and spray with zineb 1¼ lb a.i. (2 lb 65% w.p.)/100 gal H.V. *or* maneb 1 lb a.i. (1¼ lb 80% w.p.)/100 gal *or* thiram 3 lb a.i. (3·8 lb 80% w.p.)/100 gal H.V. and repeat at intervals of 10–14 days. Spray deposits may soil blooms if sprays are applied when flowering begins. Maneb gives good control but its use should be restricted to stock or non-flowering plants as the flowers may be damaged.

*****Slow wilt** (*Erwinia chrysanthemi*). A systemic bacterial disease which is now rare as a result of modern 'cultured cutting' propagating techniques. Plants wilt over considerable periods before death. To control the disease, plant in steamed soil, for no chemical eradicant is yet available. Stock plants should be tested for freedom from the disease if there is any risk of its introduction.

**Stem-rot** (*Fusarium roseum*) causes a rotting of the base of the stem and sometimes of branch stems; woody tissues are not discoloured. The pathogen often invades wounds made when removing shoots or picking flowers and may cause basal rot of cuttings taken from diseased plants. As the pathogen is not systemic, it is not detected by cultured-cutting technique. It is common in many unsterilized soils. Control is effected by spraying stock plants with captan 1 lb a.i. (2 lb 50% w.p.)/100 gal H.V. at 10–14 day intervals before and during cutting-taking, by planting in steam sterilized soil and by propagating in sterilized sand or other sterile medium. If established plants are attacked, remove infected shoots and spray with captan several times at 7–10-day intervals.

**Verticillium wilt** (*Verticillium (Phialophora) cinerescens*) causes rapid wilting, often on one side initially. Then plants soon turn grey-green and desiccate with extensive light brown discoloration of the xylem. The fungus persists for many years and penetrates deeply into soil. It may be controlled by steaming the soil in raised beds, though the normal steaming of borders is ineffective. There is evidence from trials in Britain and overseas that metham-sodium 3–4 lb a.i. (7–10 pint 33%)/100 gal, applied at 25 gal/100 ft², to the subsoil and at the same rate to the upper 12 in before planting controls the disease but more evidence is required before a firm recommendation can be made. Remove diseased plants and isolate the area from which they are removed.

CHRYSANTHEMUM (see M.A.F.F. BULL. 92)

There are marked differences in the tolerances of different chrysanthemum varieties to pesticides and their susceptibilities to diseases and pests. Year-round chrysanthemums, especially newer American varieties, are usually softer and more susceptible to damage than those grown by older systems. In case of doubt it is advisable to make a trial on a few plants of any new variety. Methods of pesticides application are an important factor.

PESTS

**Angleshades moth** (*Phlogophora meticulosa*), and other Noctuid moths. Caterpillars of these moths, particularly of the first, feed on the foliage and flowers of chrysanthemums. They are controlled by DDT atomized 3 g a.i./1,000 ft³ *or* 5% dust at 20 lb/acre *or* DDT at 1 lb a.i. (4 pints 20% e.c.)/100 gal H.V. *or* 4 oz a.i./gal L.V. at 5–7 gal/acre *or* TDE 12 oz a.i. (1½ lb 50% w.p.)/100 gal H.V. *or* DDT smoke 3·3 g/1,000 ft³. Repeat after 14 days if necessary.

Specific 'Chrysanthemum Sprays' may be used if red spider mite or mildew are also present.

Atomized concentrates of DDT/malathion, DDT/gamma-BHC, sprays of DDT/petroleum or smokes of DDT/BHC can be used where pests susceptible to the second ingredient are present.

**Spodoptora littoralis** (*Prodenia litura*) caterpillars are resistant to DDT. Young larvae are controlled by dichlorvos atomized at 1 g/1,000 ft$^3$ *or* trichlorphon 2 lb a.i. (2½ lb 80% s.p.)/100 gal H.V. *or* TDE 1 lb a.i. (2 lb 50% w.p.)/100 gal H.V.

*****Aphids.** Chrysanthemums in glasshouses are attacked by several species of aphids, the most important being the peach-potato aphid (*Myzus persicae*) whose wingless forms are green or greenish-yellow; the mottled arum aphid (*Aulacorthum circumflexum*) yellowish green with a well defined dark horseshoe shaped mark on the abdomen; the chrysanthemum aphid (*Macrosiphoniella sanborni*) blackish-red or black, and the dark green leaf-curling plum aphid (*Brachycaudus helichrysi*).

Many control methods are available but the control of *Myzus persicae* is complicated by varying patterns of resistance to insecticides of strains infesting chrysanthemums. Systemic organophosphorus compounds give the best control where aphids are not too resistant to them and the varieties are tolerant. Their application as soil drenches to moist soil is usually safer than as sprays, but less effective from budding onwards. Control should be effected before aphids enter flowers.

Recommendations for control are: to water at 2 gal/yd$^2$ *or* 1 pint/10 in pot with demeton-S-methyl 2·3 oz a.i. (4 fl. oz 58% e.c.)/gal, *or* oxydemeton-methyl 2·2 oz a.i. (4 fl. oz 57% e.c.)/100 gal; to spray H.V. with demeton-S-methyl 3·4 oz a.i. (6 fl. oz 58% e.c.)/100 gal, *or* diazinon 3·2 oz a.i. (16 fl. oz 20% e.c.)/100 gal, *or* malathion 18 oz a.i. (30 fl. oz 60% e.c.)/100 gal (not on open blooms), *or* morphothion 4 oz a.i. (1 pint 20% e.c.)/100 gal, *or* nicotine 10 fl. oz 95–98%/100 gal, *or* oxydemeton-methyl 3·4 oz (6 fl. oz 57% e.c.)/100 gal, *or* schradan 16 oz a.i. (30 fl. oz 33% s.c.)/100 gal (not on year-round varieties), *or* gamma-BHC 2 oz a.i. (10 fl. oz 20% e.c.)/100 gal (not effective against *A. circumflexum*).

Atomized formulations of dichlorvos equivalent to 1 g a.i./1,000 ft$^3$ can be used on most varieties, but information on the value of slow-release formulations is lacking. The vaporization of naled 1 fl. oz 50%/10,000 ft$^3$ from the heating pipes controls most strains of *Myzus persicae* and is safe on most chrysanthemum varieties, see Carnation—Aphids, p. 236.

Atomized solutions of diazinon 1·8 g a.i./1,000 ft$^3$ *or* malathion 2·5 g a.i./1,000 ft$^3$ are suitable before the foliage becomes dense.

Smoke generators yielding gamma-BHC 0·7 g/1,000 ft$^3$ (except for *A. circumflexum*) or parathion 0·25 g/1,000 ft$^3$ *or* nicotine 4 g/1,000 ft$^3$ (not on soft foliage) can also be used.

Thionazin, applied to control leaf and bud eelworm, see below, also controls non-resistant aphids on cuttings.

****Capsid** (Tarnished plant bug) (*Lygus rugulipennis*). The bug feeds on stems, leaves and flowers causing puckered, misshapen leaves, distorted stems and small, malformed flowers. Very occasionally *Lygocoris pabulinus* and *Calocoris norvegicus* cause similar damage. Control of the pest is achieved by: Atomized solutions of DDT 3 g a.i./1,000 ft³ *or* gamma-BHC 1 g a.i./1,000 ft³. If aphids are present, use gamma-BHC *or* malathion/gamma-BHC; *or* Smokes yielding gamma-BHC 0·7 g/1,000 ft³ *or* DDT 3·3 g/1,000 ft³ *or* combined 0·5 g BHC, 3·0 g DDT/1,000 ft³ if aphids present; *or* Sprays of DDT as for Angleshades moth, see above. Specific 'Chrysanthemum Sprays' containing DDT, petroleum oil and a fungicide such as thiram or zineb may be used if it is necessary also to control red spider mite or mildew.

***Cutworms** (A.L. 225). Caterpillars of several noctuid moths occasionally occur in glasshouses and feed on the roots and stems of chrysanthemums. If so, dust the soil surface beneath plants with DDT dust at 2 lb a.i. (40 lb 5% dust)/acre *or* water lightly with DDT at 1½ lb a.i. (1½ pints 25% e.c.)/100 gal.

****Earwigs** (*Forficula auricularia*) feed on flowers, giving them a ragged appearance and increasing their susceptibility to grey mould. To control, spray with gamma-BHC 2 oz a.i. (10 fl. oz 20% e.c.)/100 gal H.V. *or* TDE 12 oz a.i. (1½ lb 50% w.p.)/100 gal H.V. repeating at 14 day intervals if necessary. Carbaryl 2 lb a.i. (4 lb 50% w.p.)/100 gal *or* trichlorphon 0·8 lb a.i. (1 lb 80% s.p.)/100 are also effective. Where H.V. sprays are undesirable, e.g. where grey mould (Botrytis) is present, atomize gamma-BHC 1·0 g/1,000 ft³ *or* smoke with gamma-BHC 0·7 g/1,000 ft³. Repeat at 7-day intervals if necessary.

****Leaf and Bud eelworm** (*Aphelenchoides ritzemabosi*) (A.L. 339, 379). The first sign of attack is a yellowish-green or purple blotching on the basal leaves, often delineated by veins; later these blotches darken and desiccate.

Plants should be propagated from eelworm-free stock. If this is impractical, clean plants may be obtained by applying drenches of thionazin 4·6 oz a.i. (10 fl. oz 46% e.c.)/100 gal to the cuttings when well rooted and again after 14 days. This chemical treatment has largely superseded hot-water treatment of the washed, cut-back stools of stock plants by immersion for 5–5½ minutes in water kept at 115°F *or* for 20–30 minutes in water kept at 110°F, followed by immediate immersion in cold water and growing on until cutting material is obtained. Hot-water treatment may delay cutting production and is severely phytotoxic to some varieties.

****Leaf miner** (*Phytomyza atricornis*). The larvae of this small fly disfigure the leaves by eating tunnels (mines) in leaves. Sprays of gamma-

BHC 2 oz a.i. (10 fl. oz 20% e.c.)/100 gal H.V. at 14 day intervals, or atomized gamma-BHC 1 g a.i./1,000 ft³ control the pest except where BHC-resistant strains occur. Where these exist or where other pests susceptible to diazinon are present, use sprays of diazinon 3·2 oz a.i. (16 fl. oz 20% e.c.)/100 gal H.V. at 14 day intervals or atomized diazinon 1·8 g a.i./1,000 ft³ at 7-day intervals.

***Red spider mite** (*Tetranychus urticae*). For a description see Cucumber red-spider. Damage is most severe to flowers of year-round chrysanthemums; effect control before flowering. Strains of the mite resistant to acaricides are prevalent, but use sprays of dicofol or tetradifon as for carnation red-spider, or sprays of quinomethionate 2 oz a.i. (8 oz 25% w.p.)/100 gal H.V. *or* of petroleum emulsion (glasshouse grade) 1 gal/100 gal H.V. (petroleum/thiram or petroleum/zineb at equivalent petroleum concentration where mildew or rust control is required), or (if aphids present) diazinon 3·2 oz a.i. (16 fl. oz 20% e.c.)/100 gal H.V.

Atomized solutions of chlorfenson, dicofol, tetradifon or fenson or tetradifon smokes as for carnation red-spider, may be effective. Where aphids are present, atomize diazinon 1·8 g/1,000 ft³. For plants in active growth before budding, water with demeton-S-methyl *or* oxydemeton-methyl as for carnation red-spider mite, p. 237.

**Stool miner** (*Psila nigricornis*). The larvae tunnel into basal tissues and reduce cutting production. They may be controlled by drenching sprays of gamma-BHC 1·5 oz a.i. (7·5 fl. oz 20% e.c.)/100 gal in early May and again in early September.

**Symphylids** (*Scutigerella immaculata*). For description and control see Tomato—Symphylids, p. 269.

*Thrips**—see Carnation—Thrips, p. 237.

*Whitefly** (*Trialeurodes vaporariorum*) (M.A.F.F. Bull. 86)—see Cucumber—Whitefly, p. 248. For control, atomize solutions of dichlorvos 1 g a.i./1,000 ft³ (some varieties intolerant) *or* malathion 2·5 a.i./1,000 ft³ *or* DDT 3 g a.i./1,000 ft³ *or* malathion/DDT and repeat at least twice at 10–14 day intervals. Sprays of DDT 1 lb a.i. (4 pints 20% e.c.)/100 gal H.V. *or* malathion 18 oz a.i. (30 fl. oz 60% e.c.)/100 gal H.V., repeated once or twice at 14 day intervals, are effective. Also smoke with DDT 3·3 g/1,000 ft³ *or*, if earwigs or aphids are present, with DDT 3·0 g + gamma-BHC 0·5 g/1,000 ft³ and repeat several times at 7-day intervals. Fumigation with naled has given good results.


DISEASES

*Blotch** (*Septoria chrysanthemella*) causes dark grey-black spots or blotches on older leaves with defoliation if severe. To control, increase ventilation and spray with zineb 1¼ lb a.i. (2 lb 65% w.p.)/100 gal H.V. (For 'tank-mixed' zineb see Tomato leaf mould, p. 271.)

***Grey-mould** (Botrytis flower-rot) (*Botrytis cinerea*) causes spotting and rotting of buds and flowers. Infection may also be serious in the leaves and stems, especially in cuttings inserted in the propagating bench after protracted periods of cold storage. Grey mould is most common under warm, humid conditions and the mass of grey spores which are produced at the rotting stage distinguishes this disease from petal blight and ray blight. With Botrytis-susceptible varieties in humid weather, it is advisable to apply a preventive spray as a fine mist to the crop at bud burst, using either captan 1 lb a.i. (2 lb 50% w.p.)/100 gal, *or* thiram 3 lb a.i. (3·8 lb 80% w.p.)/100 gal, *or* dust with captan 1 lb a.i. (6·6 lb 15% dust)/acre. Dichlofluanid at ½–1 lb a.i. (1–2 lb 50% w.p.)/100 gal has given good results in recent trials but its safety on many varieties has yet to be established. If infection has already occurred, remove diseased blooms promptly and apply one of the above-mentioned treatments at 7–10-day intervals if humid conditions prevail.

Spraying stock plants with dichlofluanid at 1 lb a.i. (2 lb 50% w.p.)/100 gal prior to the removal of cuttings greatly decreases the incidence of Botrytis on cuttings, and additional control is obtained in mist-propagated cuttings if sprays are applied 5 and again 10 days after insertion.

Trials in America, and practical experience in Britain, indicate that dusting cuttings and the surface of the rooting medium with captan, dicloran, thiram or quintozene dusts immediately after inserting cuttings, or watering in with captan 1 oz a.i. (2 oz 50% w.p.)/3 gal reduces rotting of cuttings by Botrytis.

*****Mildew** (*Oidium chrysanthemi*). A typical powdery mildew, starting as patches on the upper side of young leaves and progressing in severe cases to cover upper and lower leaf surfaces. It may be controlled by the vaporization of commercially pure sulphur 4 g/1,000 ft$^3$ at intervals of 7–14 days, or nightly or continuously at 0·5–1 g/1,000 ft$^3$/24 hr, using apparatus designed to vaporize sulphur without risk of ignition, for fumes from burning sulphur are extremely phytotoxic. Alternatively spray with quinomethionate 2 oz a.i. (8 oz 25% w.p.)/100 gal (also controls red spider mite) *or* thiram 3 lb a.i. (3·8 lb 80% w.p.) + petroleum emulsion 2¾ pints/100 gal H.V. *or* zineb 1 lb a.i. (1·5 lb 65% w.p.) + petroleum 2¾ pints/100 gal H.V. *or* salicylanilide 12 oz a.i. (2½ pints 25% solution) + petroleum emulsion 2¾ pints/100 gal H.V. Glasshouse grade petroleum emulsion (M.A.F.F. Tech. Bull. No. 1) must be used. Repeat at 14-day intervals. Formulations of thiram or of salicylanilide with petroleum are available. Dinocap 1 oz a.i. (2 fl. oz 50% e.c.)/100 gal H.V. or dinocap atomizing solution should be used only where their safety on the varieties to be treated has been established.

****Petal blight** (*Itersonilia perplexans*) is first seen on the petals as minute, reddish-brown spots which under humid conditions coalesce to form water-soaked spots. It usually first appears on the tips of the outer petals and gradually extends inwards to affect the entire bloom. It may be confused with grey mould (*q.v.*) but the sporulating mycelium is seen as a greyish-white sheen on the rotting petals. The disease is more prevalent in unheated houses where water droplets persist on the petals overnight. To control avoid high humidity by ventilation and heating where possible. Alternatively, preventive sprays of zineb $1\frac{1}{4}$ lb a.i./100 gal prepared by the tank-mixed method (see Tomato—Leaf mould, p. 271) should be applied as a fine spray at the coloured bud stage. If humid conditions prevail, further applications should be given at 7 day intervals. The incidence of grey mould is also decreased by these sprayings. Although less effective than freshly prepared tank-mixed material, formulated zineb at $1\frac{1}{4}$ lb a.i. (2 lb 65% w.p.)/100 gal, *or* zineb dust at 3 lb (20 lb 15% dust)/acre may be used.

****Phoma root-rot** (*Phoma* spp.). The plants are stunted or killed by severe rotting of roots and, ultimately, of base of stem. Lower leaves show interveinal and chlorotic areas which become necrotic. Planting in steam-sterilized soil will avoid the disease. Pre-planting soil drenches of nabam $1\frac{1}{2}$ lb 93%/100 gal at 4 gal/yd$^2$ give good control.

***Ray-blight** (*Mycosphaerella ligulicola*). Infection starts at the centre of flowers causing them to become dark brown. The fungus also attacks stems, especially of cuttings. Sterilize the soil by steaming after an infected crop for there is no reliable information available on the value of chemical sterilants. Spray with captan 1 lb a.i. (2 lb 50% w.p.)/100 gal H.V. at 10 day intervals after planting out, and reduce to $\frac{1}{2}$ lb a.i./100 gal applied every 7 days, when the buds show colour. Practical experience suggests that maneb 13 oz a.i. (1 lb 80% w.p.)/100 gal H.V. *or* mancozeb 13 oz a.i. (1 lb 80% w.p.)/100 gal H.V. give effective control but may leave visible deposits on flowers and foliage.

***Rust** (*Puccinia chrysanthemi*) attacks the leaves under humid conditions causing yellowish spots which develop into rusty, spore-producing pustules on the undersides. Spray with zineb $1\frac{1}{4}$ lb a.i. (2 lb 65% w.p.)/100 gal H.V. *or* thiram 3 lb a.i. (3·8 lb 80% w.p.)/100 gal H.V. at 10–14-day intervals when the disease first appears. 'Chrysanthemum Spray' formulations containing thiram or zineb, petroleum oil and DDT are also effective, and give simultaneous control of caterpillars, capsids, mildew and red spider mite. Preventive spraying may be advisable if the disease is endemic.

***Verticillium wilt** (*Verticillium albo-atrum*). The lower leaves turn yellow-brown and discoloration of foliage progresses upwards; the plants are stunted and woody. Care in the selection of cuttings and planting in sterilized soil help to avoid the disease.

## CINERARIA

PESTS

**Aphids** (*Myzus persicae. Aulacorthum circumflexum, A. solani* and *Aphis gossypii*). Control, which should be effected before flowering if possible, may be obtained with gamma-BHC *or* diazinon *or* malathion *or* naled *or* nicotine *or* parathion as for Carnation—Aphids, p. 236, bearing in mind that *A. circumflexum* is resistant to gamma-BHC. Smokes of gamma-BHC 0·7 g a.i./1,000 ft³ are advised for flowering plants except for *A. circumflexum*. Drenches of demeton-S-methyl *or* morphothion, as for Chrysanthemum—Aphids, p. 240, have been successfully applied to cinerarias but should be used with care and not in bright hot weather.

**Leaf miner** (*Phytomyza atricornis*). See Chrysanthemum—Leaf miner, p. 241.

**Whitefly** (*Trialeurodes vaporariorum*). Control with DDT *or* malathion, as for Tomato—Whitefly, p. 269.

DISEASES

**Powdery mildew** (*Sphaerotheca fuliginea*). Control by spraying with dinocap as for Begonia—Powdery mildew, p. 235, avoiding, if possible, applications during hot, bright weather. There is insufficient information on the safety and effectiveness of quinomethionate.

**Grey mould** (*Botrytis cinerea*). Attacks flowers, leaves and stems; for control spray with captan *or* thiram as for Rose—Grey mould, p. 265.

## CUCUMBER AND MELON

PESTS

**Aphid** (*Aphis gossypii*). This aphid is very variable in colour—yellowish green to black, and feeds on foliage causing yellowing, and on fruitlets which are distorted or fail to develop. For control on seedlings and young plants, spray with parathion 1·6 oz a.i. (8 fl. oz 20% e.c.)/100 gal H.V., not later than 4 weeks before cutting. On planted out young plants, spray with demeton-S-methyl 3·4 oz a.i. (6 fl. oz 58% e.c.)/100 gal H.V. *or* oxy-demeton-methyl 3·4 oz a.i. (6 fl. oz 57% e.c.)/100 gal H.V. not later than 3 weeks before cutting, *or* schradan 11 oz a.i. (20 fl. oz 55% s.c.)/100 gal not later than 4 weeks before cutting. On established plants, use sprays of dimethoate 4·8 oz a.i. (12 fl. oz 40% e.c.)/100 gal H.V., allowing at least 7 days interval before next cutting, *or* nicotine 8 fl. oz (95–98%)/100 gal H.V. *or* malathion 18 oz a.i. (30 fl. oz 60% e.c.)/100 gal H.V., with at least 4 days interval before cutting *or* use parathion smokes 0·25 g/1,000 ft³ *or* nicotine smokes 4 g/1,000 ft³ *or* sulfotep smokes *or* atomized solutions of dichlorvos 1 g a.i./1,000 cu. ft *or* parathion 0·5 g/1,000 ft³.

From mid May onwards well established plants will tolerate the following additional treatments: atomized solutions of malathion 2·5 g a.i./1,000 ft³ *or* diazinon 1·8 g a.i./1,000 ft³.

**'**French-fly**' (*Tyrophagus dimidiatus*) is a glistening creamy-white, globular, sluggish mite with prominent bristles, commonly occurring in straw and undecomposed horse manure. It migrates from beds and feeds on and within the leaves of the shoot tips of young plants causing irregular perforations in the leaves, distortion and sometimes blindness of the shoots. It does not reproduce on cucumbers.

To control, spray the plants, especially growing tips and surface of beds, with parathion 1½ oz a.i. (8 fl. oz 20% e.c.)/100 gal as soon as damage is seen and not later than 4 weeks before cutting. Incorporate a wetting agent if necessary to ensure that the spray penetrates the growing tips. Repeat after 7 days if the attack continues. Parathion smokes 0·25 g/ 1,000 ft³ should be used if it is necessary to control the pest within 4 weeks of cutting.

****Fungus gnat (root) maggots** (*Sciara* spp.). The larvae of these midges are about ¼ in long with black heads and creamy translucent bodies. They are common in beds containing horse manure, and occasionally attack the large roots and stem-bases of cucumbers in beds, especially if the beds become too dry. They also destroy young plants in pots. To control, drench the beds with parathion 1 oz a.i. (5 fl. oz 20% e.c.)/100 gal at 1–1½ gal/yard run of bed. One application only per crop is acceptable under official recommendations and is usually sufficient to control an attack.

*****Root-knot eelworm** (*Meloidogyne hapla* and *M. incognita acrita*) (A.L. 307) cause swollen, gall-like growths on the roots which reduce the vigour of the plants and greatly decrease cropping. Attacks are most serious at high temperatures in light soils and, for control, the borders should be steam-sterilized or treated with 'D-D mixture' *or* methyl bromide as described under 10.1A (see p. 229).

*****Millepede (glasshouse)** (*Oxidus gracilis*) (A.L. 150). The glasshouse millepede is common in cucumber houses and, when numerous, destroys plants by feeding on the soft tissues of the stem base and larger roots. To control this pest, drench the beds, with DDT 15 oz a.i. (approx 4 pints 20% e.c.)/100 gal at ½ gal/yd run *or* gamma-BHC (liquid formulation) 2 oz a.i. (10 fl. oz 20% e.c.)/100 gal at ½ gal/yd run *or* parathion 4 oz a.i. (20 fl. oz 20% e.c.)/100 gal at 1 gal/yd run (only 1 application per crop) *or* nicotine 10 oz (95–98%)/100 gal, with a wetting agent as for H.V. spraying, at 1 gal/yd run, applied in the evening and followed the next morning by spraying the millepedes lying on the surface of the beds, paths, etc., with some of the fluid. Do not allow DDT or gamma-BHC to come into contact with the foliage of cucumbers.

*****Red spider mite** (*Tetranychus urticae*). Adult mites of this species hibernate within the glasshouse and emerge when heat and light conditions are favourable, to lay eggs on main crop cucumbers during propagation or when planted in borders. Eggs hatch within a few days, producing whitish-pink globular larvae with 3 pairs of legs which pass through nymphal stages to become greenish oval mites. The complete life cycle from egg to adult takes from 22 days at 15°C (60°F) to 5 days at 30°C (85°F). Strains of mites resistant to acaricides are selected with greater facility from populations in cucumber houses than in most other populations. It is advisable, therefore, to vary, if possible, the type of chemical used to control the pest, and to spray thoroughly at H.V. to cover both leaf surfaces.

Prevent or reduce the hibernation of adult mites by removing an old crop as soon as picking ceases, especially in autumn when mites tend to leave plants which are not producing growth on which they can feed. Remove all weeds from vacant houses.

On seedlings and young plants, spray once only with schradan 16 oz a.i. (30 fl. oz 55% s.c.)/100 gal H.V., not later than 3 days after planting *or* demeton-S-methyl 3·4 oz a.i. (6 fl. oz 58% e.c.)/100 gal *or* oxydemeton-methyl 3·4 oz a.i. (6 fl. oz 57% e.c.)/100 gal *or* parathion 1·6 oz a.i. (8 fl. oz 20% e.c.)/100 gal, observing the required intervals between application and cutting. If parathion is used repeat after 10 days but not within 4 weeks from cutting. Where mites are resistant to organophosphorus compounds, it may be necessary to use dicofol at 2 oz a.i. (8 fl. oz 25% e.c.)/100 gal *or* tetradifon at 2 oz a.i. (1¼ pint 8% e.c.)/100 gal at a slight risk of hardening the plants, in which case spray only leaves seen to be infested, and avoid as much as possible wetting the young leaves on leaders or lateral shoots ('breaks').

On cucumbers established in beds, but before mid May, spray with dicofol 2 oz a.i. (8 fl. oz 25% e.c.)/100 gal H.V., *or* tetradifon 1·6 oz a.i. (20 fl. oz 8% e.c.)/100 gal H.V., *or*, at least 7 days before cutting fruit, dimethoate 4·8 oz a.i. (12 fl. oz 40% e.c.)/100 gal H.V., *or* petroleum emulsion (glasshouse grade) 1 gal/100 gal H.V. Repeat applications at 14-day intervals if attack persists. Where mites have remained susceptible to azobenzene and organophosphorus compounds, combined smokes of azobenzene 2·5 g a.i./1,000 ft³ with parathion 0·25 g a.i./1,000 ft³ or sulfotep 1 g a.i./1,000 ft³ may be used.

On established plants from mid May onwards: spray with dicofol *or* tetradifon *or* petroleum as in preceding paragraph *or* quinomethionate as for Cucumber mildew (p. 249) *or* diazinon 3·2 oz a.i. (16 fl. oz 20% e.c.)/ 100 gal H.V. *or* dinobuton 8 oz a.i. (15 fl. oz 50% e.c.)/100 gal H.V. (not more than five applications). Where it is difficult to spare labour for H.V. spraying in the summer, atomize dicofol 2 g/1,000 ft³ *or* dicofol 1·8 gram plus diazinon 1·5 g/1,000 ft³ *or* azinphos-methyl 1·8 g plus chlorfenson 4 g/1,000 ft³ *or* tetradifon 2 g/1,000 ft³ *or* smoke with tetradifon 1 g/1,000 ft³. Many strains of red spider which are resistant to parathion are resistant to diazinon and azinphos-methyl.

Very effective control of red spider mite on cucumbers may be obtained by the introduction of the predatory mite *Phytoseiulus riegeli* into infested houses under the correct conditions, but as this mite is highly susceptible to insecticides used against other cucumber foliage pests and also to most fungicides effective against cucumber mildew, its acceptance as a standard recommendation must await further developments, see Cucumber—Powdery mildew, p. 249.

****Springtails** (*Collembola* spp.). Adults of the species most commonly occurring in cucumber beds are elongate ($\frac{1}{8}$ in) and dark grey: often the young stages are lighter in colour. They occasionally occur in large numbers in cucumber beds and may injure plants by feeding on roots. To control, water the beds with gamma-BHC 2 oz a.i. (10 fl. oz 20% e.c.)/100 gal *or* malathion 18 oz a.i. (30 fl. oz 60% e.c.)/100 gal at $\frac{1}{2}$ gal/yd run, *or* dust beds with gamma-BHC dust (0·6%) at 1 oz/yd². Watering is preferable to dusting, especially where the beds have become compact.

***Symphylid** (*Scutigerella immaculata*). This insect occasionally attacks cucumbers, checking growth and making the roots susceptible to fungal decay. To control use diazinon *or* parathion as for Tomato—Symphylids, p. 269. The application to the base soil of gamma-BHC dust at $1\frac{1}{2}$–2 oz/yd² *or* 1 pint 10% gamma-BHC e.c./100 gal at 30 gal/100 yd² is likely to be effective if uninfested soil is used for bed construction.

***Thrips** (*Thrips tabaci* and *T. fuscipennis*) sometimes damage cucumbers by feeding on leaves and flowers and are detected by small irregular white marks on the leaves. They are controlled by the methods given for aphids.

*****Whitefly** (*Trialeurodes vaporariorum*). These small, moth-like white insects normally rest on the undersides of leaves. Conical eggs are laid endwise on underside of leaves and from these minute scale-like larvae emerge and become non-motile, oval scales increasing in size to about 1/25 in, from which adults emerge. Immature and adult stages feed on leaves and excrete honeydew, on which grows a sooty fungus which soils fruits. They are controlled by parathion sprays, smokes, or atomized solutions as for red spider or by atomized solutions of malathion *or* diazinon or by H.V. sprays of malathion *or* diazinon as for aphids. Do not use atomized solutions until mid May.

****Woodlice** (*Armadillidium*, *Oniscus*, *Porcellio*, and *Trichoniscus* spp.) damage cucumbers by gnawing stems and lower leaves. For their control, water beds with DDT *or* parathion as for millepedes or bait with a mixture of Paris green 1 part and dried blood 56 parts *or* bran 28 parts, sprinkled on the beds at $\frac{1}{2}$ oz/yd² (see note on p. 52 *re* toxicity of Paris green).

DISEASES

***Anthracnose** (*Colletotrichum lagenarium*). Pale-green or reddish-brown spots which enlarge until the whole leaf withers. The fungus also attacks stems and causes pinkish depressions which develop into black spots. Spray with wettable or colloidal sulphur $1\frac{1}{2}$ lb a.i. (2 lb 75% w.p.)/100 gal H.V. *or* dust with sulphur 20 lb/acre.

***Basal stem rot** (*Erwinia carotovora*) is a slimy soft-rot of the stem at soil level. To control, dust the stems at soil level heavily with a copper dust. Keep the base of stem dry, especially with melons.

*****Black root rot** (*Phomopsis sclerotioides*). Recent work has shown that this disease is the most important cucumber root rotting fungus, and that tap root decay previously ascribed to *Fusarium*, *Pythium* and other causes is probably due to *Phomopsis*. Typical symptoms are black spotting on the small roots, black lesions on large roots and rotting of the tap root. If young plants are transplanted into infested soil, rotting of the tap root and hypocotyl develops so rapidly, especially in cold soil (below 58°F), that the plant dies before the development of the typical black lesions. Steam sterilization of the soil annually is the most effective control.

****Gummosis** (*Cladosporium cucumerinum*) causes sunken scab-like depressions on fruit from which the sap exudes and forms an amber gum. Fruits are often distorted, especially if infected when young. Infection is favoured by cool, wet conditions and inadequate ventilation. Hence raise temperature if possible, reduce humidity and remove all diseased fruits. Spraying with zineb $1\frac{1}{4}$ lb a.i. (2 lb 65% w.p.)/100 gal H.V. and repeating at 10 day intervals while the disease persists *or* dusting with zineb dust at 3 lb a.i. (20 lb 15% dust)/acre at 7 day intervals will control the disease.

*****Powdery mildew** (*Sphaerotheca fuliginea* (*Erysiphe cichoracearum*)) is a typical white powdery mildew. Spores produced on cucumber are viable only for about 10 days. To control the disease, destroy all weeds within and near houses and, at the first sign of the disease, spray with dinocap 1 oz a.i. (2 fl. oz 50% e.c.)/100 gal H.V. on young plants and 2 oz a.i./100 gal H.V. on established plants in summer repeating at 10–14-day intervals *or* atomize solutions of dinocap 2 g/1,000 ft³ *or* dust with dinocap dust at 3–6 oz a.i. (10–14 lb 20% dust)/acre at 7–10 day intervals *or*, after April, spray with quinomethionate 2 oz a.i. (8 oz 25% w.p.)/100 gal at 10–14-day intervals (3 oz a.i./100 gal for first two applications if attack is severe), *or* dinobuton as for Cucumber—red spider mite, p. 248, *or*, where petroleum oil emulsions are being used to control red spider, add copper fungicides 10–15 oz metallic copper/100 gal diluted petroleum emulsion. Sulphur, quinomethionate and zineb are highly incompatible with petroleum oil; dinocap is slightly incompatible with petroleum oil and sulphur.

A systemic, substituted pyrimidine fungicide, to which it is proposed to give the common name dimethirimol, has given good control of mildew in trials, and has the merit of not being toxic to the predator *Phytosiulus riegeli* (see Cucumber—Red spider mite, p. 248). Suggested methods of application, for which metering equipment is available, depend on plant size. The compost of pot plants should be drenched with a 1 in 125 dilution ($1\frac{1}{4}$ fl. oz/gal) of the concentrate containing $10\%$ a.i. To plants in beds (including straw bales) the following doses should be applied to the recently watered soil over an area of about 4 in diameter around the stem base: plants less than 4 ft high, 20 ml ($\frac{2}{3}$ fl. oz) of a 1 in 60 dilution of the concentrate; plants 4–6 ft high, 20 ml of a 1 in 30 dilution; plants over 6 ft high, 20 ml of the undiluted $10\%$ concentrate. Splashing the leaves should be avoided. Treatment should begin as soon as the first spot of mildew is seen and repeated if any new infection is noticed after several weeks.

***Root-rot** (*Fusarium* spp.). Cucumber roots, especially where growing in unsterilized compost or wet beds, are attacked, causing the plants to wilt and die. To control the disease, steam the border and propagating soil. Chemical sterilants are unsatisfactory for the disinfection of cucumber soil.

***Verticillium-wilt** (*Verticillium albo-atrum*). The plants wilt and the leaves yellow and desiccate from base upwards. Control is as for root-rot.

CYCLAMEN

PESTS

***Aphids** (*Aulacorthum circumflexum*, and other species). For control measures see Chrysanthemum, aphids, p. 240. *A. circumflexum* is not controlled by BHC. Do not apply parathion to cyclamen foliage.

****Tarsonemid (cyclamen) mite** (*Steneotarsonemus pallidus*). The minute, waxy, white-brown mites feed on developing foliage causing it to become puckered, distorted, stunted and brittle. When the flower buds are infested, they are either distorted or fail to open. The best control is obtained by spraying with endosulfan 6·4 oz a.i. (32 fl. oz $20\%$ e.c.)/100 gal H.V. *or* endrin 6·4 oz a.i. (32 fl. oz $20\%$ e.c.)/100 gal H.V. Dicofol 5 oz a.i. (20 fl. oz $25\%$ e.c.)/100 gal is usually satisfactory. Repeat applications once or twice at 28-day intervals.

***Thrips** (*Thrips tabaci*, *T. fuscipennis*, *Heliothrips haemorrhoidalis*) damage foliage and flowers. See Carnation thrips, p. 237, but do not use parathion on cyclamen.

****Vine weevil** (*Otiorhynchus sulcatus*). The whitish, wrinkled, legless grubs usually lie in the soil in a curved manner, and feed on the corms and roots. The admixture of aldrin *or* dieldrin to the compost at 2 oz of $1\frac{1}{4}\%$ dust/bushel prevents attacks. BHC at 8 oz $0\cdot5\%$ gamma-BHC dust/bushel may also be used. When grubs are attacking plants, drenching with gamma-BHC 1½ oz a.i. (7½ fl. oz 20% e.c.)/100 gal destroys grubs which have not eaten their way into the corms.

DISEASES

***Grey mould (Botrytis rot)** (*Botrytis cinerea*). To prevent this disease avoid very high humidity; plant in soil sterilized with steam or formaldehyde; spray with thiram 3 lb a.i. (3·8 lb 80% w.p.)/100 gal H.V. at intervals of 10 days and ensure that the corms are thoroughly wetted *or* dust with dicloran (4%) dust.

**Root rots** (*Thielaviopsis basicola, Cylindrocarpon radicicola*). Use steamed soil for propagating and potting. Overseas workers state that watering with captan 1¼ lb a.i. (2½ lb 50% w.p.)/100 gal after each repotting and at 21-day intervals after the final potting controls *Thielaviopsis*, and the admixture of zineb at 2 oz a.i. (3 oz 65% w.p.)/yd³ of compost controls *Cylindrocarpon* in unsterilized soils. Metham-sodium and dazomet appear to be suitable alternatives to steam sterilization.

FREESIA

PESTS

****Aphids** (*Aulacorthum circumflexum, Macrosiphum euphorbiae* and *Myzus persicae*) feed on freesias and cause yellow markings on the leaves and stunted growth. *M. euphorbiae* is a vector of freesia mosaic. The best control is given by H.V. sprays or soil drenches of demeton-S-methyl, *or* dimethoate, *or* formothion, *or* oxydemeton-methyl, as for Carnation— Aphids, p. 236. Gamma-BHC, *or* diazinon, *or* malathion, *or* naled, *or* nicotine, may also be used as for the latter.

***Bean seed fly** (*Delia platura*) is occasionally serious in Guernsey. The larvae of the insect enter and feed on the embryo corm of seedlings, causing them to develop a bluish tinge and lack vigour. It can be controlled by drenching with gamma-BHC 4 oz a.i. (1 pint 20% e.c.)/100 gal to wet the soil thoroughly at the first sign of damage.

***Red spider mite** (*Tetranychus urticae*) may be controlled on freesias with demeton-S-methyl, *or* dimethoate, *or* formothion, *or* oxydemeton-methyl sprays or drenches, *or* sprays of dicofol *or* tetradifon, as for Carnation—Red spider mite, p. 237.

**Slugs**—For control see Tomato—Slugs, p. 269.

DISEASES

****Fusarium yellows** (*Fusarium moniliforme, F. oxysporum*) *F. moniliforme* is seed borne and, in the absence of fungicidal treatment, spreads from infected to healthy seed during 'chitting', causing decreased germination or death of seedlings. Both pathogens are soil borne and invade freesia roots and subsequently the vascular tissue of the corm causing a reddish-brown rot. Foliage of the infected corms becomes yellow, commencing with the outer leaves and spreading inwards until, in severely infected plants, the foliage dies.

For control, seed should be treated with a captan *or* thiram seed dressing and sown in sterilized soil. Glasshouse soil in which an infected freesia crop has been grown must be sterilized, preferably with steam, but where this is impossible formaldehyde *or* metham-sodium may be used, see p. 230. Spread of the disease in a crop can be decreased by drenching with captan 1 lb a.i. (2 lb 50% w.p.)/100 gal at 1 gal/yd². The treatment of corms with suspected *Fusarium* infection, by soaking for 30 min in formaldehyde (3 fl. oz 40% formaldehyde/gal), is recommended in the U.S.A.

****Grey mould** (*Botrytis cinerea*) attacks the stems, leaves and flowers of densely grown freesias under humid conditions. Prevent by dusting weekly with captan 1½ oz a.i. (10 oz 15% dust)/1,000 ft² *or* spraying at 10-day intervals with captan *or* thiram as for Chrysanthemum—Grey mould, p. 243. Dichlofluanid ½–1 lb a.i. (1–2 lb 50% w.p.)/100 gal has given effective control but insufficient information is available concerning its safety on all varieties.


FUCHSIA

PESTS

***Aphid** (*Myzus persicae*). Control as for Chrysanthemum—Aphids, p. 240; but do not use dimethoate.

***Red spider mite** (*Tetranychus urticae*). Control as for Chrysanthemum —Red spider mite, p. 242; but do not use dimethoate.

****Whitefly** (*Trialeurodes vaporariorum*). This pest may cause serious damage to fuchsias by making the leaves fall prematurely and by dis-figurement through fungi growing on the insect's secretions. Control with DDT sprays or smokes *or* malathion aerosols or sprays, as for Tomato—Whitefly, p. 269. Two or three applications at 14-day intervals are necessary.


DISEASES

***Grey mould** (*Botrytis cinerea*) may attack flowers under very humid conditions. Control by spraying with captan *or* thiram, as for Rose—Grey mould, p. 265.


GERANIUM AND PELARGONIUM

PESTS

****Aphids** (*Myzus persicae, Aulacorthum solani*). Control as for Carnation —Aphids, p. 236; but apply demeton-S-methyl, *or* dimethoate, *or* formo-thion, *or* oxydemeton-methyl to zonals only as soil drenches.

***Leaf hopper** (*Zygina pallidifrons*). See Calceolaria—Leaf hopper, p. 235.

***Mealy bug** (*Planococcus citri*). Control with diazinon, *or* malathion, as for Grape—Mealy bug, p. 254.

***Mites.** The Tarsonemid (cyclamen) mite (*Steneotarsonemus pallidus*) and the broad mite (*Hemitarsonemus latus*) may infest glasshouse geraniums. Control as for Cyclamen—Tarsonemid mite, p. 250.

***Vine weevil** (*Otiorhynchus sulcatus*). The larvae of this beetle occasionally feed on the roots of geraniums. They may be introduced in unsterilized compost or hatch from eggs laid by adults within the glasshouse. For control see Cyclamen—Vine weevil, p. 250.

***Whitefly** (*Trialeurodes vaporariorum*). Control as for Tomato—Whitefly, p. 269.

DISEASES

***Grey mould** (*Botrytis cinerea*). Leaves, flowers and stems of geraniums, especially of 'soft' plants, are susceptible to rotting if the plants are kept under damp conditions. Infection usually occurs where plant debris, such as dead flowers, are allowed to remain on the plants. For control see Azalea —Grey mould, p. 233.

***Leaf spot** (*Alternaria tenuis*) causes small water-soaked spots with a necrotic central fleck to appear on the undersides of the leaves. The lesions increase up to about 3 mm diameter, become necrotic and may coalesce. Control is obtained by spraying with thiram 3 lb a.i. (3·8 lb 80% w.p.)/100 gal, *or* zineb 1½ lb a.i. (2½ lb 65% w.p.)/100 gal.

***Root** and **Stem rots.** Geraniums are susceptible to several soil-borne diseases. *Xanthomonas pelargonii* causes Bacterial (Black) stem rot. Fungi such as *Sclerotinia* spp. *Fusarium* spp., *Thielaviopsis basicola* and *Pythium* spp. are also responsible for the death of plants from root or stem decay. The only satisfactory preventive is the use of sterilized cutting and potting composts and propagation from healthy stock.

***Rust** (*Puccinia pelargonii-zonalis*). This disease, which attacks zonal pelargoniums causing concentric rings of raised, reddish-brown spore-producing tissue on the undersides of leaves and corresponding yellow areas on the upper sides, may occur on cuttings or plants under glass. Spraying with maneb, thiram *or* zineb, as for Chrysanthemum—Rust, p. 244, will probably control this disease if several applications are given at 10–14-day intervals.

GRAPE (VINES)

PESTS

***Brown scale** (*Parthenolecanium corni*). The adult stages are protected by a hard, reddish brown scale beneath which eggs are laid. Young scale insects emerging therefrom are active and much more readily killed. Stems mainly are attacked. Their control is as for mealy-bug but using diazinon at 6·4 oz a.i. (32 fl. oz 20% e.c.)/100 gal H.V., *or* malathion at 30 oz a.i. (2½ pints 60% e.c.)/100 gal repeating at 21-day intervals until the bunches are thinned.

***Mealy bug** (*Planococcus* spp.). Aphid-like insects which, when adult, are covered by a white, mealy protective wax. They feed on stems, berries, and leaves and disfigure berries and leaves by secreting honeydew, on which a sooty mould grows. For their control scrape dormant rods and spray in December with tar-oil winter wash conforming to the Ministry of Agriculture specification for tar-oil winter washes, 4 gal/100 gal H.V. In spring spray with diazinon 3·2 oz a.i. (16 fl. oz 20% e.c.)/100 gal H.V., *or* malathion 18 oz a.i. (30 fl. oz 60% e.c.)/100 gal H.V., and repeat several times at 14-day intervals, but not after thinning. Adequate wetting agent must be used to wet thoroughly the waxy insect.

*****Red spider mite** (*Tetranychus urticae*). For description see Cucumber —Red spider mite, p. 247, and for control, spray with diazinon *or* chlorbenside *or* chlorfenson *or* fenson *or* petroleum as for Tomato—Red spider mite, p. 268.

Atomize solutions of diazinon 1·8 g a.i./1,000 ft³, *or* chlorfenson 5 g a.i./1,000 ft³, *or* chlorfenson/parathion 5 g a.i. and 0·5 g a.i./1,000 ft³, as for Tomato—Red spider. Smoke with parathion 0·25 g a.i./1,000 ft³ at 7-day intervals *or*, from May onwards, azobenzene/parathion 3 g a.i. and 0·25 g a.i./1,000 ft³, at 14-day intervals, *or* azobenzene 3 g a.i./1,000 ft³ at 7-day intervals. Do not use smokes or atomized solutions within a week of cutting or sprays after thinning. Dicofol or tetradifon are probably safe on vines but there is insufficient information to permit a firm recommendation.

DISEASES

***Grey mould (Botrytis rot)** (*Botrytis cinerea*). Under warm, very humid conditions the fungus may attack ripening berries, causing rapid rotting. To prevent this, dust the bunches with sulphur dust at the first appearance of disease and repeat at 7–10-day intervals. The disease is unlikely to occur when sulphur is regularly vaporized to control mildew.

****Powdery mildew** (*Uncinula necator*). A typical whitish powdery mildew attacking leaves, young shoots, flowers and fruits, controlled by vaporizing sulphur as for chrysanthemum mildew or by spraying with colloidal or wettable sulphur, 2–3 lb sulphur (2½–4 lb 75% w.p.)/100 gal H.V. *or* dinocap 2 oz a.i. (8 oz 25% w.p.)/100 gal when the shoots are growing freely; repeat when flowers open and again when the berries have set, but not after thinning.

HIPPEASTRUM (AMARYLLIS)

PESTS

***Bulb scale mite** (*Steneotarsonemus laticeps*). Infestation by this mite is shown by dark redish spots on the leaves, flower stalks and bulb scales accompanied by distortion. Hot water treatment of dry, dormant bulbs for $1\frac{1}{2}$ hours at 110°F followed by gradual cooling will kill the mites. Infested growing plants should be sprayed with endosulfan *or* endrin, see Narcissus—Bulb scale mite, p. 314. Treatment with thionazin (*loc. cit.*) may be found effective.

**Mealy bug** (*Planococcus citri*). To control this pest spray thoroughly to wet leaf bases with diazinon *or* malathion as for Grape—Mealy bug, p. 254.

**Red spider mite** (*Tetranychus urticae*). Control with demeton-S-methyl, *or* diazinon, *or* dicofol, *or* dimethoate, *or* formothion, *or* oxydemeton-methyl, *or* tetradifon, as for Carnation—Red spider mite, p. 237. Azobenzene smokes 5–8 g a.i./1,000 ft³ applied several times at weekly intervals are usually effective.

DISEASES

**Leaf scorch** (*Stagonospora cutisii*) causes raised, red or brownish streaky spots on the flower stalks and leaves with bending or deformation. The fungus perennates in dormant bulbs and the disease is most severe under warm, humid conditions. To control, spray the emerging foliage from suspected bulbs and any foliage and buds at the first sign of disease, with Bordeaux mixture equivalent to 1 lb metallic copper/100 gal, *or* copper oxychloride at 6 oz metallic copper (12 oz 50% w.p.)/100 gal, *or* mancozeb $1\frac{1}{4}$ lb a.i. ($1\frac{1}{2}$ lb 80% w.p.)/100 gal, or tank-mixed zineb (see Tomato—Leaf mould, p. 271). Several fungicide applications at 14-day intervals will be required.


HYDRANGEA

Hydrangeas are very susceptible to damage by sprays applied in sunshine or if the plants are dry at their roots.

PESTS

**Red spider mite** (*Tetranychus urticae*). Control by using demeton-S-methyl, *or* diazinon, *or* dicofol, *or* formothion, *or* oxydemeton-methyl, *or* tetradifon, at the concentrations given under Carnation—Red spider mite, p. 237.

DISEASES

**Grey mould** (*Botrytis cinerea*) may occur on dense flower clusters of plants growing under excessively humid conditions. Control with captan sprays, as for Azalea—Grey mould, p. 233.

**Powdery mildew** (*Microsphaera polonica*). For description and control see Hydrangea—Powdery mildew, p. 311.

## LETTUCE

Where losses due to grey-mould (*Botrytis*) or *Corticium* occur frequently, preventive measures should be taken. With other diseases and pests, control measures should be taken immediately the disease or pest is noticed as it is almost impossible to spray or dust adequately leaves which are in contact with the ground or have folded in.

PESTS

***Aphids** (A.L. 392). Several species, of which *Myzus persicae*, *Nasonovia ribisnigri* and *Aulacorthum solani* are most prevalent, attack lettuce. Losses are usually due to planting infested seedlings, or planting in houses where aphids are present on plants or weeds. Hence, for control, raise plants in isolation or ensure other plants are aphid free. Clean up the houses 7–10 days before planting and smoke with gamma-BHC 0·7 g/ 1,000 ft$^3$ if aphids are present. Dip the young plants, before setting out, in nicotine 1 fl. oz (95–98%)/10 gal, *or* malathion 1 oz a.i. (approx. 1$\frac{1}{2}$ fl. oz 60% e.c.)/5 gal for 3 minutes. Plants raised in soil-blocks or peat-pots, etc., should be thoroughly sprayed with nicotine or malathion at these concentrations. Rubber gloves must be worn.

For attacks after planting out, spray with demeton-S-methyl *or* dimethoate *or* gamma-BHC *or* diazinon, *or* formothion, *or* malathion, *or* oxydemeton-methyl, *or* nicotine as for Tomato—Aphid, p. 267. Do not use gamma-BHC sprays until the transplants are well established. Repeat applications if necessary but strictly observe the required periods before cutting; demeton-S-methyl 21 days, dimethoate 7 days, diazinon 14 days, formothion 7 days, malathion 1 day (preferably 4 to obviate taint), nicotine 2 days.

****Springtails**—see Cucumber—Springtails, p. 248 for descriptions. For control, dust the soil surface with gamma-BHC dust (0·6%) at 1 oz/yd$^2$ or spray soil surface with gamma-BHC 2 oz a.i. (10 fl. oz 20% e.c.)/100 gal.

****Symphylids**—see Tomato—Symphylids, p. 269.

***Wireworms**—see Tomato—Wireworms, p. 270.

DISEASES

***Damping-off** (*Corticium solani*). This fungus sometimes causes the death of seedlings and may attack older leaves in contact with the soil. Propagate in sterilized soil and treat seed with captan or thiram seed dressing. Apply quintozene 20% dust at 1 oz/yd$^2$ to the seed bed before sowing. Where seedlings are being attacked, remove all seedlings seen to be affected and water with thiram 1$\frac{1}{2}$ lb a.i. (2 lb 80% w.p.)/100 gal.

****Downy mildew** (*Bremia lactucae*) is first noticed as pale green or yellow areas on lower leaves. White or greyish fungal growth soon develops on undersides of leaves. For control, avoid over-watering and a humid atmosphere; raise temperature to 15–17°C (60–65°F) and increase ventilation; spray with zineb 1¼ lb a.i. (2 lb 65% w.p.)/100 gal H.V., *or* thiram 3 lb a.i. (3·8 lb 80% w.p.)/100 gal H.V. and repeat at 14-day intervals *or* dust with zineb dust (15%) at 20 lb/acre and repeat at 7-day intervals. On seedlings thiram is safer than zineb and also more effective against grey mould.

***Grey mould** (*Botrytis cinerea*). To control this disease remove all debris from soil surface and obtain a level surface without a very fine tilth. Avoid wilting due to shortage of water, poor temperature control and inadequate ventilation, organic manure or plant debris on the surface of the soil and deep planting. Spray the superstructure of the vacant house and soil surface with formaldehyde 1 gal 38–40% formalin/50 gal at least 3 weeks before planting and ventilate freely after 2 days.

Before sowing or planting-out, dust the surface of the soil in heated or cold houses with dicloran 4% dust at ½ oz/yd² before sowing *or* 1 oz/yd² before planting, *or*, in heated houses, with tecnazene 5% dust at ½ oz/yd² before sowing *or* 1 oz/yd² before planting *or*, in cold houses, with quintozene 20% dust at 1 oz/yd² and rake into soil surface. When *Botrytis* occurs on planted-out plants, dust not later than 3 weeks before cutting with dichloran 4% dust at ¼ oz/yd² and repeat after 2–3 weeks on lettuces for Xmas cutting or at 6-week intervals on other crops, *or* with thiram dust at 3 lb a.i. (20 lb 15% dust)/acre at 7–10-day intervals, *or* spray with thiram 3 lb a.i. (3·8 lb 80% w.p.)/100 gal H.V. at intervals of 14–21 days. In all cases try to cover the undersides of the leaves and bases of stems.

Spraying with dichlofluanid at 16 oz a.i. (2 lb 50% w.p.)/100 gal or dusting with dichlofluanid at 3 lb a.i. (20 lb 15% dust)/acre at 14-day intervals from the time conditions are conducive to Botrytis infection until not later than 4 weeks before cutting, have given promising results. Plants set out in soil given pre-planting treatment should not be sprayed or dusted within 14 days of planting.

## MUSHROOM

PESTS

**Flies.** Many species of three families, Cecidomyiidae, Phoridae, and Sciaridae, are pests of mushrooms. The life histories and behaviour of these groups differ, consequently they are dealt with separately.

***Cecid midges.** The larvae are white or orange and distinguishable from other mushroom maggots by a dark brown spot on the body just behind the head. They are paedogenetic, giving birth to daughter larvae and, as this process may continue for a long time, the small delicate adult midges are rarely seen. With heavy infestations larvae ascend mushrooms and enter surface tissues and the mycelium may be destroyed.

Adequate peak-heating of the compst at 55–60°C (130–140°F) will give control but, as the larvae are dispersed in a similar manner to eelworms strict attention should be paid to hygiene. It is not possible to reduce larval populations in composts without damage to the crop, but the following casing-layer treatment acts as a barrier to larval migration to mushrooms. Damage can be reduced by mixing 2 oz gamma-BHC (e.g. 10 fl. oz 20% e.c.)/ton of peat and chalk casing material.

***Phorid flies** (*Megaselia nigra, M. halterata, M. bovista*) are small, dark, stout-bodied flies. The maggots, which are white, legless and lacking the black head of sciarid larvae, feed on mushroom mycelium and some species tunnel in mushrooms. Their attacks are usually most frequent in summer and autumn and come mainly from flies entering spawn-running rooms or cropping houses from outside.

To avoid attack, screen ventilators and keep spawn-running rooms closed as much as possible.

Phorid flies are very susceptible to the vapour of dichlorvos and the use of slow-release formulations to yield a continuing concentration of 0·006 g/1,000 ft² (0·2 $\mu$g/l) throughout spawn-running and the first few weeks of cropping is recommended. Thoroughly mixing diazinon in the compost and/or casing at the rate of 2/5 oz a.i. (2 fl. oz 20% e.c.)/ton also prevents infestation.

Alternative methods are to atomize dichlorvos 0·5 g/1,000 ft³ *or* malathion 2·5 g/1,000 ft³ *or* diazinon 1·8 g/1,000 ft³ at intervals of 1 day–4 days (depending on the number of flies) when the spawn is running and less frequently during cropping *or* use diazinon *or* malathion dusts at the same active ingredient rates and intervals.

***Sciarid flies.** Maggots of the many species of these flies, which are small, delicate insects with a tendency to run rather than fly, are white with black heads. They tunnel into the stalks and caps of mushrooms and sever the mycelium from the stalk, causing the buttons to die, and carry bacterial pathogens and eelworms. Flies are attracted to both manure and the growing crop. For their control, peak-heat the compost to 55–60°C (130–140°F).

If this temperature cannot be obtained throughout the compost, or where flies are present in spawning or cropping rooms, the use of dichlorvos slow release formulations, as for phorids, provides the most satisfactory control.

Alternatively atomize diazinon *or* dichlorvos *or* malathion *or* apply malathion dusts as for phorids *or* atomize gamma-BHC 1 g/1,000 ft³ *or* smoke with gamma-BHC 0·7 g/1,000 ft³.

If larvae are present in the casing before cropping commences, water with malathion 1·6 oz a.i. (2¾ fl. oz 60% e.c.)/100 gal, using 40 gal/1,000 ft².

***Nematodes** (*Ditylenchus* spp., *Aphelenchoides* spp.) feed on mycelium causing its disappearance in patches. Where numerous the compost may become dark, overmoist and stinking. Nematodes, under certain dry conditions, can persist for at least two years in dry compost, woodwork, etc. They can be carried from place to place on tools, boxes, etc., or by flies.

No acceptable means of controlling eelworms in spawned or cropping compost is available. Peak-heating kills eelworms in the unspawned compost, and cooking-out those in spent compost and structures. Methyl bromide fumigation also gives control in composts before spawning and may be approved when more experience has been acquired. Woodwork, structures, etc., should be cleansed thoroughly as described under Disinfection of Mushroom Houses, p. 261.

*Slugs** only damage mushrooms grown in very indifferent structures, which they can enter from outside. They may be controlled with metaldehyde bait as for Tomato—Slugs, p. 269, but do not place the bait on the beds.

*Springtails** (*Collembola* spp.). For description see Cucumber—Springtails, p. 248. Controlled by peak-heating, followed by gamma-BHC as for sciarids.

*Woodlice** are likely to occur when mushrooms are grown in floor beds in glasshouses. Before cropping begins, apply 5% DDT dust to the floor and crevices in which woodlice hide.


DISEASES

Mushroom yields are reduced both by parasitic organisms and by fungi which grow on the compost and compete for nutrients with the mushroom mycelium. Most if not all losses from either cause can be prevented by strict attention to hygiene, correct compost fermentation, adequate pasteurization of the compost by 'peak-heating' before spawning, and 'cooking-out' before emptying or removing old trays or beds. There are no satisfactory control measures for several pathogenic and competitive fungi.

**Bacterial blotch** (*Pseudomonas tolaasi*) appears as slightly sunken yellowish-brown blotches and spots on the pilei, and coalesce making the sporophore discoloured and sticky. The disease is favoured by high temperatures, moist conditions and inadequate ventilation, therefore ventilate after watering. If the disease appears, a routine watering with sodium hypochlorite equivalent to 2½ oz available chlorine/100 gal (1 pint 10% available chlorine in 80 gal) *or* with chlorinated water containing that amount (160 ppm) of chlorine, will reduce losses.

*Bacterial pit.** The cause of this trouble is unknown but bacteria appear to be associated. It starts as small cavities beneath the skin of the cap, the skin eventually collapsing to leave open pits. Treatment as for bacterial blotch has reduced its incidence.

****Bubbles  (white-mould)** (*Mycogone perniciosa*). Mildly infected mushrooms usually show wavy gills covered by a white velvety fungal growth. More severely attacked mushrooms are deformed, covered by the white mould from which exudes a clear golden liquid, and rapidly become brown, wet and foul-smelling. The fungus is derived usually from infected casing soil and it attacks mushrooms most readily when the temperature is high. Strict hygiene must be observed. The casing soil should be sterilized with steam or formaldehyde. Spray small patches of infected mushrooms and 6 in around them with formaldehyde 1 pint 38–40% formalin/2 gal, and then remove diseased mushrooms. Next, spray the beds once between each flush, with zineb 5·2 oz a.i. ($\frac{1}{2}$ lb 65% w.p.)/100 gal at 1 gal/50 ft² at least three times *or* dust with zineb dust, 1–2 oz a.i. (6–12 oz 15% dust)/ 1,000 ft² at weekly intervals. Where the disease was present on a previous crop, start treatment with zineb after casing and repeat between first and second, and second and third flushes.

Watering with hypochlorite or chlorine as for bacterial blotch has given promising results but more information is necessary before it can be firmly recommended.

***Cobweb** (*Dactylium dendroides—Hypomyces rosellus*). The pinkish-white, loose cobwebby mycelium of fungus envelops entire mushrooms and spreads on surrounding casing. For control, treat with zineb as for bubbles *or* spray the beds with quintozene 2 oz a.i. (4 oz 50% w.p.)/100 gal to wet casing to a depth of $\frac{1}{2}$ in immediately after casing, and $\frac{1}{2}$–1 lb a.i./ 100 gal after the first flush has been picked *or* dust, after picking, with quintozene dust at 0·8 oz a.i. (4 oz 20% dust)/1,000 ft² at weekly intervals beginning after the first flush. Spray the bed-boards and woodwork with quintozene $\frac{3}{4}$ lb a.i. (1$\frac{1}{2}$ lb 50% w.p.)/100 gal H.V.

***Dry bubble (verticillium)** (*V. malthousei* or *V. psalliotae*) appears first as light brown blotches on caps. When severe the mushrooms are distorted, often with swollen and bulbous stalks and very small cups. Mushrooms become covered with grey-white mould but do not decay to produce the offensive odour associated with bubbles (*Mycogone*). Apply zineb or sodium hypochlorite as for bubble disease. Disinfect house at end of cropping—see below.

***Gill mildew** (*Cephalosporium lamellicola*) causes symptoms similar to a mild attack of dry bubble but the blotches are darker. The gills may become covered with a fine mycelial growth. The control is as for dry bubble (*Verticillium*).

***Red geotrichum** (*Sporendonema purpurascens*) is a competing fungus which develops in the compost and casing. Young mycelium is white but at the sporulating stage the fungus turns buff and then bright pink. It is favoured by poor ventilation and high temperature and may be avoided by using sterilized casing soil and attending to humidity and temperature during spawning. Use zineb or quintozene as for bubbles (*Mycogone*).

**Truffle (false truffle)* (*Dichliomyces microsporus—Pseudobalsamia microspora*). Spores of this fungus withstand high temperatures and may survive 'peak-heating' and 'cooking-out'. The mycelium, which runs rapidly through compost, is cream and makes the compost dark brown and soggy. Fruiting bodies are up to 1 in in diameter with a convoluted surface. Water the compost at the last turn with copper sulphate $1\frac{1}{2}$ lb/ton, using a suitable volume of water in relation to the dryness of the compost. If the fungus develops in shelf beds, remove the compost to form a trench extending at least 2 ft beyond the visible limit of the fungus.

### DISINFECTION OF MUSHROOM-HOUSES

To destroy spores of pathogenic and competing fungi on walls, floors, woodwork, etc., the houses should be treated with a fungicide after the diseased crop has been removed, even though the old compost has been 'cooked-out' by heating to 135–160°F. This is most important if brown plaster-mould (*Papulaspora byssina*), olive green mould (*Chaetomium* spp.), or truffle were present, for the spores of these fungi are heat tolerant.

Suitable sprays for this purpose are formaldehyde 5 gal 38–40% formalin/100 gal H.V. *or* sodium pentachlorophenate 10 lb/1,000 gal *or* sodium-DNOC 1 lb/100 gal H.V. (for safety regulations where sodium-DNOC is used see p. 000) *or* emulsified cresylic acid 20 lb cresols/100 gal H.V. Tools and equipment on which diseases or pests can be carried, should be cleansed with one of these chemicals, but do not use sodium-DNOC for this purpose.

### ORCHID

After the application of sprays to orchids with sheath type leaves, it is necessary to drain accumulations of spray retained in the leaf axils. Spray under quick drying conditions but not in bright sunshine.

PESTS

***Aphids.* Several species of aphid including *Aulacorthum circumflexum*, *Cerataphis lataniae* and *Myzus persicae*, breed on orchids. *C. lataniae*, when apterate, is reddish-brown and scale-like. Control with smokes of gamma-BHC, *or* gamma-BHC plus DDT, *or* nicotine, *or* parathion, *or* sulfotep, *or* fumigation with naled, *or* atomized solutions of malathion, *or* sprays of gamma-BHC *or* malathion, as for Carnation—Aphids, p. 236. Some genera appear tolerant to sprays and drenches of dimethoate, *or* oxydemeton-methyl, but further experience is necessary before systemic organophosphorus compounds can be recommended.

*Cattleya fly (*Eurytoma orchidearium*), **Cattleya midge** (*Parallelodiplosis cattleyae*) and **Cattleya weevil** (*Cholus cattleyae*). These, and several other less important dipterous and coleopterous pests, have almost been completely eliminated from orchid houses by insecticides used for other pests. Should they occur, applications of DDT in H.V. sprays at 12 oz a.i. ($2\frac{1}{2}$ pints 20% e.c.)/100 gal, *or* dust at 1 lb 5% dust/200 yd², *or* gamma-BHC sprays at 2 oz a.i. (10 fl. oz 20% e.c.)/100 gal will give control.

**Mealy bugs (*Planococcus* spp.). Control with sprays of malathion at 18 oz a.i. ($1\frac{1}{2}$ pints 60% e.c.)/100 gal, *or* DDT at 12 oz a.i. ($2\frac{1}{2}$ pints 20% e.c.)/100 gal. Several applications at 14–21-day intervals may be required.

***Red spider mite (*Tetranychus urticae*). These small greenish or red mites are frequently a pest on orchids and, if not controlled, will, by their puncturing of the leaf cells, cause the leaves to become hard and the plants unthrifty. In the absence of other pests control should be obtained by using sprays of dicofol *or* tetradifon, *or* atomized solutions of these *or* fenson, *or* smokes of tetradifon, as for Carnation—Red spider mite, p. 237. If other pests are present, sprays or atomized solutions of diazinon at 10–14-day intervals, *or* fumigation with naled several times at 4-day intervals may be advantageous. Several applications at 7-day intervals of smokes of azobenzene, 3 g a.i./1,000 ft³ will give control with safety to most varieties.

**Scales. Many species of scale insects occur on orchids, most having been introduced on imported plants. Recommended control measures are to spray 2–3 times at 14–21 day intervals with diazinon $6\frac{1}{2}$ oz a.i. (32 fl. oz 20% e.c.)/100 gal, *or* malathion 30 oz a.i. ($2\frac{1}{2}$ pints 60% e.c.)/100 gal, *or* DDT as for Cattleya fly.

*Thrips. Of the species occurring on orchids, *Heliothrips haemorrhoidalis*, *Hercinothrips femoralis* and *Anothrips orchidaceous* appear the commonest. Control as for aphids (above) *or* with DDT sprays as for Cattleya midge (above).

*Vine weevil, see Cyclamen—Vine weevil, p. 250.

*Woodlice. The species of woodlice which damage orchids may be controlled with DDT, *or* parathion, *or* Paris green baits, as for Cucumber—Woodlice, p. 248. Apply drenches sparingly to orchid composts.


DISEASES

*Anthracnose. Two fungi cause diseases commonly termed anthracnose. *Glomerella cincta* causes dark, decayed areas on the leaves, usually starting on the tips. The spots caused by *Gloeosporium* spp. are yellowish—light brown at first—becoming soft and sunken later. The recommended control is to spray several times at 10–14-day intervals with a copper fungicide, e.g. copper oxychloride at 6–8 oz metallic copper/100 gal.

***Bacterial soft rot** (*Pectobacterium carotovorum*). A slimy rotting of the pseudobulbs and basal stems. Plants may be saved by cutting out all diseased tissue and dusting the wounds and surrounding areas heavily with a copper dust before rotting is extensive.

***Black rot** (*Pythium* spp.). Seedlings may damp-off due to attack. The fungus also enters the roots of older plants and spreads to the pseudobulbs which shrivel. Leaves may be attacked if the humidity is high. The immersion of seedlings for 1 hour in captan at $\frac{1}{2}$ oz a.i./gal on their removal from the culture flasks is recommended as a preventive. To prevent its spread on older plants, the humidity should be reduced and the plants sprayed as for anthracnose (above).

### POINSETTIA

The bracts of poinsettia are very susceptible to injury by pesticides from the time they begin to colour.

PESTS

***Mealy bugs** (*Pseudococcus citri, P. adonidum*) and

***Scales** (several spp. including *Coccus hesperidum* and *Aspidiotus hederae*) may infest poinsettia, particularly those grown with other hosts in private glasshouses. To control them spray with DDT at 12 oz a.i. ($2\frac{1}{2}$ pints 20% e.c.)/100 gal, several times at 14-day intervals.

****Red spider mite** (*Tetranychus urticae*). Control with H.V. sprays or atomized solutions of dicofol *or* tetradifon, or tetradifon smokes as for Carnation—Red spider mite, p. 237.

DISEASES

Apart from damping-off of seedlings and young plants due to soil-borne fungi such as *Corticium* spp. and *Thielaviopsis basicola* (preventable by using sterilized composts), the only fungus likely to cause trouble is

***Grey mould** (*Botrytis cinerae*), which attacks leaves and bracts under excessively moist conditions. If control is not obtained by removing the diseased parts and giving adequate ventilation, adopt the control measures given under Cineraria—Grey mould, p. 245.

### PRIMULA

#### (*Primula malacoides, P. obconica*)

PESTS

****Aphids** (*Aulacorthum circumflexum, Myzus persicae*) Control, if possible before flowering, by spraying with or dipping in gamma-BHC (not *A. circumflexum*), *or* malathion, *or* nicotine, at concentrations as for

Carnation—Aphids, p. 236, or use atomized solutions of gamma-BHC, or malathion, or smokes of gamma-BHC, or vaporize naled, as for Carnation—Aphids. Although some cultivars may be sensitive to systemic organophosphorus compounds, demeton-S-methyl, or dimethoate, or formothion, or oxydemeton-methyl, have been used successfully as sprays and drenches, as for Carnation—Aphids, on several cultivars of both species.

***Leaf hopper** (*Zygina pallidifrons*). An active, yellowish-green insect which feeds on the undersides of the leaves, causing chlorotic spotting of the upper surfaces. Easily controlled by DDT, or gamma-BHC, or malathion, as for Carnation—Thrips, p. 237.

****Red spider mite** (*Tetranychus urticae*). To control, spray at 10–14-day intervals with dicofol, or tetradifon, at concentrations as for Carnation—Red spider mite, p. 237, or dip the plants in these. The systemic compounds mentioned under Aphids (above) may be used on most cultivars.

***Vine weevil** (*Otiorhynchus sulcatus*). For description and control with gamma-BHC see Cyclamen—Vine weevil, p. 250. There are reports of growth retardation of primulas where dieldrin has been added to the compost.

ROSE

PESTS

*****Aphids.** Several spp., of which *Macrosiphum rosae* is the most important, are troublesome and their control is usually associated with the control of red spider. Where red spider is absent, atomize solutions of malathion, or gamma-BHC, or use smokes of nicotine, or parathion, or spray with malathion, or nicotine, as for Chrysanthemum—Aphids, p. 240. If caterpillars are also present atomize malathion/DDT, or gamma-BHC/DDT as for Chrysanthemum—Aphids and Caterpillars, p. 240.

Where red spider is present atomize diazinon, or spray with demeton-S-methyl, or dimethoate, or formothion, or oxydemeton-methyl, as for red spider. Rose growers consider that atomized solutions are less likely to spoil rose foliage than are H.V. sprays.

***Caterpillars.** Several species damage leaves, shoots and buds; some tortrices spin leaves together. Their control is as for Carnation—Tortrix, p. 237. Atomized solutions or dusts are preferable.

***Leaf hoppers.** For description and control see Tomato—Leafhopper, p. 268.

***Mealy bug.** For description and chemical control see Grape—Mealybug, p. 254.

*****Red spider mite** (*Tetranychus urticae*). To control the mite, atomize solutions of diazinon 1·8 g/1,000 ft³ or dicofol 2 g/1,000 ft³ or dicofol

1·8 g plus diazinon 1·6 g/1,000 ft³ *or* spray, when shoots are growing freely, with dimethoate 4·8 oz a.i. (12 fl. oz 40% e.c.)/100 gal, *or* formothion 7 oz a.i. (16 fl. oz 43% e.c.)/100 gal H.V., *or* oxydemeton-methyl 3·4 oz a.i. (6 fl. oz 57% e.c.)/100 gal H.V. *or*, when roses will not be cut for 3–4 days, demeton-S-methyl 3·4 oz a.i. (12 fl. oz 58% e.c.)/100 gal H.V.

In the absence of aphids or thrips, sprays of dicofol 1·6 oz a.i. (8 fl. oz 20% e.c.)/100 gal H.V. *or* tetradifon 2 oz a.i. (25 fl. oz 8% e.c.)/100 gal H.V. *or* smokes of tetradifon 1 g a.i./1,000 ft³ can be used.

***Rose (scurfy) scale** (*Aulacaspis rosae*). The older branches become covered with a scurfy mass of scales and colonies of young yellowish-white scales may appear on shoots and leaves. Their control is as for Scale on grapes.

DISEASES

***Black spot** (*Diplocarpon rosae*) occasionally occurs under glass and forms black or purple spots on the upper surfaces of leaves, which shed leaflets prematurely. Spray, at the first sign of disease, with captan 1 lb a.i. (2 lb 50% w.p.)/100 gal H.V. and repeat at 10-day intervals until the disease is eradicated. If the disease was present previously, spray after pruning and again when the new leaves unfold and continue at 10-day intervals. Trials in America and elsewhere have shown folpet 1 lb a.i. (2 lb 50% w.p.)/100 gal H.V. *or* dodine ¾ lb a.i. (approx 18 oz 65% w.p.)/100 gal H.V. *or* maneb 1 lb a.i. (20 oz 80% w.p.)/100 gal H.V. are also effective but more experience of their value on glasshouse roses in Britain is required before a firm recommendation can be made.

***Downy (black) mildew** (*Peronospora sparsa*) produces irregular yellow-grey or purplish spots on leaves and occasionally on flower stems. Whitish-grey downy fungal growth appears on undersides of leaves which fall prematurely. Spray with copper/petroleum emulsion, 6 lb metallic copper (e.g. 12 lb copper oxychloride containing 50% Cu) and 4 pints petroleum oil emulsion (glasshouse grade)/100 gal H.V., *or* zineb 1¼ lb a.i. (2 lb 65% w.p.)/100 gal, H.V., *or* tank-mixed zineb as for Tomato—Leaf mould, p. 271, at 14-day intervals.

***Grey mould** (*Botrytis cinerea*). Under excessively humid conditions, flowers and occasionally stems may be attacked. Germinating spores may produce spotting of the petals and subsequent fungal growth causes rotting of the blooms. The fungus can also invade stem wounds causing die-back of more woody tissue. To check the disease, remove infected blooms and decrease the humidity. Sulphur fumigation as for powdery mildew (below), *or* spraying with captan 1 lb a.i. (2 lb 50% w.p.)/100 gal, *or* thiram 3 lb a.i. (3·8 lb 80% w.p.)/100 gal, will give control. Dipping cut flowers in thiram at the above concentration will prevent spotting of flowers in cold storage.

***Powdery mildew** (*Sphaerotheca pannosa*). A typical white powdery mildew attacking young leaves, stems, buds and petals, and causing new growth to be dwarfed and distorted. To control the disease, vaporize sulphur as for Chrysanthemum—Mildew, p. 243, avoiding as far as possible its use on dark red varieties when at height of flower production. The variety Super Star appears to be somewhat sensitive to sulphur. Spray with dinocap 1 oz a.i. (2 oz 50% e.c.)/100 gal H.V., *or* copper/petroleum emulsion as for downy mildew. Results of overseas work suggest that triamiphos may be of value for the control of powdery mildew but additional evidence is required before it can be recommended. Some rose varieties are sensitive to dinocap.

### SAINTPAULIA (AFRICAN VIOLET)

Do not apply sprays or dips in bright sunshine or under humid, slow drying conditions or if the pot compost is dry. Use tepid water for sprays and preferably drain the plants on their sides.

PESTS

*****Mealy bug** (*Planococcus citri*). For control see Orchid—Mealy bugs, p. 262.

****Mites**—Broad mite (*Hemitarsonemus latus*), Tarsomenid (Cyclamen) mite (*Steneotarsonemus pallidus*) and *Tarsonemus confusus*. The effects of these mites on plants are similar. The plants become stunted with down-curling of the leaf edges, twisted and brittle leaves and stems. The leaves are often smooth and glossy, and the flower buds may be distorted and fail to open. For control see Cyclamen—Tarsonemid mite, p. 250.

DISEASES

*****Grey mould** (*Botrytis cinerea*). The conditions conducive to attack, and the control measures are as for Azalea—Grey mould, p. 233.

### SWEET PEA

The flowers are likely to be blemished if sprayed with any pesticide.

PESTS

****Aphids** (*Myzus persicae*). This and occasionally other species feed on the leaves and flower buds of sweet peas under glass. Control is as for Carnation—Aphids, p. 236, but use malathion with care and only on well grown plants.

****Red spider mite** (*Tetranychus urticae*). This mite is frequently a serious pest on sweet pea and control measures should be taken as soon as it is noticed. H.V. sprays of demeton-S-methyl, *or* diazinon, *or* dicofol, *or* dimethoate, *or* oxydemeton-methyl, *or* tetradifon, as for Carnation—Red spider mite, p. 237.

DISEASES

***Powdery mildew** (*Erysiphe polygoni*). A typical powdery mildew, first appearing as a whitish powdery growth on the undersurface of the leaves. Flower buds and stems may also be attacked. Avoidance of wide temperature and humidity fluctuations may be sufficient to stop the development of this disease, but it is often necessary to use a fungicide. Dinocap at 2 oz a.i. (4 fl. oz 50% e.c.)/100 gal, applied several times at 7–10-day intervals is recommended. Good results have also been obtained by vaporizing sulphur as for Chrysanthemum—Mildew, p. 243.

TOMATO

PESTS

***Aphid.** The most serious is the glasshouse-potato aphid (*Aulacorthum solani*), which feeds on the foliage causing yellowing and increasing its susceptibility to grey mould; on immature fruits it causes raised, paler areas which remain yellow when the fruit ripens. The aphid is controlled on young plants by spraying with demeton-S-methyl 3·4 oz a.i. (6 fl. oz 58% e.c.)/100 gal H.V., *or* dimethoate 4·8 oz a.i. (12 fl. oz 40% e.c.)/100 gal H.V., *or* oxydemeton-methyl 3·4 oz a.i. (6 fl. oz 57% e.c.)/100 gal H.V., *or* nicotine 8 oz 95–98%/100 gal H.V., *or* parathion 1·6 oz a.i. (8 fl. oz 25% e.c.)/100 gal H.V., *or* by smoking with parathion 0·25 g/ 1,000 ft³. Repeat if necessary after 10 days.

On established plants, atomize solutions of diazinon 1·8 g/1,000 ft³ *or* dichlorvos 1 g/1,000 ft³, *or* gamma-BHC 1 g/1,000 ft³, *or* malathion 2·5 g/1,000 ft³, *or* smoke with gamma-BHC 0·7 g/1,000 ft³, *or* nicotine 4 g/1,000 ft³; *or* spray with gamma-BHC 2 oz a.i. (10 fl. oz 20% e.c.)/100 gal H.V. allowing 4 days before picking, *or* malathion 18 oz a.i. (30 fl. oz 60% e.c.)/100 gal H.V. allowing 1 day before picking, *or* if red spider is also present, diazinon 3·2 oz a.i. (16 fl. oz 20% e.c.)/100 gal H.V., allowing 2 weeks before picking. Demeton-S-methyl, *or* oxydemeton-methyl, *or* dimethoate can be used as for young plants if the necessary periods of 21 days and 7 days respectively before picking can be observed.

A second application of atomized fluids or smokes may be required after an interval of 10 days.

***Leaf miner** (*Liriomyza bryoniae*). The larvae tunnel cotyledons and stems of seedlings early in the year, often killing the seedlings. Later generations sometimes mine the leaves of established plants sufficiently to warrant control measures. The larvae pupate and overwinter in the soil. The first indications of its attack are small rounded pits on cotyledons, caused by adult flies feeding. Avoid propagating tomatoes in houses where the previous crop was infested unless the soil in the house is steamed to kill the pupae. Spray seedlings and small plants with diazinon 3·2 oz a.i. (16 fl. oz 20% e.c.)/100 gal H.V. parathion 1·6 oz a.i. (8 fl. oz 20% e.c.)/100 gal H.V. *or* and repeat at 14-day intervals if fresh pit-marks are seen on leaves *or* use dimethoate 4·8 oz a.i. (12 fl. oz 40% e.c.)/100 gal on young plants. On established plants, when attack is serious, use sprays or atomized solutions of gamma-BHC or diazinon as for aphids.

***Leafhopper** (*Zygina pallidifrons*) may attack young tomatoes propagated in houses containing other infested hosts. If so, spray the seedlings with parathion as for aphids. Dust young plants with DDT dust *or* smoke with DDT smokes as for tomato moth. Atomized solutions of DDT *or* diazinon *or* malathion may be used if attacks occur on established plants in summer.

*****Potato root eelworm** (*Heterodera rostochiensis*) (A.L. 284). Tomato plants attacked by the eelworm are retarded, the leaves become dark green with a purplish under-surface. The plants wilt during hot bright weather and their roots may become invaded by fungi and rot. Propagate in steamed soil and avoid introducing eelworm into glasshouses. Sterilize borders with steam *or* with dazomet *or* with metham-sodium, *or*, where root-knot eelworm is present and the use of dazomet or metham-sodium is inadvisable, with methyl bromide or D-D mixture, see p. 229.

****Red spider mite** (*Tetranychus urticae*). For description see Cucumber—Red spider mite, p. 247. To control on seedlings and young plants, spray with dimethoate *or* demeton-S-methyl as for aphids, *or* spray with dicofol 2 oz a.i. (8 fl. oz 25% e.c.)/100 gal H.V., *or* tetradifon 2·0 oz a.i. (25 fl. oz 8% e.c.)/100 gal H.V., and repeat after 14 days.

On established plants before mid-May, spray with dicofol *or* tetradifon *or* demeton-S-methyl *or* dimethoate, *or* oxydemeton-methyl, as for Cucumber—Red spider, p. 247, *or* with diazinon 3·2 oz a.i. (16 fl. oz 20% e.c.)/100 gal H.V., *or* smoke with azobenzene 3 g and parathion 0·25 g/1,000 ft³ *or* tetradifon 1 g/1,000 ft³, if the temperature can be maintained above 21°C (70°F) for at least 4 hours, *or*, on normal plants exceeding 3 ft in height, atomize azinphos-methyl 1·5 g/1,000 ft³ allowing at least 2 days before picking and limiting the applications to not more than three per season.

On established plants after mid-May, in addition to the substances recommended above for established plants, the following can be used: chlorfenson 5 g/1,000 ft³ *or* fenson 5 g/1,000 ft³, *or* dicofol 2 g/1,000 ft³ *or* tetradifon 1 g/1,000 ft³. Where an immediate kill of adults is required, diazinon can be used together with one of the above four compounds. Repeat applications at intervals of 10–14 days.

***Root-knot eelworm** (*Meloidogyne* spp.) (A.L. 307). For control see Cucumber—Root-knot eelworm, p. 246.

Where plants have to be planted in soil known to be infested by root-knot eelworm, watering with parathion 6 oz a.i. (30 fl. oz 20% e.c.)/100 gal at 1 gal/yd$^2$ within 3 days of planting will protect plants from early attack and increase their yields. In view of the relatively high concentration of parathion used, the glasshouses must be well ventilated during application, rubber boots must be worn when entering the glasshouse, and rubber gloves worn when handling the soil for the following 14 days; these precautions are additional to other statutory requirements (see Chapter 4).

*****Slugs** (A.L. 115) occasionally enter houses from weedy land outside or in dung, and are controlled by watering or spraying the soil, in the area of attack, with liquid metaldehyde preparations at 2 oz a.i. (e.g. 10 fl. oz 20% suspension)/100 yd$^2$ or by baiting with metaldehyde bait at $\frac{1}{2}$–1 oz a.i. (e.g. 16–33 oz 3% pellets)/100 yd$^2$.

*****Springtails**—see Lettuce—Springtails, p. 256.

****Symphylids** (*Scutigerella immaculata*) (A.L. 484) are active, white centipede-like insects which descend too deep into the soil to be destroyed by soil sterilization in autumn and winter, but return to the surface in spring and stunt growth of plants by damaging young roots. To control, water the plants within 3 days of planting with parathion 2 oz a.i. (10 fl. oz 20% e.c.)/100 gal *or* diazinon 2 oz a.i. (10 fl. oz 20% e.c.)/100 gal. Thorough incorporation of gamma-BHC 2$\frac{1}{2}$ lb a.i./acre as dusts into the top 4 in of soil within 4 weeks prior to planting has proved satisfactory.

*****Thrips** (*Thrips tabaci*) seldom cause direct damage but are vectors of spotted wilt virus and thus of importance where different species of plants are grown together. For their control, see Carnation—Thrips, p. 237.

****Tomato moth** (*Lacanobia oleracea*). The green or brown caterpillars of this moth, when young, skeletonize leaves and, when larger, eat foliage fruit and stems. For their control, use DDT either as dust at $\frac{3}{4}$–1 lb (15–20 lb 5% dust)/acre *or* as atomized solution at 2·5 g a.i./1,000 ft$^3$ *or* as spray at 1 lb a.i. (4 pints 20% e.c.)/100 gal H.V. *or* 6 oz a.i. (1$\frac{1}{2}$ pint 20% e.c.)/1$\frac{1}{2}$ gal/acre L.V. *or* as smokes 3·5 g a.i./1,000 ft$^3$. On seedlings use the dust only.

Allow at least 14 days between spraying or dusting and picking; 2 days in the case of aerosols or smokes.

****Whitefly** (*Trialeurodes vaporariorum*) (A.L. 86). For description, see Cucumber—Whitefly, p. 248. For their control on seedlings or small plants, use sprays of DDT 1 lb a.i. (4 lb 20% w.p.)/100 gal *or* malathion 18 oz a.i. (30 fl. oz 60% e.c.)/100 gal, *or* smokes of DDT 3·5 g/1,000 ft$^3$, *or* DDT 3 g plus gamma-BHC 0·5 g, *or* parathion 0·25 g/1,000 ft$^3$.

On established plants, use atomized solutions of dichlorvos 1 g a.i./

1,000 ft³ *or* malathion 2·5 g/1,000 ft³, or sprays and smokes as for seedlings, but using e.c. DDT formulations. Malathion aerosols have more effect than DDT aerosols on the scales. Repeat treatment at 14 days in cool weather or 10 days in hot weather as necessary.

***Wireworms** (*Agriotes obscurus*) (A.L. 199) usually occur only in the first 3–4 years after building glasshouses on grassland but may enter under walls from heavily infested land and attack adjacent plants. Before planting in new houses apply to the soil gamma-BHC dust ¾ lb a.i. (e.g. 125 lb 0·6% dust)/acre, and work into top 4 in of soil. One application is sufficient to eradicate wireworms. Where plants are being attacked, water with gamma-BHC, 1½ oz a.i. (7·5 fl. oz 20% e.c.)/100 gal at ½ pint/plant provided that the soil is well drained and the plants set out so that no fluid remains in a pool around the stems; otherwise use parathion 1½ oz a.i. (7·5 fl. oz 20% e.c.)/100 gal.

***Woodlice** (see Cucumber—Woodlice, p. 248) occasionally destroy seedlings and small plants during propagation. Attacks can be usually prevented by attention to hygiene, but, if troublesome, dust the soil surface, staging and places infested with woodlice with 5% DDT dust at 1–2 oz of dust/yd² or water infested soil with DDT 1 lb a.i. (e.g. 4 pints 20% e.c.)/100 gal at ½ gal/yd², or use Paris green bait as for Cucumber—Woodlice, p. 248.


DISEASES

*****Brown root-rot,** see Root-rots.

***Buck-eye rot** (*Phytophthora parasitica*). The lower fruits show grey brownish-red patches with a series of concentric dark brown rings and have become infected by the fungus being splashed onto the fruit from contaminated soil or the fruit being in contact with the soil. (See also Foot-rot.) The trusses should be tied up out of contact with soil and splashing when watering should be avoided. The disease may be prevented by propagating in soil and containers sterilized by steaming or formaldehyde, and by planting in sterilized borders. Where the disease occurs, spray the lower part of the plants and the surface of the soil with copper sprays at 6 lb metallic copper (e.g. 12 lb copper oxychloride containing 50% Cu)/100 gal H.V. *or* zineb 1¼ lb a.i. (2 lb 65% w.p.)/100 gal.

*****Colletotrichum root-rot,** see Root-rots.

*****Corky root,** see Root-rots.

****Foot-rot (Damping-off)** (*Phytophthora parasitica, P. cryptogea, Corticium solani*). These fungi may cause damping-off of seedlings or foot-rot of small plants. Infection arises from the soil or occasionally from contaminated boxes, pots or water. (See also Buck-eye rot.) Prevent attack by propagating in sterilized soil, etc., as for buck-eye rot. Unsterilized soil should be watered at seed sowing, potting or planting out with

Cheshunt compound $\frac{1}{2}$ oz/gal or other copper fungicides equivalent to 1 oz metallic copper/8 gal, or zineb 1 lb a.i. ($1\frac{1}{2}$ lb 65% w.p.)/100 gal H.V. Where seedlings or young plants are being attacked, healthy plants can be protected by watering with these fungicides but great care must be taken to reject all doubtful plants when potting or planting.

***Grey mould** (*Botrytis cinerea*). This fungus may invade stems from infections established on damaged or senescent leaves, on petiole stumps or deleafing wounds, producing stem lesions which may ultimately kill the plant. Masses of grey spores may be produced on the infected tissue and distinguish this disease from stem rot (*q.v.*). Under humid conditions *Botrytis* spores will germinate on green, immature fruits causing the small 'water' or 'ghost' spots which persist to disfigure the ripe fruit, or may infect flowers causing premature drop. Infected flowers may produce further infections of fruit or foliage.

To control the disease, reduce humidity by adequate ventilation. Remove all plant debris from borders, and decaying leaves and fruit from plants. Trim to allow circulation of air through the lower parts of plants and remove leaves and shoots cleanly.

Lesions on the stems should be cut out and the wound painted with a paste of liver of sulphur *or* with lime sulphur concentrate *or* a paste of dicloran (4%) dust. Painting the lesions with creosote is commonly practised and is satisfactory if care is taken to limit the treated area to the actual lesion.

To protect foliage and fruit from infection, spray with thiram 3 lb a.i. (3·8 lb 80% w.p.)/100 gal H.V. at 7–10-day intervals, *or* dust with thiram dust 6 lb a.i. (40 lb 15% dust)/acre at 4–7-day intervals *or* smoke with tecnazene. Particular care must be taken to adequately cover the lower parts of the plants. Dichlofluanid $\frac{1}{2}$ lb a.i. (1 lb 50% w.p.)/100 gal at 14-day intervals has proved very effective but must not be used until recommendations for its safe use are available.

***Leaf mould (mildew)** (*Cladosporium fulvum*) (A.L. 263) is first noticed as pale yellow patches on the upper surface of older leaves, the reverse surfaces of which become covered with a pale grey mould-like growth which darkens to brownish violet. The disease seldom occurs before early June, when the temperature and overnight humidity are high. Adequate ventilation controls the disease but, where the conditions are such that the disease appears every year, grow varieties which are resistant or of low susceptibility, or adopt one of the following routine preventive measures, starting when the disease usually appears or weather conditions are conducive to attack:—zineb prepared from nabam as stated below, *or* as a w.p. 2 lb a.i. (3 lb 65% w.p.)/4–8 gal/acre L.V., *or* zineb dust 3 lb a.i. (20 lb 15% dust)/acre, *or* salicylanilide 2 lb a.i. ($6\frac{1}{2}$ pints 25% solution)/4–8 gal/acre L.V. at 5–7-day intervals.

If control measures are delayed until the disease appears, spray (H.V.) at 10-day intervals with tank-mixed zineb prepared by adding, with agitation, a solution of 20 oz zinc sulphate (white vitriol) to a solution of 28 oz nabam (93%) and making the final volume to 100 gal, *or* zineb 1¼ lb a.i. (2 lb 65% w.p.)/100 gal H.V., *or* maneb 1¼ lb a.i. (1½ lb 80% w.p.)/100 gal H.V., *or* 3 lb zinc sulphate and 4 lb 3 oz nabam (4½ lb 93%)/5–10 gal/acre L.V. A wetting agent must be used with H.V. tank-mixed zineb sprays and with H.V. sprays prepared from w.ps when used on infected plants. When petroleum emulsions are used to control red spider mite, add copper fungicides equivalent to 10–15 oz metallic copper/100 gal diluted emulsion.

***Root rots.** Tomato roots are attacked and rotted by several fungi, the most important being a grey sterile fungus (G.S.F. or *Mycelia sterilia*) responsible for the swollen, suberized root condition termed corky root, and *Colletotrichum atramentarium* which destroys the cortex of the roots and stem below soil level. The effect of these fungi is to reduce plant vigour and crop production; plants tend to wilt in bright hot weather. For their control, sterilize the soil by steaming or with chemicals as described on p. 229. Formaldehyde is the least effective chemical but, except for methyl bromide permits the shortest interval before planting.

Recent experiments have shown that the use of a 60 ppm solution of nabam (1 oz 93%/100 gal) for watering plants on all occasions after planting gives increased crops in unsterilized, disease-contaminated soils. Increased yields are obtained, especially where Rhizoctonia is present, when zineb (1¼ lb a.i. (2 lb 65% w.p.)/100 gal) is applied at ½ pint/plant soon after planting and repeated at 3–4-week intervals at 2–4 pint/plant under U.K. conditions, or as a trench watering at about the time of set of the 5th truss under Channel Isles conditions.

****Stem rot** (*Didymella lycopersici*). Infection very occasionally originates from seed saved from diseased fruits, and then appears as spots on the leaves and proceeds to attack the main stem. Stem-base attacks 3 weeks or more after planting are much more usual and result from planting in infected soil. Later in the year, attacks on other parts of the stem may result from secondary infections. The lesions are dark brown and sunken, a grey woolly fungal growth is not produced as with *Botrytis* (see Tomato— Grey mould, p. 271). Most serious attacks occur on steamed soils which have become re-infected. Where disease was present on a crop, remove debris carefully and spray the superstructure and soil surface with formaldehyde 2 gal 38–40% formalin/100 gal H.V. Sterilize the soil with chemicals, or steam if necessitated by other considerations, and avoid risks of re-contamination. As a preventive against infection, spray the bottom 4–6 in of stem and the soil surrounding the stem to a radius of 2–3 in with captan 2½ lb a.i. (5 lb 50% w.p.)/100 gal, using 3 fl. oz/plant, within 2 days after planting and repeat 2–3 weeks later using 4–5 fl. oz/plant. Do not exceed these quantities or apply to plants which are dry at their roots or to pot plants. A slight temporary check to growth may result. Treatment at the 5 fl. oz rate may prevent further infections if applied immediately an attack occurs. Maneb is more effective than captan against stem-rot but may injure tomatoes in some soils, particularly steamed soils. Good results have been obtained, however, on some soils without injury by spraying the planting holes 24 hours before planting with maneb 1 lb a.i. ($1\frac{1}{4}$ lb 80% w.p.)/100 gal at 5 fl. oz/hole and spraying the stem bases with 3 fl. oz about 6 days and again 4 weeks after planting.

Painting stem lesions with maneb 1 lb a.i. ($1\frac{1}{4}$ lb 80% w.p.)/10 gal will eradicate the fungus if the lesions are treated in the very early stage.

***Wilt diseases** (*Verticillium albo-atrum*, *V. dahliae*, *Fusarium oxysporum*, *F. lycopersici*) (A.L. 53). Laboratory techniques are required to establish the organism responsible for wilting. Verticillium wilts are usually most severe when temperatures are low, whereas Fusarium wilt is more common at higher temperatures and is widespread in the Channel Isles but rare in Britain. Roots are rotted by Fusarium but may appear healthy in plants attacked by Verticillium. With Verticillium a light brown discoloration of the woody stem tissue extends practically to the growing point whereas, with Fusarium, the dark brown discoloration (almost black within the root) is less extensive. Their control rests in soil disinfection (see p. 229). Steaming is the preferred method. There is some evidence that chloropicrin is the most effective chemical sterilant against Verticillium.

Where Verticillium wilt occurs, avoid planting in cold, wet soil. Raise the temperature of the greenhouse to 25°C (77°F), shade lightly if necessary to reduce wilting, and damp plants overhead.

## 10.3 MISCELLANEOUS COMMERCIAL GLASSHOUSE PLANTS

In addition to the major glasshouse crops already dealt with, many other species of plants are grown or propagated commercially in glasshouses. Because of their relatively low economic importance or, in some cases, their novelty, little reliable information is available regarding the tolerance of such plants to pesticides. There are many conflicting reports. Factors such as stage of growth, climatic conditions, nutrition and the particular cultivar have a big influence on the susceptibility to injury. In general, seedlings and young plants, particularly when growing under adverse light conditions, are most susceptible. Damage by sprays is most likely to occur if they are applied in hot, bright sunshine or to plants suffering from water stress. The latter condition is also conducive of damage by soil

U

drenches. On the other hand conditions should be such that H.V. sprays dry within a few hours.

Most pests of minor crops have been mentioned in connection with the major crops and, providing the host is tolerant, can be controlled with the same pesticides.

# CHAPTER II

# PESTS AND DISEASES OF VEGETABLE CROPS

The direct fungicidal control of many vegetable diseases is either not possible or quite uneconomic. For this reason the scope of this chapter is restricted and must not be thought to present a balanced picture of vegetable diseases likely to be met. The measures necessary for the control of many of these diseases include efficient field sanitation, adequate crop rotations and the use of resistant varieties, as discussed in general terms in Chapter I. More specific information can be obtained from the M.A.F.F. Bull. 123 'Diseases of Vegetables'.

## 11.1 GENERAL PESTS AND DISEASES

PESTS

**Cutworms** (*Noctuidae*). Cutworms are caterpillars, chiefly of the turnip moth (*Agrotis segetum*) and the garden dart moth (*Euxoa nigricans*), which inhabit the surface layers of the soil, feeding on plants just above or below ground level. Beet, brassicae, carrots, celery, lettuce, swedes and turnips are attacked, plants being either cut off at ground level or large holes eaten into the roots. Most damage occurs in June and July and to crops just after transplanting or thinning out.

Cutworms are usually much more common in crops grown after a heavy weed cover, than in those on land that has been kept free of weeds. Dusts and sprays containing DDT are effective, especially if applied when the caterpillars are young and under moist conditions during the late afternoon or in the evening. As a dust, use DDT at $2\frac{1}{2}$ lb a.i. (50 lb 5% dust)/acre. As a spray, use DDT at 1 lb a.i. (3 pints 25% e.c.)/acre; apply, in 40–100 gal water/acre, along the rows; it is beneficial to hoe or harrow afterwards.

Poison bait of bran mixed with DDT broadcast during the later afternoon or evening will check cutworm attacks. To prepare the bait mix 4 oz DDT (8 oz 60% w.p.) with 28 lb bran; add $1\frac{1}{2}$ gal water and mix thoroughly; broadcast at 30–40 lb/acre. A bait of Paris green, 1 lb/28 lb bran, can also be used.

**Leatherjackets** (*Tipula* spp.) are the larvae of crane flies. They are normal inhabitants of the soil in damp grasslands but will feed on and damage a wide range of agricultural and horticultural crops. Most damage occurs after the ploughing-up of grassland or leys and the worst attacks usually occur in spring. Thinly sown or transplanted crops are usually those most affected. The larvae feed at or just below the soil surface. Early ploughing of grassland or leys (July or early August) will prevent egg-laying. Consolidation of the soil and the preparation of a good tilth will assist the crop to resist attack. For chemical control, see Chapter 5, p. 98.

**Millepedes.** Several kinds of millepede damage crop plants; they feed on live or dead vegetable matter in the soil and the roots, tubers, etc., of most kind of plants are eaten. Beans, particularly the germinating seeds, carrots, peas, potatoes and strawberries are especially subject to injury. Millepedes occur in most soils but, where they are numerous and causing damage, they are difficult to control. Surface applications of pesticides are apt to be ineffective as the millepedes rarely come above the soil surface. DDT *or* gamma-BHC dust or spray controls millepedes and should be applied to the open seed drills or worked into the soil before planting. Gamma-BHC should not be applied when there is a taint hazard (see Chapter 2 and also Chapter 6).

*Swift moths* (*Hepialus* spp.). The caterpillars of the garden swift moth and the ghost swift moth feed on the roots and rhizomes of many plants and may cause much damage to nursery and market garden crops. Strawberries, anemones, flax, Michaelmas daisies and hops are perennials particularly affected, and a wide range of annuals, including lettuce, carrots, parsnips, swedes and red beet may be damaged, especially if planted on infested land. Periodic digging of the soil, control of perennial weeds and the lifting and replanting of bulbous and herbaceous plants at fairly frequent intervals are all means which reduce caterpillar populations in the soil. The chemical control of the pest in established crops is difficult for the larvae may be deep in the soil. Promising results have been obtained with an emulsion containing DDT at $0.05\%$ ($1\frac{1}{2}$–2 pints $25\%$ e.c.)/100 gal, applied as a drench at $\frac{1}{4}$ pint per plant.

**Wireworms** (*Agriotes* spp.) See Chapter 5, p. 100.


DISEASES

***Damping-off diseases.*** Germinating seedlings of many vegetables can be attacked by species of *Pythium* and *Phytophthora*, either before or after emerging above the soil. The damage caused by these fungi is usually most severe under conditions when germination is slow, for example, early in the season or in cold wet soils. Considerable protection against such soil-borne pathogens can be obtained by dusting the seed with captan *or* thiram, either alone or mixed with insecticidal seed dressings. The dressings normally contain 50–75 per cent a.i. and are used at a rate sufficient to give a good cover of the seed. Recommended rates vary, according to the vegetable, from $2\frac{1}{2}$–9 oz/cwt seed.

*****Seedborne fungal diseases.** Several fungal pathogens of vegetables and other crops are carried within the seeds and cannot be eliminated by dust applications. Previously, treatment was by immersing the seeds in hot water but a new 'Thiram soak treatment' has several advantages and is being widely used on certain vegetable seeds (e.g. celery, beet, brassicas – *q.v.*). The seeds are enclosed in mesh bags and immersed for 24 hours in a suspension of thiram containing 2 lb a.i. ($2\frac{1}{2}$ lb 80% w.p. or 3 pints 50% colloidal suspension)/100 gal maintained at 30°C (86°F). The seeds imbibe water and should be dried after treatment in a stream of cool or slightly warmed air. Up to 50 lb seed can be treated in 100 gal suspension but for small quantities (up to $1\frac{1}{2}$ lb) a 2 gal bucket can be used, the temperature being maintained by an aquarium heater and thermostat. In the larger baths the suspension is circulated and the thiram thereby deposited on the seeds also gives protection against damping-off diseases.

The thiram soak treatment does not affect germination of most vegetable seeds but some varieties or samples of brassicas and lettuce may be sensitive. Before treatment of large bulks of seeds whose reaction is unknown, a pilot test should be made. Further information can be obtained from the National Vegetable Research Station, Wellesbourne, Warwick.

## 11.2 SPECIFIC CROPS

### BROAD BEAN

PESTS

*****Black bean aphid** (*Aphis fabae*). This aphid is a common and often serious pest of field and broad beans. Heavy infestations may also occur on sugar and red beet, spinach, mangolds and French and runner beans. On wild hosts it occurs on fat hen and thistles and it overwinters in the egg stage on the spindle tree (*Euonymus europaeus*). On beans, heavy infestations on the leaves and stems cause stunting, leading to a marked reduction in yield.

Autumn or early spring sowing of beans allows the plants to become well-grown and usually in flower before the attack starts. The removal of the plant tips when infestation has just begun usually reduces the subsequent attack. Crops may become heavily infested in June, July or August and should be treated with insecticides before this occurs. For spraying use demeton-S-methyl at 3·8 oz a.i. (6 fl. oz 58% e.c.)/acre; *or* dichlorvos at 8 oz a.i. (16 fl. oz 50% e.c.)/acre; *or* dimethoate at 5 oz a.i. (12 fl. oz 40% e.c.)/acre *or* formothion at 7 oz a.i. (16 fl. oz 43% e.c.)/acre *or* malathion at 18 oz a.i. (30 fl. oz 60% e.c.)/acre *or* mevinphos at 2–3 oz

(99%)/acre *or* nicotine at 8 fl. oz (95–98%)/acre *or* oxydemeton-methyl at 3·4 oz a.i. (6 fl. oz 57% e.c.)/acre *or* phosphamidon at 3·2 oz a.i. (16 fl. oz 20% s.c.)/acre. Apply in 40–100 gal of water as a fairly coarse spray along the rows. Nicotine is not effective at low temperature; 65–70°F is preferred. Observe the periods: 1 day for malathion and dichlorvos, 2 days for nicotine, 3 days for mevinphos, 7 days for dimethoate and formothion and 21 days for demeton-S-methyl, oxydemeton-methyl and phosphamidon which must elapse between application and harvest when the crop is for direct human consumption. Sprays containing derris and pyrethrum may be preferred to nicotine for small-scale use as they are less hazardous.

Nicotine dusts, if applied under warm dry conditions, can give satisfactory results. Depending on plant size and density, 50 lb to 200 lb/acre of a dust containing 4% nicotine is required.

Alternative treatments include: menazon at 3 oz a.i. (8 fl. oz 40° e.c.)/40–100 gal/acre may be used on broad beans, at least 21 days before harvest; schradan at 12 oz a.i. (30 fl. oz 40% s.c.)/40–100 gal/acre is satisfactory for use on field beans only but must not be used after mid-September (if applied from April to July, the interval before harvest must be at least 4 weeks, if applied August to mid-September at least 6 weeks). Disulfoton granules (7·5% a.i.) at 14 lb/acre *or* phorate granules (10% a.i.) at 10 lb/acre on broad beans before flowering are most effective. They are more persistent than sprays and are less toxic to pollinating insects. The granules are applied along the rows as a band over the plants; there must be an interval of at least 6 weeks before the crop is harvested.

FRENCH AND RUNNER BEANS

PESTS

****Bean seed fly** (*Delia platura*). The larvae of this insect tunnel in the germinating seeds and in the young stems of French and runner beans and may cause serious crop losses by killing or stunting the plants. The damage is usually worst on crops sown early and when germination is slow. The fly is attracted to decaying plant material and it is becoming increasingly common in market gardening areas.

Seed dressings based on dieldrin are very effective and economical to use. Use 1 oz dressing (containing 75% dieldrin)/28 lb seed. Mix dressing and seed thoroughly before sowing.

****Black bean aphid** (*Aphis fabae*). Heavy infestations of this aphid often develop on French and runner beans during July and August and the black smothering colonies cause stunting, flower drop and malformation of the pods. Crops should be treated with insecticides before they become heavily infested and to avoid harm to bees should not be sprayed when in flower. For chemical control see p. 277 excepting schradan and phorate which are not recommended for use on French or runner beans. Take into account crop maturity when choosing chemicals so that the necessary period between treatment and harvesting occurs. Disulfoton granules at the rate given for broad beans must be applied at sowing time, 1 in below seed. They should not be used on light sandy soil.

BEETROOT (RED BEET)

PESTS

***Mangold fly** (*Pegomya betae*). The larvae of the mangold fly feed and cause large blisters in the leaves of sugar beet, red beet, mangolds and spinach. The damage is most serious when the plants are small but larger plants may be quite heavily attacked with little effect on the final yield of roots. On spinach, however, considerable reduction in the market value of the crop can follow even a moderate attack. Crops can be helped to grow away from an attack by providing a good tilth for germination and by top dressing with a nitrogenous fertilizer. If an attack develops in the seedling stage, singling should be delayed. Good control may be obtained by spraying with DDT at 12 oz a.i. (3 pints 20% e.c.)/acre *or* trichlorphon at 6 oz a.i. (8 oz 80% s.c.)/acre. Apply in 20–100 gal water and direct spray on the rows.

The above insecticides are not advised when both mangold fly and aphids are present. A satisfactory control of both pests with demeton-S-methyl at 4·6 oz a.i. (8 fl. oz 58% e.c.)/acre *or* dimethoate 3 oz a.i. (7 fl. oz 40% e.c.)/acre *or* formothion 2 oz a.i. (5 fl. oz 43% e.c.)/acre *or* oxydemeton-methyl at 4·6 oz a.i. (8 fl. oz 57% e.c.)/acre applied in about 50 gal water. Although these materials are probably not quite as effective against mangold fly as trichlorphon when the pest is abundant and the larvae are large, they are more effective than DDT.

If one of the above insecticides is used on spinach, account must be taken of the time which should elapse between application and harvest.

**Black bean aphid** (*Aphis fabae*). This aphid is frequently common on beet causing malformation of the leaves and stunting of the crop. The aphid also spreads sugar beet virus yellows and certain viruses of beet and mangolds. With the exception of nicotine, dichlorvos and malathion (which are unlikely to be as effective as systemic materials), use any of the insecticides given on p. 277 for this pest and observe the same periods between application and harvest.

DISEASES

**Black leg** (*Pleospora betae*). The stems of young seedlings become blackened and shrivelled. The disease is seed-borne and may be controlled by organomercury seed treatment with dusts containing 1–1·5% metallic mercury at 12 oz/cwt seed (dry) *or* 4 fl. oz/cwt seed (wet) *or* with slurries containing 0·6–2·0% metallic mercury (generally 4 fl. oz/cwt seed). The thiram soak treatment (p. 277) gives complete control of seed-borne infection accompanied by an improvement of germination.

***Silvering** (*Corynebacterium betae*). This bacterial disease affects the seed crop, especially of Cheltenham Green Top. The leaves on seeding plants become silvery and the plants then wilt and die. The pathogen is seed-borne and may be controlled by soaking the seed for 24 hours in a solution containing 200 ppm streptomycin as the sulphate. Dry the seed and store until required. Less effective yet giving an appreciable control is seed treatment with an organomercury dressing at the rate of 1 oz/5 lb seed.

BRASSICA CROPS

PESTS

*****Cabbage aphid** (*Brevicoryne brassicae*) is a mealy aphid which infests the leaves and shoots of many cruciferous crops and frequently causes severe damage to Brussels sprouts, cauliflowers, cabbage, kale and swedes. The aphid overwinters as eggs and sometimes as adults on brassica plants which, if kept for seed, may become heavily infested by early summer and serve as the main source of aphids for spread to spring and summer sown crops.

Methods of control vary according to the stage of development and the variety of the crop. Insecticidal sprays are usually more effective than dusts and should be applied at 60–150 gal/acre, depending on crop size and density. Enough spray should be applied just to wet the foliage and an additional wetting agent may be required to ensure wetting of the waxy aphids and foliage. On Brussels sprouts, in August and September, pendant lances should be used to control aphids on the lower leaves and sprouts. The following insecticides have proved effective: demethion at 4·8 oz a.i. (16 fl. oz 30% e.c.)/acre; *or* demeton-S-methyl at 4·6 oz a.i. (6 fl. oz 58% e.c.)/acre; *or* dimethoate at 6 oz a.i. (15 fl. oz 40% e.c.)/acre; *or* formothion at 10 oz a.i. (25 fl. oz 43% e.c.)/ acre; *or* malathion at 18 oz a.i. (30 fl. oz 60% e.c.)/acre; *or* mevinphos at 2–3 oz (99%)/acre *or* nicotine at 8 fl. oz (95–98%)/acre; *or* phosphamidon at 4 oz a.i. (20 fl. oz 20% s.c.)/acre; *or* schradan, on Brussels sprouts at 32 oz a.i. (4 pints 40% s.c.)/acre, on other brassicae at 24 oz a.i. (3 pints 40% s.c.)/acre. Demeton-S-methyl and schradan appear to be the most effective of these insecticides. Edible crops must not be sprayed with schradan after mid-September and the least period between application and harvest for schradan is 4 weeks for crops sprayed between April and July and 6 weeks for crops sprayed in August or early September. For the other insecticides this time period is, for malathion 1 day, nicotine 2 days, mevinphos 3 days, dimethoate and formothion 7 days, demethion, demeton-S-methyl and phosphamidon 21 days; demeton-S-methyl should not be applied after October.

Nicotine dusts can give satisfactory results if applied under warm dry conditions; 50–100 lb dust (4% nicotine) is required per acre. Disulfoton granules provide good control on Brussels sprouts, cabbages and cauliflower at $1\frac{1}{4}$ lb a.i. (17 lb 7·5% granules)/acre, either as a foliar dressing *or* a soil application at planting time. A second application (foliar) may follow the first (foliar or soil) at not less than 4 weeks. The minimum interval between application and harvest is 6 weeks. An alternative treatment is the use of a soil drench of menazon, 1 lb 70% w.p. in 50 gal water applied at $\frac{1}{10}$ pint to the base of the plant within 4 days of transplanting. Menazon may also be used as a root dip, but not for broccoli or cauliflower; the plants are dipped, for 10 seconds before planting out, in a paste of 1 lb 70% w.p. in 2 gal water, enough for 10,000 plants. Menazon can be combined with dieldrin as a drench or root dip for the control of both aphid and cabbage root fly. For continuous protection follow up the menazon treatment with a spray of menazon at 6 oz a.i. (15 fl. oz 40% e.c.)/100 gal water/acre, 8 weeks after transplanting and again, if necessary, 3 weeks later. At least 21 days must elapse between the last spray and harvest.

**Cabbage moth** (*Mamestra brassicae*). Caterpillars of this moth may cause damage to brassica crops over the period June to October. Cabbage suffer most for the caterpillars eat into the hearts which they spoil. The treatments recommended for the control of cabbage white caterpillars (see p. 283) give control. Mevinphos at 4 fl. oz (99%) in 50–100 gal water/acre is the most effective treatment but the crop should not be harvested until at least 3 days after spraying.

***Cabbage root fly** (*Erioischia brassicae*) occurs throughout Great Britain and is a serious pest of cabbage, cauliflower and Brussels sprouts. The larvae feed on the roots which they may partially or completely destroy. Attacked plants are stunted or they may collapse and die. The larvae also tunnel into fleshy roots of swedes, turnips and radishes, causing loss of yield and quality. There are two or three overlapping generations in a year and plants may be attacked at any stage of growth. Methods of control vary according to the stage of development and type of crop. Plants and seedlings may be attacked at any time between mid-April and September.

The development of strains of cabbage root fly resistant to aldrin, dieldrin and BHC has largely limited the usefulness of these insecticides to the northern parts of the country. In areas where this resistance occurs, organophosphorus insecticides give good control and should be used. Control measures using aldrin, dieldrin and BHC are dealt with first.

To give control in seedbeds, either seed treatment or soil treatment may be used. For the former, add 1 fl. oz paraffin (kerosine) to 1 lb seed, mix well and shake thoroughly with 1 oz gamma-BHC seed dressing (75%). This method should not be used on radish or turnip because of the risk of taint. For soil treatment, apply aldrin dust ($1\frac{1}{4}$% a.i.), at $\frac{1}{2}$ oz per 10 yds of row, to the seed drill before covering the seed. Alternatively, aldrin, *or* dieldrin can be worked into the nursery seedbed before drilling; aldrin at $2\frac{1}{4}$ lb a.i. (6 pints 30% e.c.) in 20–50 gal water/acre *or* dieldrin at 1 lb a.i. ($5\frac{1}{4}$ pints 15% e.c. *or* 2 lb 50% w.p.) in 20–50 gal water/acre. For post-emergence treatment, the rows should be sprayed or dusted at the second or third true leaf stage with either aldrin at 1 lb dust (1·25%)/60-yd row *or* spray at 0·6 oz a.i. (2 fl. oz 30% e.c.)/10 gal water *or* dieldrin at 0·45 oz a.i. (3 fl. oz 15% e.c.)/10 gal water *or* 1 oz (50% w.p.)/10 gal water. The sprays should be applied at 10 gal/400-yd row.

To plants in the field, the following treatments are effective: at transplanting, dip the roots and lower part of the stem for about 10 seconds immediately before transplanting in a dieldrin suspension 1 oz a.i. (2 oz 50% w.p.)/10 gal water. Do not let the roots dry out. This dip treatment should give protection for at least 1-month unless the attack is severe; it may cause some check, especially to cauliflowers, but the use of the wettable powder is less likely to cause a check than the use of e.c. formulations.

After transplanting, a drench applied to the soil at the base of the plant gives excellent control. The drench may be of aldrin at 6 oz a.i. (20 fl. oz 30% e.c.)/100 gal water *or* dieldrin at $4\frac{1}{2}$ oz a.i. (30 fl. oz 15% e.c.) *or* 10 oz (50% w.p.)/100 gal water. Each plant should receive about $\frac{1}{8}$ pint of the drench.

For potted plants, such as early cauliflower, the pots may be watered with aldrin or dieldrin suspensions, at the concentrations given in the above paragraph. Do not use more than 2 gal per 250 plants and plant out as soon as possible.

No seed treatment with an organophosphorus insecticide can be recommended but chlorfenvinphos at 16 oz a.i. (4 pints 24% e.c.) in 20–50 gal water/acre sprayed on the soil surface and worked into 4 in will control the pest in nursery seedbeds. For direct sown crops drill through a 3–6 in band of chlorfenvinphos granules at 1 oz a.i. (9·2% granules)/44 yd row *or* spray a band of chlorfenvinphos at 0·8 oz a.i. (3·2 fl. oz 24% e.c.)/1 gal/ 300 yd of row; with swedes and especially turnips these treatments carry some risk of phytotoxicity.

For transplants from seedbeds, root dips of organophosphorus insecticides cannot be safely recommended but these materials can be very effective when applied either as liquid or granules to the soil at the base of the plant within 4 days of transplanting. For liquid application, use azinphos-methyl at 8·8 oz a.i. (2 pints 22% e.c.)/ 100 gal *or* chlorfenvinphos at 7·2 oz a.i. (1½ pints 24% e.c.)/100 gal *or* diazinon at 6·4 oz a.i. (1 lb 40% w.p.)/100 gal, giving each plant ⅛ pint of drench, *or* thionazin at 5 oz a.i. (½ pint 46% e.c.)/100 gal, giving each plant 1/10 pint of the drench. When applying granules, use chlorfenvinphos at 1 lb granules (9·2% a.i.)/2,500 plants *or* diazinon at 1 lb granules (5% a.i.)/500 plants *or* mecarbam at 1 lb granules (4% a.i.)/500 plants *or* thionazin at 1 lb granules (10% a.i.)/2,000 plants. There is no recommended treatment with organophosphorus insecticides for cauliflowers or other brassicas in pots or soil blocks.

Cabbage root fly larvae sometimes occur in the buttons of Brussels sprouts and for this, trichlorphon at 1 lb a.i. (1¼ lb 80% s.p./50–100 gal/ acre gives a good control. Apply three sprays at 7-day intervals, the first 1 month before the anticipated harvest. This may be as late as mid-September for crops harvested up to the end of October. Obtain a good cover of the bottom half of the plant by using pendant lances.

**Cabbage stem weevil* (*Ceutorhynchus quadridens*). This insect infests spring sown brassica plants, especially those in seed beds. Attacks occur from April to July and the larvae mine in the petioles and stems. Plants are killed or stunted. The stems of infested plants may be spongy and snap readily on transplanting. Good control has been obtained with gamma-BHC applied either as a seed dressing at 1 oz (containing 75% a.i.)/lb seed or by applying a gamma-BHC spray at ⅓ oz a.i. (⅔ oz 50% w.p.)/10 gal water in a band along the row at 1 gal/30 yd of row when the seedlings are in the second or third true leaf stages.

****Cabbage white butterflies* (*Pieris* spp.). The caterpillars feed on many kinds of brassicas, on stocks and nasturtiums and on cruciferous weeds. Most damage is caused to brassica crops from July to September, the caterpillars eating the leaves and fouling the crop with excrement. Dusts and sprays containing one of the following insecticides, applied to the leaves and hearts of the plants, have proved effective. As the foliage of brassica crops is difficult to wet, it may be necessary to use additional wetter. The dusts are DDT at 2½ lb a.i. (50 lb 5% dust)/acre *or* derris dust (0·2–0·5% rotenone) at 50 lb/acre. The sprays are DDT at ½ lb a.i. (2 pints 20% e.c. *or* 1 lb 50% w.p.)/acre; *or* mevinphos (99%) at 4 fl. oz/acre; *or* nicotine (95–98%) at 8 fl. oz/acre; *or* trichlorphon 12 oz a.i. (1 lb 80% s.c.)/acre. When spraying, the insecticide should be applied in 40–100 gal water/acre, the higher volumes usually giving the better results. DDT emulsions are usually more effective and persistent than DDT dusts or w.p. formulations. The nicotine spray requires the addition of a wetter and is effective only against young caterpillars. At least 14 days should elapse between the application of DDT and harvest, 3 days in the case of mevinphos and 2 days in the cases of nicotine and trichlorphon.

***Diamond-back moth* (*Plutella xylostella*). The caterpillars of this moth feed on the leaves of many brassicae, including cabbages, cauliflowers, Brussels sprouts, swedes and turnips. Although they normally cause little damage, outbreaks occasionally occur during July and August especially in coastal districts. These are associated with the migration of moths from the Continent. Dusts or, preferably, sprays of DDT, applied principally to the undersides of the leaves, give effective control. For dusting, use either DDT at $2\frac{1}{2}$ lb a.i. (50 lb 5% dust)/acre *or* derris at 2–3 oz a.i. (28–56 lb 0·2–0·5% rotenone)/acre; the latter is effective but is less persistent than DDT. As a spray, use DDT at 1 lb a.i. (4 pints 20% e.c. or equivalent w.p.) in 40–100 gal water/acre. Trichlorphon at 18 oz a.i. ($1\frac{1}{2}$ lb 80% s.c.)/40–100 gal water/acre also gives good control. The minimum time intervals between insecticide application and harvesting are as for cabbage white butterflies.

***Flea beetles** (*Phyllotreta* spp.) often cause serious damage to newly-emerged seedlings of cabbage, cauliflower, kale and other brassica crops. They eat holes in the leaves and stems and may check or even destroy the young plants. Most damage is caused in late April and in May; crops sown before early April or after the end of May usually suffer only slight damage. They may be controlled by seed dressings and post-emergence sprays and dusts.

A seed dressing based on gamma-BHC is effective and economical to use. It protects the seedlings during emergence and in the cotyledon stage; if an attack develops later a dust or spray can be applied. For moderate or heavy attacks, use $\frac{3}{4}$ oz dressing (containing 75% gamma-BHC)/lb seed; for light attacks use $\frac{1}{2}$ oz dressing/lb. When the higher rate of dressing is used, mix $\frac{1}{4}$ fl. oz paraffin (kerosine) with each 1 lb seed before mixing with the dressings. Seed treated with the paraffin sticker should be sown within 1 week. To avoid injury to the seedlings, gamma-BHC dressings should not be used if the seeding rate exceeds 10 lb/acre when drilled or 20 lb/acre when broadcast. Radish and turnips for early bunching should not be treated with gamma-BHC because of the risk of taint.

Post-emergence sprays and dusts should be applied when the seedlings begin to emerge through the soil; re-apply, if necessary, 7–10 days later. Use gamma-BHC at 4–6 oz a.i. (1 lb 25% w.p. or 50 lb 0·65% dust)/acre. The gamma-BHC spray may be applied to the drills in a band or as an overall spray at 20–80 gal water/acre. The use of gamma-BHC is restricted by risks of taint (see Chapter 2, p. 31). DDT, used at $\frac{1}{2}$–1 lb a.i. (1 lb 50% w.p. or 3 pints 20% e.c. or 50 lb 2% dust)/acre is also very effective.

**Turnip gall weevil** (*Ceutorhynchus pleurostigma*). The rounded galls of this weevil may occur on all cultivated brassicae and on mustard and charlock. They occur on the root just below soil level and can be distinguished from the swellings caused by clubroot as each gall contains a maggot or the cavity caused by it. Most of the damage arises in the seedbed and badly galled plants may be seriously checked. Gamma-BHC seed dressing 75% a.i. used at ¾ oz/lb seed will reduce the attack. Well-established plants usually suffer little but swedes and turnips lose value if they are galled. Effective control on transplants can be obtained with gamma-BHC at 3–4 oz a.i. (¾ lb 25% w.p.)/100 gal water. Apply $\frac{1}{10}$ pint to the base of the plant in May within 4 days of setting out. The use of aldrin as for the control of cabbage root fly (see p. 282) will also control the weevil. A promising control of the weevil has been obtained with dusts containing a low percentage of gamma-BHC, applied as a band about 2 in wide along the rows of seedlings when they have two rough leaves. Similar dusts applied to newly transplanted cabbages at the rate of 1 lb/100 plants have proved effective, but both recommendations are tentative. There is a risk of root crops following this treatment being tainted.


DISEASES

*Canker (*Phoma lingam*) is a seed-borne disease which causes light brown or purple lesions on the stems of Brussels sprouts and other brassicas resulting in a stunting of the plants. Seed infection can be controlled by the thiram soak treatment (p. 277).

***Clubroot (*Plasmodiophora brassicae*) (A.L. 276), appears as swellings on the roots which eventually decay and the plants often wilt and die. It is particularly severe on summer crops and may be confused with turnip gall weevil (round swellings containing larvae) or with damage by 2,4-D or MCPA (galls on stem base). Dipping the roots of transplants into a suspension of 2 oz calomel in 1 pint of water immediately before planting effects control; some insecticidal chemicals may be added for cabbage root fly control (see p. 252). Alternatively, the dibble holes may be treated with a heaped teaspoonful of 4% calomel dust *or* with corrosive sublimate. The latter is extremely poisonous but to use it dissolve 1 oz in a little methylated spirit and add water to make 12½ gal; add ¼ pint per dibble hole twice before planting. Calomel may also be used by applying ⅙ oz (4 oz 4% dust) per 5-yard row.

In addition to applying one of these chemical treatments, ensure that the ground is well drained and has a high lime status.

**Damping-off and wirestem (*Rhizoctonia solani*) is a dark brown or black rot on the stem base of seedlings, especially of cauliflower, usually causing a pronounced constriction of the stem. The seedlings often die but may survive as stunted plants with a typical wirestem appearance. Control is achieved by raking into the soil surface before sowing the seed, quintozene at 2 oz a.i. (9–12 oz 20% dust)/10 sq. yd. Do not treat soil where cucumber, marrows, melons or solanaceous plants, including tomato but not potato, are to be grown.

**Downy mildew** (*Peronospora parasitica*). This disease may be trouble-some in the seedling stage, especially on cauliflower in Dutch lights, causing stunting or death of young plants. Spray with dichlofluanid at 12 oz a.i. (1½ lb 50% w.p.)/100 gal/acre, starting as soon as the first seedlings emerge and subsequently every 3–4 days for 10 days. Continue at weekly intervals until the seedlings have fully-grown true leaves, using higher volumes if necessary.

***Leaf spot** (*Alternaria brassicicola*) can cause a severe disease of brassica seed crops in wet seasons when the fungus readily invades the developing pods and causes considerable loss of seed. The disease is seed-borne and also gives rise to a damping-off of young seedlings. Seed infection can be largely controlled by the thiram soak treatment (see p. 277).

CARROT

PESTS

*****Carrot fly** (*Psila rosae*) is a widespread and often serious pest of carrots and, in some districts, causes damage to celery. Parsnips and parsley are also often attacked. The larvae feed on the roots and young plants may be stunted or killed. The mature roots of carrots and parsnips may be made unmarketable by the tunnelling of the larvae. There are two generations in a year, the first attacking crops in May and June, the second in August and September. Growing crops away from hedges or other shelter, sowing carrots in June to escape from the first generation, not growing susceptible crops frequently in the rotation, are all ways of reducing attack.

On carrots, a seed dressing containing gamma-BHC usually gives adequate protection on mineral soils where only light or moderate attacks are usually experienced. Use at the rate of ⅔ oz gamma-BHC (75%)/lb of seed. On soils of high organic content such as fen soils the treatment is less effective but where carrot fly attack is light sufficient protection on early and late sown carrot may be obtained by using the dressing at ⅔ oz/lb of seed. On maincrop carrots grown under similar conditions use 1 oz of gamma-BHC (75%)/lb of seed. These dressing rates should not be exceeded or off-flavour of carrots may occur.

Aldrin and dieldrin are not recommended for the control of carrot fly but several organophosphorus insecticides can be used effectively, though complete control of late attacks on fen soils may not always be achieved. For carrots grown on mineral soils, chlorfenvinphos granules or spray, diazinon or mecarbam granules control this pest and granules of disulfoton or phorate give combined control of carrot fly and carrot-willow aphid (see below). For fen soils use diazinon, disulfoton or phorate granules.

Chlorfenvinphos can be broadcast either as granules at 2 lb a.i. (20 lb 9·2% granules)/acre *or* sprayed at 1¾ lb a.i. (7 pints 24% e.c.)/acre and harrowed into the soil before drilling the crop. Diazinon granules at 2 lb a.i. (40 lb 5% granules)/acre on mineral soils and 2½–3 lb a.i. (50–60 lb 5% granules)/acre on fen soils can be broadcast and rotavated into the top 3 in but are more effective when drilled as a 1-in band about 1 in below the soil surface directly beneath the seed. The application rate for this treatment should be at 1½ lb a.i. (30 lb 5% granules)/acre on mineral soils and 2 lb a.i. (40 lb 5% granules)/acre on fen soils for 15-in rows (11,616 yds/acre) and should be correspondingly adjusted for other widths of row. Alternatively, for this placement treatment on mineral soils, use mecarbam at 1¼–2 lb a.i. (32–48 lb 4% granules)/acre according to row width (*i.e.* 1 lb 4% granules/300 yd row) on main crop carrots. Reduce the rate to 1 lb mecarbam 4% granules/375 yd row on early and late carrots.

Disulfoton granules at 16·8 oz a.i. (14 lb 7·5% granules)/acre on mineral soils and 31·2 oz a.i. (26 lb 7·5% granules)/acre on fen soils *or* phorate granules at 1½ lb a.i. (15 lb 10% granules)/acre on mineral soils and 2½ lb a.i. (25 lb 10% granules)/acre on fen soils, should be applied as a 3-in band immediately in front of the seed coulter. These rates, which are for 15-in rows, should be adjusted proportionately for other row widths. For carrots grown in close rows, the insecticides should be broadcast and worked in, disulfoton granules at 2 lb a.i. (26 lb 7·5% granules)/acre on mineral soils only *or* phorate granules at 2 lb a.i. (20 lb 10% granules)/acre on mineral soils and 3 lb a.i. (30 lb 10% granules)/acre on fen soils.

For parsnips on mineral soils, use disulfoton *or* phorate granules at the same rates and in the manner as for carrots.

On parsley, grown on mineral soils, an effective control has resulted from the use of a seed dressing containing 75% gamma-BHC at the rate of 1½ oz/lb seed.

**Carrot-willow aphid** (*Cavariella aegopodii*). Heavy infestations of this aphid often occur from late May to early July and cause considerable loss of yield in early and mid-season crops. The aphids infest carrots in the cotyledon stage, as well as older plants, and spread the carrot motley dwarf virus which produces a yellow mottling of the leaves and stunts the plants. Parsnips and celery are also attacked by the aphid. As soon as any aphids are seen on the plants the rows should be sprayed with either demeton-S-methyl at 3·8 oz a.i. (6 fl. oz 58% e.c.)/40–100 gal water/acre; *or* dimethoate at 5 oz a.i. (12 fl. oz 40% e.c.)/40–100 gal water/acre; *or* oxydemeton-methyl at 3·4 oz a.i. (6 fl. oz 57% e.c.)/40–100 gal water/acre. If, on carrots for bunching, it is necessary to control the aphid shortly before harvest, malathion or nicotine (see cabbage aphid) should be used, as the requisite interval between application and harvest is short with these two insecticides. Alternative recommendations are the application of phorate granules (10% a.i.) at 15 lb/acre *or* disulfoton granules (7·5% a.i.) at 14 lb/acre as soon as the crop has germinated, *or* 7·5 lb/acre of the disulfoton granules may be applied then followed by 7·5 lb/acre 3 weeks later. Both disulfoton and phorate also give a good control of the carrot fly on mineral soils.

DISEASES

***Black rot** (*Stemphylium radicinum*) causes black sunken lesions on the mature roots and can be troublesome in store. It can also cause loss of seed in the seed crop and a damping-off of seedlings. Seed-borne infection can be controlled by the thiram soak treatment (see p. 277).

***Leaf blight** (*Alternaria dauci*) can cause lesions on the leaves in wet seasons but is more important (especially in precision drilled crops) as a cause of seedling damping-off. Eradication of infection from the seed by the thiram soak treatment (see p. 277) gives a clean stand of seedlings.


CELERY

PESTS

****Carrot fly** (*Psila rosae*), for biological details see under carrots, p. 286. Several treatments are possible against this pest. The potting soil for soil blocks or boxes can be sprayed with diazinon at 1·2 oz a.i. (3 oz 40% w.p.)/4 pints water/4 cwt soil; mix thoroughly. Good results have been obtained by drenching seedlings in soil blocks or boxes with diazinon at 1 lb 40% w.p./1 gal/250 plants; this method also succeeds with seedlings in boxes when enough soil remains adhering to the roots at planting time. A root dip with gamma-BHC at 5% ($\frac{1}{2}$ pint 80% suspension/1 gal water) can control the pest but the roots must remain in the dip no more than 10 seconds and should not dry before planting; even so some check of growth may occur and, because of the risk of taint, avoid growing potatoes on the same land in the following year. Alternatively, the soil on the roots of seedling celery can be treated at the time of transplanting with insecticide granules from an applicator mounted (outside the operator's cab) on a mechanical planter. For this purpose, use diazinon at 40 lb 5% granules *or* disulfoton at 14$\frac{1}{2}$ lb 7·5% granules *or* mecarbam at 50 lb 4% granules to every 30,000 self-blanching plants or every 14,000 late celery plants. The last two treatments will also reduce damage by celery fly. Another method, which also gives some celery fly control, is to apply disulfoton granules at 16·8 oz a.i. (14 lb 7·5% granules)/acre for celery in wide rows and at 31·2 oz a.i. (26 lb 7·5% granules)/acre for self-blanching celery in narrow rows, to the furrow. The granules should be at a depth of not less than 3 in and the celery planted into this band. Mecarbam can be used instead of disulfoton at 2 lb a.i. (50 lb 4% granules)/acre on wide rows and double this rate on narrow row widths. The pest can also be effectively controlled by applying a liquid spot treatment within a few days of planting

out. For this purpose, use diazinon at 8 oz a.i. ($1\frac{1}{4}$ lb 40% w.p.)/100 gal
*or* mecarbam at 8 oz a.i. ($\frac{1}{2}$ pint 80% e.c.)/100 gal, applied at 100 gal/3,500
plants (about $\frac{1}{4}$ pint/plant). Alternatively disulfoton 7·5% granules *or*
mecarbam 4% granules can be applied to the plant bases within a week
of transplanting at the same rates as for the furrow application.

****Celery fly** (*Philophylla heraclei*). The larvae of this fly feed and cause
large blisters in the leaves of celery and parsnips. The damage is most
severe on celery, especially when the plants are small, although large plants
may be quite heavily attacked. Attacks may occur from May until October.
Sprays of DDT at $1\frac{1}{3}$ lb a.i. (4 pints 25% e.c.)/acre; *or* malathion at 18 oz
a.i. ($1\frac{1}{2}$ pints 60% e.c.)/acre, applied in 40–100 gal water/acre, have
been found effective. Dimethoate at 6 oz a.i. (15 fl. oz 40% e.c.)/40–100 gal
water/acre; *or* trichlorphon at 6 oz a.i. (8 oz 80% s.c.)/40–100 gal water/
acre have also been found effective.


DISEASES

*****Leaf spot** (*Septoria apii-graveolentis* and *S. apii*) (A.L. 241) causes
brown rusty spots on leaves and stems and spreads rapidly in cool damp
weather, causing much damage. The fungi are carried on the seed which
should receive the thiram soak treatment (p. 277) or be immersed in water
at 122°F for 25 minutes. If the disease is seen, the plants should be
sprayed with one of the following fungicides, repeated every 2–3 weeks:
Bordeaux mixture (10 : 12 : 100) or a copper fungicide (w.p. or colloidal)
at a rate giving $2\frac{1}{2}$ lb metallic copper/100 gal, H.V. or L.V.; *or* zineb at
21 oz a.i. (2 lb 65% w.p.)/100 gal *or* tank mix zineb (5 pints 22% nabam
solution *or* $1\frac{3}{4}$ lb 93% nabam plus $1\frac{1}{4}$ lb zinc sulphate to 100 gal water). If
water is not readily available, L.V. sprays or dusts may be used. The latter
should be applied more frequently than sprays and preferably in the early
morning when there is dew on the plants.


LETTUCE

PESTS

*****Aphids.** At least five species of aphids live on the leaves of lettuce
which they stunt, malform and contaminate by their cast skins and excre-
tions. All are found throughout the year on lettuce grown under glass but
only three species are serious pests of outdoor lettuce. Certain of these
aphids, besides directly damaging the plants, spread viruses which cause
severe stunting. Control measures should be applied early, not only so
that the crop is not infested when it begins to heart, but to reduce the
spread of virus when the plants are young. Sprays of demeton-S-methyl
at 3·4 oz a.i. (6 fl. oz 58% e.c.)/acre; *or* diazinon at $3\frac{1}{4}$ oz a.i. (16 fl. oz 20%

W

e.c.)/acre; *or* dimethoate at 5 oz a.i. (12 fl. oz 40% e.c.)/acre; *or* formothion at 8 oz a.i. (1 pint 43% e.c.)/acre; *or* malathion at 18 oz a.i. (30 fl. oz 60% e.c.)/acre; *or* nicotine at 8 fl. oz 95–98%/acre; *or* oxydemeton-methyl at 3·4 oz a.i. (6 fl. oz 57% e.c.)/acre; *or* TEPP at 1½ oz a.i. (4 fl. oz 40% conc.)/acre with wetter, have proved effective, though nicotine is less satisfactory at temperatures below 65–70°F. On summer lettuce, apply the insecticide in 80–100 gal water/acre, using a fairly coarse spray directed along the rows. Winter lettuce become infested in the seed-bed during autumn and, if treated in late October or November, will remain substantially free of aphids until cutting. The period which must elapse between application and harvest is, with malathion, 1 day; with nicotine and TEPP, 2 days; with dimethoate and formothion, 7 days; with diazinon, 14 days and with demeton-S-methyl and oxydemetonmethyl, 21 days.

Another method of control is to dip the seedlings, before transplanting, in a wash either of malathion, 1 fl. oz 60% e.c. in 3 gal water, *or* nicotine 1 fl. oz 95–98% in 9 gal water. Ten gallons of wash is enough to treat 6,000 plants, dipped in batches of 1,000 or less. Each batch is dipped in the wash for about 3 minutes, then removed and planted immediately. If the lettuce is not transplanted, spray with a malathion spray in November.

The lettuce mosaic virus is seed-borne and the risk of aphids spreading the virus can be much reduced by using virus tested seed.

**Lettuce root aphid** (*Pemphigus bursarius*). This aphid feeds on the roots causing yellowing of the foliage, stunting and, when the soil is dry, the death of the plant. It overwinters on Lombardy poplar, and in June and July, winged migrants fly to lettuce where their progeny infest the roots, increasing rapidly in numbers especially in hot weather. After mid-August, most of the aphids return to poplar but a few remain in the soil to infest subsequent crops of lettuce planted on the same ground in the autumn and following spring. Certain varieties, such as Salad Bowl and White Favourite, are resistant to attack. For lettuce sown between mid-April and early July, a satisfactory control can be obtained by treating the soil with diazinon at 1 lb a.i. (4 pints 20% e.c.)/acre. Spray the insecticide on to the soil in 30–50 gal water/acre and cultivate into a depth of 3 in before sowing. Lettuce sown at other periods usually escape attack and treatment is unnecessary.

DISEASES

**Downy mildew** (*Bremia lactucae*) appears as pale green or yellow angular areas on the older leaves which may bear numerous whitish spores of the fungus, especially on the lower surface. The areas later die and become brown. Under cool moist conditions, copious sporing may occur on fairly normal leaves. The disease is of most importance in the early autumn on outdoor lettuce and in late autumn on frame lettuce but also attacks overwintering lettuce, especially in frames. It may be followed by Botrytis or soft rot. For control of downy mildew, use zineb at 1¼ lb a.i. (2 lb 65% w.p.)/100 gal/acre to check the disease on field plantings and repeat every 3–4 weeks, *or* use thiram at 5 oz a.i. (6 oz 80% w.p.)/10 gal on seedlings in frames, a method which will also check Botrytis.

****Grey mould** (*Botrytis cinerea*). This disease is most serious under cool damp weather conditions and is therefore most prevalent in the spring on overwintered lettuce. The first sign of infection is often a complete collapse of the plant, caused by a basal stem rot, which may occur at any stage. On young plants it is known as 'red-leg' but sometimes the most severe attack occurs as the plants are approaching maturity. The fungus produces copious grey spores on decaying leaves and stems and also forms sclerotia which persist in the soil. The disease may start on the seedlings, sometimes following an attack of Rhizoctonia damping-off, and remain quiescent for many weeks before causing serious damage to the plant. Protectant treatments must therefore be applied from the seedling stage onwards and careful attention should be paid to efficient culture. Quintozene, applied at 1 oz 20% dust/yd², before sowing the seed or putting out transplants, checks attack by Rhizoctonia and has been shown to reduce Botrytis infection in frames, Dicloran, at 1 oz 20% dust/yd², applied before planting out in frames and followed by one or two further dustings at ¼ oz/yd² at 6-weekly intervals provides control but the last application must be made not less than 3 weeks before harvest. Seedlings grown in frames for spring planting may be sprayed with a 0·3% thiram suspension (6 oz 80% w.p./10 gal) every 3–4 weeks throughout winter; adequate wetting of the stem bases is essential. This treatment also checks downy mildew (*q.v.*). Promising results have also been obtained with dichlofluanid at 1 lb a.i. (2 lb 50% w.p.)/20–100 gal/acre, applied every 10 days to seedlings or after planting out.

***Ring spot** (*Marssonina panattoniana*) forms circular spots, ⅛–¼ in diameter, on the outer leaves and elongated spots resembling slug injury on the leaf midribs. The disease is rarely serious although rather disfiguring but is occasionally severe. The fungus is carried on seed and in plant debris. Bordeaux mixture (1% metallic copper), thiram (0·2% a.i.) *or* captan (0·2% a.i.) are reported to give some control if applied from the seedling stage onwards.

MINT

DISEASES

*****Rust** (*Puccinia menthae*). The first sign of infection is the production in spring of swollen distorted shoots bearing yellow pustules of spores.

Later infections appear as brown pustules on the undersides of leaves and the affected leaves eventually dry up and fall. No spray treatment will cure the disease but reductions in infection of second and third cuts from outdoor crops have been obtained by spraying the stubble of earlier cuts with sulphuric acid (B.O.V. full strength) at the rate of 24 gal/acre.

## ONION

PESTS

***Onion fly** (*Delia antiqua*). Maggots of the onion fly cause serious damage to seedling onions and plants which are bulbing-up; onion sets, leeks and shallots are sometimes infested. Treatment of the seed with a dieldrin seed dressing is an effective control measure. First mix the seed thoroughly with $\frac{1}{4}$ fl. oz paraffin (kerosine)/lb seed; then add the seed dressing at the rate of $\frac{1}{2}$ oz dieldrin ($\frac{3}{4}$ oz 75% dust)/lb seed and mix until the seeds are uniformly covered. Starch paste at 2 oz/lb seed *or* the methyl cellulose sticker described below under smut (p. 293) can be used instead of paraffin as the sticker.

Some control may be obtained by the use of gamma-BHC dust at 6 oz a.i. ($\frac{1}{2}$–2 cwt 0·2–0·65% dust)/acre applied as a band, about 3 in wide, along the rows. The first application should be made when the seedlings are in the 'loop' stage and a second two weeks later.

DISEASES

**Downy mildew** (*Perenospora destructor*). Pale oval areas appear on the leaves or the tips of the leaves may become pale and die back. The leaves often fold downwards at the infected area on which grey, later brown-purple, spores of the fungus may develop. These spread extensively in cool moist conditions. Other leaf moulds also grow on the lesion and may obscure the pathogen. The fungus grows internally and can infect the bulbs which, if kept for seed raising, produce stunted leaves. The flower stalks of seeding plants may bear a conspicuous pale oval lesion at which point the stalk often breaks. Oospores of the fungus can persist in the soil and infect new crops. The disease is also carried over in perennial onions and shallots and possibly in wild Allium species.

The chief control measures are to avoid contaminated land and to grow onions, where possible, on warm well-drained soils in sites with good air circulation. Overwintering hosts should be eradicated. Some control of leaf infection has been obtained by the application of zineb at $1\frac{1}{4}$–2 lb a.i. (2–3 lb 65% w.p.)/100 gal/acre, preferably before the appearance of the disease and then continued at intervals of 7–10 days.

***Smut** (*Urocystis cepulae*) A.L. 261). This disease is soil-borne. Germinating mycelium from the spore balls in the soil infects the bases of the leaves of young seedlings causing dark lead-coloured spots or streaks beneath the skin. These leaves later become thickened and twist or curl backwards, and further leaves or bulb scales may bear dark areas which later split exposing black spores of the fungus. Soil which has carried an infected crop should not be used for onion growing for as long a time as possible. When it is used, the disease can be reduced by applying form-aldehyde to the drill just after sowing but before covering the seed, using 1 pint commercial 40% formalin/16 gal water/800 yd row. In very wet weather the concentration of formalin should be doubled. When soil contamination is not heavy, good control can be obtained by pelleting the seed with thiram (80% w.p.) using a resin-alcohol or methyl cellulose sticker and applying 10 oz 80% thiram/lb seed. The resin sticker is prepared by dissolving 6 oz resin in 1 pint methylated spirit. Mix 1 fl. oz/lb seed thoroughly before adding the fungicide and then shake vigorously for 5 minutes. Alternatively, dissolve 1 oz methyl cellulose in 1 pint of warm water and use 3 fl. oz/lb seed. Stringent precautions should be taken against spreading infested soil on boots, implements, etc. The disease also occurs on leeks, shallots, chives and garlic.

**White rot** (*Sclerotium cepivorum*) (A.L. 62). The plants are stunted with yellow leaves, and the base of the plant is rotten and often covered with a white fungal growth in which the black resting bodies (sclerotia) of the pathogen may be embedded. The fungus is soil-borne and its sclerotia may persist for many years. The disease can be controlled to some extent by applying calomel at 1 lb 4% dust/50 yd of drill to the seed drill *or*, more effectively, by seed treatment with calomel. A methyl cellulose sticker should be used as in the smut treatment described above. The calomel is best applied at the rate of 1 lb undiluted calomel/lb seed, but a fair degree of control can be obtained by using 4 oz calomel/lb seed.

## PEA

PESTS

**Pea aphid** (*Acyrthosiphon pisum*). Periodically heavy infestations of this aphid occur on peas. It lives throughout the year on leguminous plants, overwintering as eggs, or occasionally as adults, on clovers, lucerne, trefoil and sainfoin. Pea crops can be attacked at any time from early May to autumn but those growing in June and July are generally the most seriously affected. Azinphos-methyl and especially DDT sprays used against pea moth will also control the pea aphid but if circumstances demand the use of separate control measures, DDT at 1 lb a.i. (3 pints 25% e.c.)/acre should be applied in 40–100 gal water as soon as 10% of the flower heads are aphid infested. Alternatively, any one of the insecti-cides, with the exception of phosphamidon and schradan, recommended for black bean aphid control (p. 277) may be used for controlling pea aphid. Apply at the same strength and rate as for black bean aphid and observe the same period between spraying and harvesting.

***Pea midge** (*Contarinia pisi*). Outbreaks of damage by pea midge sometimes occur, and although the small whitish, jumping larvae cause more injury by feeding on the flowers and growing points of pea plants they also infest and malform the pods. DDT spraying against pea moth or pea aphid usually also controls pea midge; but should the midge alone be troublesome spray DDT at 1 lb a.i. (3 pints 25% e.c. *or* 2 lb 50% w.p.) in 40–100 gal water/acre.

*****Pea moth** (*Laspeyresia nigricana*). The caterpillars of this moth feed inside the pods on developing peas greatly reducing their value as food or seed. The pest is most prevalent in the southern part of England, particularly in East Anglia and Kent. All varieties of peas are attacked but those with a long growing period suffer most. Quick maturing varieties sown early and those sown after the middle of June usually escape attack. Peas which come into flower during the flight period of the moth, between mid-June and mid-August, normally suffer most damage. This damage can be much reduced by spraying with azinphos-methyl at 4·4 oz a.i. (20 fl. oz 22% e.c.)/acre applied in no less than 60 gal water *or* DDT at $2\frac{1}{2}$ lb a.i. (1 gal 25% e.c. *or* 5 lb 50% w.p.)/acre applied in 120–140 gal water at medium to high pressure. With DDT, the emulsion formulation usually gives the better results. A careful timing of the spray is necessary. For crops which come into flower before the middle of June, spray in the third week of June; for those which flower between mid-June and mid-August, spray 7–10 days after the beginning of flowering: spraying is usually not necessary on crops flowering after mid-August. When using azinphos-methyl or when a higher level of control is required with DDT, as for peas for quick-freezing or canning, a second spray may be needed about 14 days after the first. The crop should not be harvested until 14 days have elapsed since the last spraying. Two sprays, timed as above, are usually necessary on dry harvested peas because they are exposdd to attack for a long period. Weather conditions may affect the time of appearance and the activity of the moths; advice on timing of sprays should be sought of the local N.A.A.S. officer.

***Pea and bean thrips** (*Kakothrips robustus*). This small insect feeds on the surface tissues of the young pods and foliage of peas and beans, causing silvery mottled patches and malformation of the pods. Heavy attacks may lead to severe stunting. Peas are particularly affected and the main attacks occur in June and July. An effective control is obtained by the application of a DDT spray at 1–2 lb a.i. (3–5 pints 25% e.c.)/120–140 gal water/acre. A DDT spray applied to control pea moth (see above) will also give adequate protection against pea thrips.

***Pea and bean weevils** (*Sitona* spp.). These weevils feed on a wide range of leguminous crops, the adults on the leaves, the larvae on the root nodules. The adults cause the greater damage and eat semi-circular notches in the leaf edges and, when large numbers are present, defoliation or destruction of the crop may occur. The adult weevils are active in April and May, most damage being done when conditions for plant growth are poor, as in cloddy soil or cold dry weather. Excellent control of the weevils is obtained with either DDT sprays at 10 oz a.i. (20 oz 50% w.p. *or* 2 pints 25% e.c.)/20–40 gal water/acre, *or* with DDT dust at 2½ lb a.i. (50 lb 5% dust)/acre; *or* with gamma-BHC dust at 6 oz a.i. (56 lb 0·65% dust)/acre. DDT emulsion persists longer on the foliage than w.p., which in turn persists longer than the dust.


### SPINACH

PESTS

****Black bean aphid** (*Aphis fabae*). This aphid causes the same type of damage as that on beet (see p. 277) and can be controlled by the same insecticides. Observe particularly the time which should elapse between application and harvest.
*****Mangold fly** (*Pegomya betae*)—see p. 279.


DISEASES

****Downy mildew** (*Peronospora effusa*) forms yellow patches on the upper surfaces of leaves accompanied by grey or violet-grey mould beneath the leaves. Badly affected leaves cease growth and may curl downwards at the edges. The disease can be very severe under moist conditions, on badly drained land and at low temperatures. The life history of the pathogen and the control of this disease have not been satisfactorily worked out and it is uncertain whether oospores carried on the seed or present in the soil serve as sources of infection of new crops. Such infections can arise early and the closeness of the affected leaves to the ground makes adequate protective spraying difficult. Protection of upper leaves can be obtained by fungicides such as zineb at 1¼ lb a.i. (2 lb 65% w.p.)/100 gal/acre. Some control has been reported from other countries by the use of copper fungicides.


### TOMATO (outdoor)

DISEASES
*****Blight** (*Phytophthora infestans*) can be severe in cool wet seasons, causing grey-brown lesions on the leaves and russet-brown marbled areas on the fruits, which become unusable. The disease is contracted from

infected potatoes and the control is the same as for potato blight (see Chapter 6, p. 117). Copper and dithiocarbamate fungicides all give good protection and one of these fungicides should be applied at 3-week intervals from the end of July, or earlier if potato blight is seen in the neighbourhood.

**Didymella stem rot** (*Didymella lycopersici*). A dark brown shrunken canker appears near the base of the stem and if this girdles the stem, the plant wilts suddenly. Later in the season similar cankers may appear on other parts of plants. There may also be brown spots on the leaves and black encrusted lesions on the stem ends of the fruits. Affected plants should be removed and burned as soon as possible and, at the end of the season, all wirework, canes, poles, etc., should be sterilized by immersion for 24 hours in 2% formalin (40%), or for 15 minutes in 5% formalin followed by wrapping in sacks for 1 week. The remaining plants in an infected area should be sprayed at the stem base with captan at 2·5 lb a.i. (5 lb 50% w.p.)/100 gal at 4–5 fl. oz per plant. In Jersey, a reduction in infection has been obtained by spraying plants with maneb (as for blight control) using $1\frac{1}{2}$–3 lb 70% w.p./acre L.V. Maneb has also been used satisfactorily, on outdoor tomatoes only, for stem base application using 4–5 fl. oz of a 1% suspension; this suspension is prepared by adding 1·4 lb of 70% w.p. to 10 gal water. This treatment can be applied to plants immediately adjacent to infected ones.

WATERCRESS

DISEASES

**Crook root** (*Spongospora subterranea* f. sp. *nasturtii*). The roots of the watercress are attacked by the swimming spores of the causal fungus which then grows in the cells of the roots. Affected roots are stunted, swollen and distorted and eventually decay. The plants become stunted and their leaves turn bluish or yellow. The disease is particularly severe from October to March. For control apply a solution of zinc sulphate, usually 20%, to the borehole water by a constant drip-feed apparatus to give a zinc concentration of $\frac{1}{10}$ ppm. The rate of flow of the water must be measured by means of a graduated weir and the application rate of the solution adjusted accordingly. Advice on procedure should be obtained from the local N.A.A.S. officer.

Alternatively, a powdered glass frit containing zinc may be used. Apply $\frac{1}{3}$ lb of the zinc frit per square yd to the top third or quarter of the cress bed near the water entrances after draining the bed. Wash the frit off the plants onto the floor of the bed. Re-admit water slowly. This method is for use on beds fed from several water sources and in which the more preferable zinc sulphate treatment cannot be applied.

# CHAPTER 12

# PEST AND DISEASE CONTROL IN OUTDOOR ORNAMENTALS AND IN TURF

The term 'outdoor ornamentals' is intended to mean those plants grown for their ornamental value whether trees, shrubs, herbaceous perennials or annuals. Although the number of different ornamental plants is large, not many can be considered of economic importance through selling regularly in large numbers. Many are purchased by amateur gardeners in small quantities, but they need to be raised by nurserymen on a fairly large scale and many require protection from pests and disease when in the nursery. Therefore, the following advice is meant primarily for nurserymen and large-scale growers, but it can also apply to small gardens where amateurs may meet with similar troubles. In any case, many of the smaller plants are grown from seed produced in Britain and it is important for people growing a commercial seed crop to know how to protect the seed heads from infection by various diseases. For this reason, some general advice on chemical control measures is given, based on experience with those used to combat similar pests and diseases affecting commercially-important crops.

Of the very large number of ornamental plants which are grown for sale to the public and for seed production it is only possible to mention a small proportion and to deal with the control of only the more important and destructive of the pests which attack them.

A number of important soil pests has been omitted here since they may attack almost every plant mentioned. These include cutworms, swift moth caterpillars, wireworms, chafer grubs, leatherjackets, millepedes, woodlice and slugs, all of which are dealt with in other chapters. The control of chafer grubs and leatherjackets in turf is, however, described, since the methods employed differ from those used on arable land.

The value of certain chemicals for preventing or controlling some kinds of diseases common to a wide range of host plants has been established by detailed experiments and large-scale usage under a variety of conditions; a summary of the present position is given in the following general guide:—

1. *Grey mould*—captan; thiram.
2. *Damping-off, foot rots and root rots*—soluble copper (Cheshunt compound) for treatment of soil used in seed boxes, etc., captan,

thiram; quintozene (PCNB) is very useful for controlling specifically *Rhizoctonia solani*.
3.  *Downy mildews and leaf spots*—copper; zineb.
4.  *Powdery mildews*—copper with white oil added; dinocap; sulphur.
5.  *Rusts*—maneb; thiram; zineb.

Detailed information on the propagation and culture of many of the plants mentioned in this Chapter may be found in the appropriate Bulletins issued by the M.A.F.F., e.g. Bulletin No. 190—'Outdoor Flowers for Cutting'. References to certain aspects included in these Bulletins are included under the host plant in the following pages.

## ACER

### DISEASES

***Tar spot** (*Rhytisma acerinum*). During the summer large black spots with a bright yellow edge develop on the leaves. This disease is more unsightly than harmful, but leaves may fall prematurely and these should be raked up and burnt. If necessary spray in spring with Bordeaux mixture containing $1–1\frac{1}{2}$ lb copper/100 gal.

## ANEMONE

### PESTS

****Aphids,** e.g. peach-potato aphis (*Myzus persicae*), mottles arum aphid (*Aulacorthum circumflexum*) and other species cause direct damage by feeding on young growth and also transmit virus disease. Control as for Dahlia (p. 305).

***Caterpillars,** e.g. angleshades moth (*Phlogophora meticulosa*) and cutworms (*Agrostis* spp.) eat stems, leaves and buds. For control measures see under Dahlia.

### DISEASES

****Downy mildew** (*Peronospora ficariae*) causes a fine whitish mould on the undersides of the leaves which tend to roll upwards. Controlled by maneb at 32 oz a.i. ($2\frac{1}{2}$ lb 80% w.p.)/acre *or* zineb spray at 21 oz a.i. (2 lb 65% w.p.)/100 gal. Application should be given at 10-day intervals.

****Grey mould** (*Botrytis cinerea*). This fungus may cause rotting of flowers and flower buds in wet weather, especially in winter months. A captan spray at 1 lb a.i. (2 lb 50% w.p.)/100 gal H.V. *or* $2\frac{1}{2}$ lb a.i./acre L.V. *or* thiram at 3 lb a.i. (3·8 lb 80% w.p.)/acre L.V. *or* $25\frac{1}{2}$ oz a.i. (2 lb 80% w.p.)/100 gal H.V. gives a worthwhile degree of protection.

## ANTIRRHINUM

PESTS

***Aphids,** e.g. peach-potato aphid (*Myzus persicae*) and potato aphid (*Macrosiphum euphorbiae*), green, yellow or pinkish aphids causing stem distortion and stunting. They may be controlled by the sprays listed under Dahlia (p. 305) but do not use malathion.

DISEASES

****Downy mildew** (*Peronospora antirrhini*). A grey mealy growth develops on under surfaces of affected leaves which are curled and dull. Growth is severely checked. Spray early with thiram at $25\frac{1}{2}$ oz a.i. (2 lb 80% w.p.)/100 gal *or* zineb at 21 oz a.i. (2 lb 65% w.p.)/100 gal.

****Grey mould** (*Botrytis cinerea*). On seed crops the seed spikes may be continually destroyed by the fungus especially in damp summers and, in some cultivars, the yield of seed can be reduced to negligible proportions. It may be necessary to spray every 7–14 days with captan at 1 lb a.i. (2 lb 50% w.p.)/100 gal *or* thiram at 3 lb a.i. (3·8 lb 80% w.p.)/acre L.V. *or* $25\frac{1}{2}$ oz a.i. (2 lb 80% w.p.)/100 gal H.V.

****Leaf spot and stem rot** (*Phyllosticta antirrhini*). On stems dark brown spots develop on which can be seen small black fruiting bodies of the fungus. When the main stem of the young plant is attacked, dieback occurs. In older plants lesions develop on the stems causing them to split and when girdling occurs the plant is killed. Pale brown circular spots develop on the leaves, flowers and fruits. Tentative recommendations are to treat the seed with captan 75% seed dressing at 2 oz/28 lb and to spray early with thiram at $25\frac{1}{2}$ oz a.i. (2 lb 80% w.p.)/100 gal *or* zineb at 21 oz a.i. (2 lb 65% w.p.)/100 gal.

## AZALEA

PESTS

***Azalea whitefly** (*Pealius azaleae*) are very localized, on evergreen azaleas. The pale green 'scales' infest the underside of leaves which become disfigured by sooty moulds. Their control is as for rhododendron whitefly (p. 323).

For other pests of azaleas outdoors see under Rhododendron, p. 322. For pests of azaleas under glass see p. 233.

## AZALEA (Deciduous types)

The growth of algae and lichen on branches may be corrected by a caustic wash made by dissolving caustic soda 10 lb/100 gal water and adding a good spreader. This spray must be applied during December or early January while the flower buds are still quite dormant. The operator should wear old clothes, goggles and gloves, but the spray is quite safe if used with care. An alternative is lime sulphur at $6\frac{1}{2}$ gal/100 gal.

AZALEA (Evergreen types, e.g. Kurume, etc.)

DISEASES

****Gall or false bloom** (*Exobasidium vaccinii*). The young, developing leaves or flower buds turn reddish and swell into small galls which then turn waxy white. Pick off the galls and spray with a copper-containing fungicide, such as Bordeaux mixture, equivalent to $1-1\frac{1}{2}$ lb metallic copper in 100 gal water, *or* copper oxychloride at 6 oz metallic copper (12 oz 50% w.p.)/100 gal.

BEECH

PESTS

****Beech aphid** (*Phyllaphis fagi*). These are yellowish-green aphids covered with long, flocculent, white 'wool'. They feed on the underside of leaves, covering them with honeydew on which grow sooty moulds. The leaves may shrivel in severe infestations. They are controlled by the measures recommended for aphids on fruit trees (see Chapter 9, p. 160).

****Beech scale** (*Cryptococcus fagi*) is a small yellowish scale covered with a white, felted mass of 'wool', which forms conspicuous colonies on the bark. They are controlled by the measures recommended for scale insects on fruit trees (see Chapter 9).

BEGONIA (Outdoor bedding types)

DISEASES

****Powdery mildew** (*Oidium begoniae*) causes white powdery spots or patches on leaves and stems. The damage can be extensive and control may be difficult. Several fungicides, such as copper oxychloride at 6 oz metallic copper (12 oz 50% w.p.)/100 gal *or* dinocap at 2 oz a.i. (8 oz 25% w.p.)/100 gal *or* thiram at 3·2 lb a.i. (4 lb 80% w.p.)/100 gal, at 3–7 day intervals, applied three times, are reported to be effective; although dinocap may cause slight scorch of the leaf edges of certain cultivars.

CAMELLIA

DISEASES

***Leaf spot** (*Pestalotia guepini*) appears as yellowish patches on the leaves and these later turn greyish white with pinpoint size black dots on the surface. The disease is only likely to be of serious import in the case of cuttings and young plants in glasshouses and frames.

It is often possible to cut out the affected part of a leaf and the remainder should be sprayed with copper oxychloride at 6 oz metallic copper (12 oz 50% w.p.)/100 gal.

## CARNATION (BORDER)

PESTS

**Aphids,** e.g. the peach-potato aphid (*Myzus persicae*), green, yellow or pinkish aphids on shoots, causing stunting and distortion. They may be controlled by the methods given for aphids on Dahlia (p. 305). The summer forms of the elder aphid (*Aphis sambuci*) infest the roots and prevent normal growth. The plants wilt in dry weather. Control is difficult but the latest methods of controlling lettuce root aphid should be tried, i.e. water soil before planting with diazinon 1 lb a.i. (4 pint 20% e.c.)/acre in at least 30 gal water. Established plants found infested should be drenched with diazinon, 3·2 oz a.i. (16 fl. oz 20% e.c.)/100 gal. For pests of carnations under glass, see p. 236.

DISEASES

***Ring spot** (*Didymellina dianthi* syn. *Heterosporium echinulatum*) causes round, grey spots in which the skin erupts into pustules arranged in roughly concentric rings. Infected leaves wither, affected stems snap and the plants flower poorly (see M.A.F.F. Bull. 151). Control may be difficult but repeated spraying with Bordeaux mixture, equivalent to $1-1\frac{1}{2}$ lb copper in 100 gal water, gives the best control.

**Rust** (*Uromyces dianthi*) appears as brown spore clusters on both sides of the leaves and stems, usually starting on the lower leaves. Many sprays have been recommended, such as lime sulphur at $2\frac{1}{2}$ gal/100 gal, repeated at 10–14 day intervals *or* thiram at $2\frac{1}{2}$–4 lb a.i. (3–5 lb 80% w.p.)/100 gal, applied at 10–14 day intervals after the disease is first seen, to keep the new foliage and stems free from severe attack. Zineb sprays at 21–32 oz a.i. (2–3 lb 65% w.p.)/100 gal H.V. have also been successfully used. Sprays should not be applied on open blossoms for fear of staining.

### CHINA ASTER

PESTS

**Aphids,** e.g. potato aphid (*Macrosiphum euphorbiae*), leaf-curling plum aphid (*Brachycaudus helichrysi*), are green, pink or greenish-black aphids causing general stunting of growth. For their control see aphids on Dahlia (p. 305).

**Capsid bugs,** e.g. tarnished plant bug (*Lygus rugulipennis*) cause tattering of leaves and distorted growth. Control measures are given under Dhalia (p. 305).

DISEASES

*Foot rot** (*Phytophthora cryptogea*) causes a brownish or blackish rot of roots and stem bases. The seed may be dressed with captan 75% *or* thiram 50% seed dressings at 2 oz to 28 lb seed. The soil can be watered with captan at 2 oz a.i. (4 oz 50% w.p.)/10 gal *or* with Cheshunt compound equivalent to 1 oz metallic copper/2 gal *or* with zineb at 2 oz a.i. (3 oz 65% w.p.)/10 gal.

## CHRYSANTHEMUM

### PESTS

****Aphids,** e.g. chrysanthemum aphid (*Macrosiphoniella sanborni*), peach-potato aphid (*Myzus persicae*) and leaf-curling plum aphid (*Brachycaudus helichrysi*). These aphids feed on shoots and foliage, causing stunting and distortion of new growth and poor quality blooms and may also transmit virus diseases. They are controlled by the methods given for aphids on Dahlia (p. 305) but do not use dimethoate or formothion. Certain varieties are also damaged by demeton-S-methyl and oxydemeton-methyl (see product labels). Morphothion 4 oz a.i. (20 fl. oz 20% e.c.)/100 gal may also be used.

***Capsid bugs**—see Dahlia (p. 305).

****Chrysanthemum leaf miner** (*Phytomyza atricornis*)—see Chapter 10, p. 241.

****Earwig** (*Forficula auricularia*). These dark brown, elongate insects with leathery wing cases and stout pincers on the hind end of the body feed at night on petals making them ragged and unsightly. During the day they hide deep among the petals or under debris on the ground. To control them, clear away all debris. If possible, shake open blooms to dislodge insects and then spray the ground and lower parts of plants with DDT 15 oz a.i. (3 pints 25% e.c.)/100 gal *or* gamma-BHC 2 oz a.i. ($\frac{1}{2}$ pint 20% e.c.)/100 gal. Good control has been given by trichlorphon at 0·8 lb a.i. (1 lb 80% s.p.)/100 gal *or* TDE $\frac{3}{4}$ lb a.i. ($1\frac{1}{2}$ lb 50% w.p.)/100 gal.

****Leaf and bud eelworm** (*Aphelenchoides ritzemabosi*)—see Chapter 10, p. 241.

### DISEASES

***Blotch or leaf spot** (*Septoria* spp.). Round brown or black spots or blotches up to 1 in in diameter develop on the lower leaves. Fruiting bodies of the fungus show on the blotches as pinpoint size black dots. The disease is especially severe in wet weather, when the spots coalesce, the leaves wither and fall prematurely. Spray at 7–10-day intervals with copper sprays containing $\frac{3}{4}$ lb metallic copper/100 gal, but avoid blooms, *or* zineb at 22 oz a.i. (2 lb 65% w.p.)/100 gal. Some cultivars may be damaged by copper.

****Grey mould** (*Botrytis cinerea*) is favoured by warm, moist weather when this disease may cause much damage to the blooms (damping) which become covered with the greyish mass of spores. For control apply captan at 1 lb a.i. (2 lb 50% w.p.)/100 gal *or* thiram at 3 lb a.i. (3·8 lb 80% w.p.)/acre L.V. or 25½ oz a.i. (2 lb 80% w.p.)/100 gal H.V. and remove regularly all affected blooms.

**Petal blight** (*Itersonilia perplexans*) appears as small oval water-soaked spots on the outer petals, rapidly spreading to spoil the bloom. Control can be achieved by spraying with zineb at 21 oz a.i. (2 lb 65% w.p.)/100 gal *or* tank-mix zineb (5 pints 22% nabam solution or 1¾ lb 93% nabam soluble powder, 1¼ lb zinc sulphate to 100 gal water). Commence when buds first show colour and repeat at 5–7-day intervals.

***Powdery mildew** (*Oidium chrysanthemi*) (see M.A.F.F. Bull. 92) shows as a superficial white powdery deposit on the leaves.

It can be controlled by copper-oil sprays containing 12 oz metallic copper and ½–1% white oil/100 gal H.V. *or* dinocap at 1 oz a.i. (4 oz 25% w.p. or 2 fl. oz 50% e.c.)/100 gal applied when the disease is first seen and repeated at 10–14 day intervals, *or* sulphur sprays containing 3 lb a.i./100 gal *or* thiram 3 lb a.i. (3·8 lb 80% w.p.) + petroleum 2¾ pint/100 gal. Care is necessary as some cultivars are copper-sensitive or may be damaged by dinocap.

**Rust** (*Puccinia chrysanthemi*) causes brown erumpent pustules on the undersides of the leaves and on stems, and may be serious under moist conditions. It appears first on the lower leaves which are often removed when potting on. To control, use copper sprays at ¾ lb metallic copper/ 100 gal at 2 or 3-weekly intervals, but keep these sprays off the blooms, *or* zineb at 32 oz a.i. (3 lb 65%)/100 gal with wetter H.V. Certain cultivars are sensitive to copper.

CLARKIA

DISEASES

***Grey mould** (*Botrytis cinerea*) attacks the bases of the stems; for control dust with tecnazene dust (5%) at 1 lb for each 60 yd².

CLEMATIS

PESTS

***Clay-coloured weevil** (*Otiorhynchus singularis*)—see Rhododendron (p. 323).

***Earwig** (*Forficula auricularia*)—see Chrysanthemum (p. 302).

DISEASES

**Powdery mildew** (*Erysiphe polygoni*) occasionally causes damage in southern counties. The disease can be checked by spraying with dinocap at 2 oz a.i. (8 oz 25% w.p.)/100 gal.

****Wilt** (is now believed to be due to *Ascochyta clematidina*). One or more shoots die back rapidly, often to the base of the plant, but new shoots usually develop in due course. Spray the developing shoots with a copper spray such as Bordeaux mixture equivalent to $1-1\frac{1}{2}$ lb copper/100 gal *or* copper oxychloride at 6 oz metallic copper (12 oz 50% w.p.)/100 gal.

## CONIFER SEEDLINGS

See Spruce, p. 327.

## CORNFLOWER

DISEASES

****Petal blight** (*Itersonilia perplexans*). Small oval water-soaked spots develop on the outer petals and spread rapidly thus spoiling the bloom. Commence spraying when the buds first show colour and repeat at weekly intervals using zineb at 21 oz a.i. (2 lb 65% w.p.)/100 gal *or* tank-mix zineb (5 pints 22% nabam solution or $1\frac{3}{4}$ lb 93% nabam soluble powder, $1\frac{1}{4}$ lb zinc sulphate/100 gal water).

***Powdery mildew** (*Oidium* spp.) shows as a white powdery coating on leaves, stems and flowers. Spraying with dinocap at 2 oz a.i. (8 oz 25% w.p.)/100 gal should control this disease.

***Rust** (*Puccinia cyani*) appears as brown spore masses on the undersides of the leaves. Spraying with thiram at $25\frac{1}{2}$ oz a.i. (2 lb 80% w.p.)/100 gal as soon as the disease is first seen, should check it.

## COTONEASTER

PESTS

****Brown scale** (*Parthenolecanium corni*)—see Chapter 9 (p. 193).

****Hawthorn webber** (*Scythropia crataegella*)—see Hawthorn (p. 309).

****Winter moths** (*Operophtera brumata, Alsophila aescularia*, etc.)—see Chapter 9 (p. 176).

****Woolly aphid** (*Eriosoma lanigerum*)—see Chapter 9 (p. 177).

## CROCUS

PESTS

***Tulip bulb aphid** (*Dysaphis tulipae*)—see Tulip (p. 329).

## DAHLIA

PESTS

****Aphids,** e.g. bean aphid (*Aphis fabae*) (A.L. 54) and leaf-curling plum aphid (*Brachycaudus helichrysi*) are greenish-black aphids infesting the foliage and young shoots, causing stunting and distortion and transmitting virus diseases. They may be controlled by spraying as necessary with demeton-S-methyl 3·4 oz a.i. (6 fl. oz 58% e.c.)/100 gal *or* diazinon 3·2 oz a.i. (16 fl. oz 20% e.c.)/100 gal *or* dimethoate 4·8 oz a.i. (12 fl. oz 40% e.c.)/100 gal *or* formothion 7·5 oz a.i. (16 fl. oz 43% e.c.)/100 gal *or* gamma-BHC 2 oz a.i. (½ pint 20% conc.)/100 gal *or* malathion 18 oz a.i. (1½ pints 60% e.c.)/100 gal *or* oxydemeton-methyl 3·4 oz a.i. (6 fl. oz 57% e.c.) *or* nicotine 8 fl. oz 95–98%/100 gal *or* parathion 1·2 oz a.i. (6 fl. oz 20% e.c.) *or* schradan 11 oz a.i. (1 pint 55% conc.)/100 gal. Demeton-S-methyl, formothion and oxydemeton-methyl may also be used as soil drenches.

****Capsid bugs,** e.g. tarnished plant bug (*Lygus rugulipennis*) and common green capsid (*Lygocoris pabulinus*). Capsids are active sap-sucking bugs, about ¼ in long. They feed on young leaves and shoots and on flower buds. Injured flower buds produce malformed, lop-sided blooms or the bud may be killed. Attacked shoots are scarred, stunted and twisted. On foliage, the feeding punctures appear initially as small brown spots of calloused tissue. As the leaves grow they become deformed and puckered. Damage may occur from June onward but is usually most serious later in the summer. To control this pest ensure that ground is kept clear of weeds and rubbish and spray the plants and ground underneath 2–3 times at 14-day intervals starting as early as possible with DDT at 15 oz a.i. (3 pints 25%)/100 gal *or* diazinon at 3·2 oz a.i. (16 fl. oz 20%)/100 gal *or* gamma-BHC at 2 oz a.i. (½ pint 20%)/100 gal *or* parathion 1·2 oz a.i. (6 fl. oz 20% e.c.)/100 gal *or* TDE ¾ lb a.i. (1½ lb 50% w.p.)/100 gal.

***Catepillars,** e.g. angleshades moth (*Phlogophora meticulosa*). Various species of caterpillar feed on foliage and flowers. They may be controlled by spraying as necessary with DDT at 15 oz a.i. (3 pints 25% e.c.)/100 gal *or* TDE at 12 oz a.i. (1½ lb 50% w.p.)/100 gal *or* dusting with a 5% DDT dust *or* 0·2% rotenone dust. Trichlorphon might also be effective against these pests.

***Earwig** (*Forficula auricularia*)—see Chrysanthemum (p. 302).

****Red spider mite** (*Tetranychus urticae*). This mite is primarily a pest of glasshouse plants but also attacks dahlia and certain other ornamentals grown outdoors. Breeding is continuous during the summer and severe infestations build up rapidly in dry, warm weather. At first leaves are lightly speckled but bronzing and death may follow.

The chemical sprays detailed in Chapter 10 (p. 237) may be used, but strains of red spider mite have developed resistance to some of these and will probably soon become resistant to others.

x

DISEASES

***Grey mould** (*Botrytis cinerea*). This fungus can be a nuisance on flowers and flower buds in moist summer weather and should be checked by spraying with captan at 1 lb a.i. (2 lb 50% w.p.)/100 gal *or* thiram at 25½ oz a.i. (2 lb 80% w.p.)/100 gal H.V. Affected flowers, leaves, etc., should be removed.

***Smut** (*Entyloma calendulae* f. *dahliae*) produces circular brown spots on the lower leaves spreading upwards and, where spots coalesce, large areas may be killed and the leaf wither. The disease is only likely to be serious on the small bedding cultivars planted close together.

A copper spray, such as Bordeaux mixture equivalent to 1–1½ lb copper/100 gal *or* copper oxychloride at 6 oz metallic copper (12 oz 50% w.p.)/100 gal gives control.

## DAPHNE

PESTS

****Aphids,** e.g. peach-potato aphid (*Myzus persicae*)—for control, see aphids on Dahlia (p. 305).

DISEASES

****Leaf spot** (*Marssonina daphnes*) causes spotting on leaves, especially at the base and on the petiole, resulting in appreciable defoliation. Spraying with copper oxychloride at 6 oz metallic copper (12 oz 50% w.p.)/100 gal is advised.

## DELPHINIUM

DISEASES

****Black Blotch** (*Pseudomonas delphinii*). This disease causes large, black or brownish-black blotches on leaves, stems or even flower buds, but it seems to be very selective as some cultivars are badly attacked while others nearby are not affected. It is a difficult disease to check, but repeated spraying with Bordeaux mixture equivalent to 1–1½ lb metallic copper/100 gal H.V. will protect young plants, if applied as soon as the shoots appear above the soil; the surface of the soil should also be sprayed.

***Powdery mildew** (*Erysiphe polygoni*) appears as the typical white, powdery coating on leaves, stems and even flowers. Spray with copper oxychloride at 6 oz metallic copper (12 oz 50% w.p.)/100 gal *or* dinocap at 2 oz a.i. (8 oz 25% w.p.)/100 gal *or* lime sulphur 1 pint in 10 gal.

## EUONYMUS

PESTS

****Bean aphid** (*Aphis fabae*) (A.L. 54) is a greenish-black aphid which uses Euonymus as winter host plant, laying eggs on the bark of the shoots in autumn and infesting the young growth in spring causing the foliage to curl. Apply sprays as for dahlia (p. 305) as soon as aphids appear on new growth.

**Spindle ermine moth** (*Yponomeuta cagnagella*) (A.L. 40). The small, grey caterpillars of this moth live in colonies, binding the shoots and foliage together in a silken 'tent' and feeding inside. They are most commonly seen in May or June and are effectively controlled by spraying with trichlorphon at 13 oz a.i. (1 lb 80% s.p.)/100 gal *or* carbaryl at 8 oz a.i. (1 lb 50% w.p.)/100 gal.

**Willow scale** (*Chionaspis salicis*)—see Willow (p. 333).

DISEASES

**Powdery mildew** (*Oidium euonymi-japonicae*). The attack of the fungus is often severe and the foliage becomes covered in a whitish powdery deposit. The disease is said to be controlled fairly easily by spraying with either dinocap at 2 oz a.i. (8 oz 25% w.p.)/100 gal *or* colloidal sulphur (40–50%) at 2 pints/100 gal H.V., or wettable sulphur (75%) 2½–4 lb/100 gal H.V.

FORSYTHIA

PESTS

**Common green capsid** (*Lygocoris pabulinus*)—see Pelargonium, p. 318.

FUCHSIA

PESTS

**Common green capsid** (*Lygocoris pabulinus*)—see Pelargonium, p. 318.

**Peach-potato aphis** (*Myzus persicae*)—control as for aphids on Dahlia (p. 305).

GLADIOLUS

PESTS

**Aphids,** e.g. potato aphid (*Macrosiphum euphorbiae*), peach-potato aphid (*Myzus persicae*), shallot aphid (*Myzus ascalonicus*), etc., may infest foliage and can be controlled in the same way as aphids on Dahlia (p. 305). Grey bulb aphid (*Dysaphis tulipae*) infests dry corms. Controls are the same as those given under Tulip (p. 329).

**Caterpillars,** e.g. angleshades moth (*Phlogophora meticulosa*) and cabbage moth (*Mamestra brassicae*) attack the foliage and are controlled as for Dahlia (p. 305).

**Gladiolus thrips** (*Taeniothrips simplex*) are tiny, elongate insects, yellow when young and dark brown when adult, which suck the sap of the stems and leaves leaving pale yellow or silvery streaks which later turn brown. They also attack the flowers, producing small white flecks on the petals.

The thrips overwinter on corms in store, feeding under the scales, and will continue to breed if the temperature is over 50°F. Infested corms have rough greyish-brown patches on the surface. As soon as any damage is seen, apply gamma-BHC dust *or* spray with diazinon at 3·2 oz a.i. (16 fl. oz 20% e.c.)/100 gal *or* gamma-BHC at 2 oz a.i. ($\frac{1}{2}$ pint 20% e.c.)/100 gal *or* malathion 18 oz a.i. (1$\frac{1}{2}$ pint 60% e.c.)/100 gal *or* nicotine at 8 fl. oz (95–98%)/100 gal.

Corms should be stored at a temperature not exceeding 50°F and dusted with DDT (5%) 1 oz/bushel.

### DISEASES

****Core rot** (*Botrytis gladiolorum*). This fungus attacks the central core of gladiolus corms and then spreads outwards to destroy the corm with a moist (but not wet) rot. In storage corms are soon destroyed and the disease can spread quickly especially if moisture is present. For control the corms should be dried thoroughly and dusted with quintozene dust at 2 lb a.i./ ton corms. 10 lb of the 20% dust suffice for 1 ton of corms. This treatment is particularly effective for rots caused by Penicillium and also gives useful measure of control of core rot.

****Dry rot** (*Sclerotinia gladioli*) causes small black spots on the flesh of the corms and these may coalesce to cause larger black areas so that badly infected corms are reduced to worthless 'mummies'.

****Hard rot** (*Septoria gladioli*) causes larger black spots which are somewhat sunken and these by enlarging will mummify the corms.

***Scab** (*Pseudomonas marginata*) shows as round, sunken craters towards the corm base and each crater is often covered by a shiny varnish-like covering. All the above corm diseases need to be controlled by a dip treatment. At planting time the corms are roughly cleaned by removal of the outermost rough scales, the badly affected ones are discarded and the others dipped in an aqueous solution of either an organomercury compound usually at 1 lb to 40 gal, *or* calomel suspension containing 2 oz/1 gal water, *or* in captan 2 lb 50% w.p./10 gal for 1 hour (treats 80 dozen corms).

## GODETIA

### DISEASES

****Grey mould** (*Botrytis cinerea*) see Clarkia, p. 303.

***Stem rot** (*Alternaria godetiae*). This fungus causes brown spots to appear on the stems and these extend and kill the stems. The disease is seed-borne and seed should be steeped for $\frac{1}{2}$ hr in a 0·25% solution of an organomercury fungicide. Seed crops should also be sprayed with Bordeaux mixture, equivalent to 1–1½ lb metallic copper/100 gal H.V. or 2–3 lb/acre L.V.

HAWTHORN

PESTS

***Hawthorn webber** (*Scythropia crataegella*). These small reddish- or yellowish-brown caterpillars feed in colonies on foliage and form a delicate silk webbing covering a considerable part of the tree. A local pest, prevalent in May and June, which may be controlled by destroying the webbing by 'fanning' with a blow-lamp and spraying with DDT 15 oz a.i. (3 pint 25% e.c.)/100 gal *or* TDE at ¾ lb a.i. (1½ lb 50% w.p.)/100 gal. Trichlorphon applied as for spindle ermine moths on Euonymus (p. 307) may also be effective.

Other caterpillars, e.g. **lackey moth** (*Malacosoma neustria*), **small ermine moth** (*Yponomeuta* spp.) and **vapourer moth** (*Orgyia antiqua*)— see Chapter 9, p. 176. Trichlorphon at 12·8 oz a.i. (1 lb 80% s.p.)/100 gal gives good control of ermine moths.

DISEASES

****Powdery mildew** (*Podosphaera oxyacanthae*). The leaves and shoots become covered by the typical powdery white coating and in the case of newly-planted hedges or young soft shoot growth the disease can be very injurious. To control spray with dinocap at 2 oz a.i. (8 oz 25% w.p.)/100 gal H.V. *or* colloidal sulphur (40–50%) 2 pints/100 gal H.V. *or* wettable sulphur (75%) 2½–4 lb/100 gal H.V.

HELLEBORE

DISEASES

****Leaf spot** (*Coniothyrium hellebori*). The leaves become spotted with round or elliptical black blotches marked by concentric zones which coalesce to form large patches until eventually the leaves wither. Smaller spots disfigure the flowers. Remove and burn diseased leaves and flowers and spray regularly with Bordeaux mixture equivalent to 1–1½ lb metallic copper/100 gal *or* copper oxychloride at 6 oz metallic copper (12 oz 50% w.p.)/100 gal.

HOLLY

PESTS

***Holly leaf miner** (*Phytomyza ilicis*). The small yellowish-white grubs tunnel in the leaf tissue. The mines are linear at first and may run along the midrib, later broadening out to form a blister which turns yellowish-brown. The mines are usually evident from early June onwards.

Their control is difficult but possible by spraying three times at 14-day intervals with gamma-BHC 2 oz a.i. ($\frac{1}{2}$ pint 20% e.c.)/100 gal, adding a wetting agent. Applications of diazinon at 3·2 oz a.i. (16 fl. oz 20% e.c.)/100 gal may be more effective.

***Holly leaf-tying moth** (*Rhopobota naevana*). The small greyish-green caterpillars of the moth are particular pests on clipped holly bushes, feeding on the young, tender leaves which they bind together with silk. They are active from May to July and may be controlled by spraying, in early May and again 2 weeks later, with DDT 15 oz. a.i. (3 pints 25% e.c.)/100 gal *or* TDE $\frac{3}{4}$ lb a.i. (1$\frac{1}{2}$ lb 50% w.p.)/100 gal. Tied leaves should be crushed or removed when seen.

## HOLLYHOCK

DISEASES

****Rust** (*Puccinia malvacearum*). On the undersurfaces of the leaves and on stems raised pustules develop which are at first orange and later turn brown. In severe cases shrivelling of leaves or even the whole plant may occur. Raise new plants at least every other year and spray from the seedling stage onwards at fortnightly intervals with thiram at 25$\frac{1}{2}$ oz a.i. (2 lb 80% w.p.)/100 gal.

## HYACINTH

PESTS

****Stem eelworm** (*Ditylenchus dipsaci*). Symptoms of stem eelworm attack on hyacinth are similar to those seen on narcissus (p. 316) except that definite 'spickels' do not form on the leaves. Hot water treatment may be applied as for narcissus.

DISEASES

****Grey bulb rot** (*Sclerotium tuliparum*). The fungus infects the nose portion of the bulb causing a dry greyish decay thus preventing the shoot getting far above ground. The bulb is soon destroyed by the dense fungal growth which quickly produces large resting bodies (sclerotia) which contaminate soil and infect other bulbs the next season. The result is that bare patches are seen in the beds. Rake 20% quintozene dust into the soil at 2 oz/yd$^2$ or 5$\frac{1}{2}$ cwt/acre.

## HYDRANGEA

PESTS

****Common green capsid** (*Lygocoris pabulinus*)—see Pelargonium (p. 318) but do not use BHC.

**Red spider mite** (*Tetranychus urticae*)—see Chapter 10, p. 268.

DISEASES

**Powdery mildew** (*Microsphaera polonica*) shows on the leaves as brown spots on which the mildew shows as a whitish coating. The control measures given under the powdery mildew of hawthorn (p. 305) apply.


## IRIS

PESTS

**Aphis,** e.g. potato aphid (*Macrosiphum euphorbiae*) and tulip bulb aphid (*Dysaphis tulipae*)—for control see aphids on Dahlia (p. 305) and on Tulip (p. 329).

**Iris sawfly** (*Rhadinoceraea micans*), appears as leaden grey, smooth bodied 'caterpillars' with black heads, eating away longitudinal strips from the margin of the leaves and active from late May to July. For their control apply, as necessary, DDT 15 oz a.i. (3 pint 25% e.c.)/100 gal *or* gamma-BHC 2 oz a.i. ($\frac{1}{2}$ pint 20% e.c.)/100 gal *or* derris 0·65 oz a.i. (13 fl. oz 5% rotenone)/100 gal.


DISEASES OF BULBOUS IRIS

**Leaf spot** (*Didymellina macrospora*) usually appears only in second year when it causes brown oval-shaped spots on the leaves which soon wither. Control is obtained by spraying with maneb at 2 lb a.i. ($2\frac{1}{2}$ lb 80% w.p.)/100 gal *or* zineb at 2 lb a.i. (3 lb 65% w.p.)/100 gal. A wetter should be added.


DISEASES OF RHIZOMATOUS IRIS

**Leaf spot** (*Didymellina macrospora*). Attacks by this fungus vary according to locality, but in some places (e.g. Western counties) the foliage is often attacked in early spring by brown oval-shaped spots which coalesce so that the leaves are badly injured or killed. The control is as for the leaf spot of bulbous iris (see above).

**Rhizome rot** (*Erwinia carotovora*) causes the rhizome to rot with a yellowish slimy rot so that the fan of leaves above shows browning of tips followed by collapse and death. To control the disease cut out affected parts and dust the wounds, rhizomes and soil with a copper-containing dust.


## JUNIPER

PESTS

**Conifer spinning mite** (*Oligonychus ununguis*)—see Spruce (p. 327).

**Juniper scale** (*Carulaspis* spp.). These small dirty-white scales disfigure stems and foliage. Their control is as for scale on Willow (p. 333) but do not use tar oil or DNOC-petroleum.

***Juniper webber** (*Dichomeris marginella*). These small brown caterpillars feed in colonies on the foliage, tying it up with webbing. The injured foliage dies and turns brown. A local pest, active in May and June. They may be controlled by the methods given for hawthorn webber (p. 309).

## LARCH

DISEASES

****Leaf cast** (*Meria laricis*) can be a serious disease of young larch trees in nurseries. The symptoms appear in early May when the leaflets turn brown from the tip downwards. The needles fall prematurely soon after becoming completely brown. Control is by application of sulphur sprays at end of March and 2–3 week intervals until end of July or earlier in a dry season, with colloidal sulphur equivalent to $1\frac{1}{2}$ lb sulphur/100 gal H.V. *or* lime sulphur 1 gal/100 gal H.V.

## LILAC

PESTS

**Lackey moth** (*Malacosoma neustria*) and **winter moths** (*Operophtera brumata, Alsophila aescularia* etc.) see Chapter 9, p. 176.

****Lilac leaf miner** (*Caloptilia syringella*). The small, whitish, legless caterpillars tunnel in the leaf tissue forming large blisters which turn brown. Older caterpillars feed externally but conceal themselves in leaves which they roll from the tip and tie with silk. They are active from May to June and from August to September and may be controlled by spraying 3 times, at 14-day intervals, with diazinon 3·2 oz a.i. (16 fl. oz 20% e.c.)/ 100 gal *or* gamma-BHC 2 oz a.i. ($\frac{1}{2}$ pint 20% e.c.)/100 gal. Treatment should begin as soon as the first signs of damage are seen.

***Privet thrips** (*Dendrothrips ornatus*)—see Privet (p. 321).

***Willow scale** (*Chionaspis salicis*)—see Willow (p. 333).

DISEASES

***Blight** (*Pseudomonas syringae*) causes small angular brown spots on leaves but the serious damage is that young shoots blacken and wither away (both flower and foliage shoots). Cut back the affected portions to a healthy bud and spray with Bordeaux mixture equivalent to $1-1\frac{1}{2}$ lb metallic copper/100 gal H.V. *or* copper oxychloride at 6 oz metallic copper (12 oz 5% w.p.)/100 gal.

## LILY

PESTS

****Aphids.** Tulip bulb aphid (*Dysaphis tulipae*) can build up rapidly in store and, together with other species (e.g. mottled arum aphid, *Aula-*

*corthum circumflexum*, shallot aphid, *Myzus ascalonicus*, etc.) will infest growing plants, causing distortion and also transmitting virus diseases. Control on bulbs in store is as for Tulip (p. 329) and, for outdoor infestations, controls may be applied as for Dahlia (p. 305).

***Lily beetle** (*Lilioceris lilii*). A red and black beetle, $\frac{1}{4}$ in long, locally common in Surrey and Berkshire. The larvae are reddish but usually covered with black, slimy fæces. Both stages chew away the tissues of stems, leaves and seed-pods. To control, spray with DDT 15 oz a.i. (3 pints 25%)/100 gal *or* apply DDT (5%) dust as necessary.

### DISEASES

***Leaf blight** (*Botrytis elliptica*). The leaves develop oval-shaped water-soaked spots and these spread rapidly in humid weather conditions. Flower stems can 'topple' if the fungus spreads from the leaves and enters the stalk. Spray as soon as the flower buds can be detected with a copper-containing fungicide, e.g. Bordeaux mixture equivalent to $1-1\frac{1}{2}$ lb metallic copper/100 gal *or* copper oxychloride at 6 oz metallic copper (12 oz 50% w.p.)/100 gal.

## LOBELIA

### DISEASES

****Leaf spot and stem rot** (*Alternaria tenuis*). Pale spots develop on leaves of seedlings often giving a scorched effect. The fungus can also cause the seedlings to damp off. Tentative recommendations are to treat the seed with captan 75% seed dressing at 2 oz/28 lb *or* to soak seed in a 0·2% aqueous thiram suspension for 24 hours at 30°C and to spray early with thiram at $25\frac{1}{2}$ oz a.i. (2 lb 80% w.p.)/100 gal *or* zineb at 21 oz a.i. (2 lb 65% w.p.)/100 gal.

## MAHONIA

### DISEASES

***Powdery mildew** (*Microsphaera berberidis*) shows as a white powdery coating on the leaves especially on young growth. Spraying with dinocap at 2 oz a.i. (8 oz 25% w.p.)/100 gal should control the disease.

***Rust** (*Cumminsiella mirabilissima* and *Puccinia graminis*). Small red spots appear on the upper surfaces of leaves and on the lower surfaces brownish, somewhat powdery, masses of spores are produced. Spraying with thiram at $25\frac{1}{2}$ oz a.i. (2 lb 80% w.p.)/100 gal may give some control of this disease.

## MALUS

### PESTS

These trees are afflicted by most of the pests recorded for apple, Chapter 9, p. 160.

DISEASES

***Scab** (*Venturia inaequalis*). On many species the apple scab fungus regularly causes brown to greenish blotches on the leaves which spoil the appearance. To control the disease, spray with either captan at 1 lb a.i. (2 lb 50% w.p.)/100 gal *or* lime sulphur 1 pint in 12½ gal at green bud and after petal fall. Dodine at 10 oz a.i. (1 lb 65% w.p.)/100 gal may also be used.


## MAPLE

See Acer, p. 298.


## MECONOPSIS

DISEASES

 **Downy mildew** (*Peronospora arborescens*) shows as furry, greyish-white patches on the undersurfaces of the leaves or even on flower stalks and seed pods. Seedlings are often attacked and sterilized soil should be used for seed pans and boxes. Control is also obtained by a copper-containing spray, e.g. Bordeaux mixture equivalent to 1–1½ lb metallic copper in 100 gal *or* copper oxychloride at 6 oz metallic copper (12 oz 50% w.p.)/100 gal. A good spreader is essential to wet these leaves.


## MICHAELMAS DAISY

DISEASES

 **Powdery mildew** (*Sphaerotheca fuliginea*) gives a white, powdery deposit covering the leaves. Spray dinocap at 2 oz a.i. (8 oz 25% w.p.)/ 100 gal.


## NARCISSUS

PESTS

 ***Bulb mite** (*Rhizoglyphus echinopus*). These are shiny, yellowish-white, globular mites which infest bulbs injured by handling, by disease or by other pests. They extend the original damage and prevent recovery from slight cuts and bruises. Infested bulbs should be examined for evidence of disease or damage by other pests and, if necessary, the appropriate control measures taken. The mites are destroyed by hot-water treatment as recommended for bulb scale mite, narcissus flies or eelworm. Alternatively fumigate for 5 days in airtight container containing paradichlorobenzene at rate of 4 oz per ft³. Cover the crystals with hessian and fill with bulbs.

 ****Bulb scale mite** (*Steneotarsonemus laticeps*) (A.L. 456) are microscopic mites which feed between bulb scales, leaving on their surface elongate, brown streaks which can be seen on cutting open the bulb and parting the scales. In store, infested bulbs become soft and light in weight.

On the growing bulb the effects on the vigour are best seen in forcing houses. Leaves are distorted and abnormally green with yellow streaks. Flowers may be malformed or killed in the bud. In the open, plants merely show lack of vigour, poor blooms and premature death of the foliage.

Hot-water treatment of fully dormant bulbs will reduce infestations. One hour at 110°F will give adequate control on bulbs for forcing during the season after treatment and, for stocks intended for growing on outdoors, 4 hours at 110°F or 3 hours at 112°F will eradicate the mites. Dipping bulbs for $2\frac{1}{2}$ hours in an emulsion of thionazin (4 pints 46% e.c./100 gal cold water) will also give good control.

Overnight frosting followed by a drench of 0·1% endrin *or* endosulfan (4 pints of 20% endrin e.c. or of 20% endosulfan e.c./100 gal, plus wetter) will check the development of mites on bulbs brought indoors for forcing.
***Large narcissus fly** (*Merodon equestris*) (A.L. 183). The larva, a yellowish-white, legless grub, up to $\frac{3}{4}$ in long, tunnels in the interior of bulbs, which become filled with wet brown material. The initial entry is in May or June after the eggs hatch but damage is not easily seen until the bulbs are replanted in September or October. Infested bulbs are softer than normal, especially around the neck, and, if replanted, will produce only small leaves ('grass') or may fail to make any growth at all. Where bulbs are lifted annually, narcissus fly larvae within infested bulbs can be killed by either a 3-hour dip in gamma-BHC ($\frac{1}{4}$ pint 20% conc./10 gal water) immediately after lifting *or* by hot water treatment (1 hour at 110°F) during the dormant period. The BHC dip may be phytotoxic and the latter method is therefore preferred, if practicable.

After hot water treatment a cold dip of 15 minutes in 0·2% aldrin ($\frac{1}{2}$ pint 30% e.c./10 gal) *or* 0·2% dieldrin (1 pint 15% e.c./10 gal) will prevent reinfestation during the following growing seasons, and may give some protection in the second or third seasons after planting.

An alternative to bulb dipping is the application of aldrin *or* gamma-BHC to the soil after planting. The use of such soil applications is biologically unsound and may cause harmful side-effects.
**Small narcissus flies** (*Eumerus tuberculatus* and *E. strigatus*) (A.L. 183). Whereas only one larva of the previous species is normally present in a bulb, there may be several of the larvae of the small narcissus fly. They are greyish-white, legless and grow to $\frac{1}{3}$ in in length.

Small narcissus flies may occasionally cause primary damage to narcissus but are generally secondary pests, following and extending damage done by eelworm, large narcissus fly or slugs. Control measures taken against these pests should therefore reduce infestations by small narcissus flies.

**Root lesion eelworms** (*Pratylenchus* spp.) feed on the roots, making small, brown or black, slit-like lesions which allow entry of fungi, etc. Roots decay and rot away and bulbs make poor growth, producing dwarfed plants. At present severe damage occurs only in the Isles of Scilly and in Cornwall. Chemical treatment of infested soil is possible though expensive and may be worthwhile where land is valuable.

***Stem eelworm** (*Ditylenchus dipsaci*) (A.L. 460). This eelworm infests the bulb and upper growth. The foliage of attacked plants may be paler than usual, also stunted and twisted with small yellow swellings (spickels) on the surface. Flower production is late and eventually ceases altogether as the bulb rots in the ground. Infested bulbs are soft to the touch, with dull, unhealthy-looking outer scales, and, when cut across, rings of brown tissue can be seen where the scales have decayed. Rogue out infested bulbs in the field. After lifting inspect bulbs and destroy those which are badly affected. The rest should be given hot-water treatment when dormant (3 hours at 112°F). A wetter should be added to the water for better penetration or the bulbs may be soaked in cold water containing a wetter before hot-water treatment is given.

Damage by hot-water treatment may be avoided by storing bulbs for a week at a temperature of 85–86°F, both before and after treatment. As an alternative to hot water treatment bulbs may be dipped for $2\frac{1}{2}$ hours in cold water containing thionazin at 0·23% a.i. (4 pints 46% e.c./100 gal). Heavily infested bulbs must be graded out before dipping.

On infested land the interval between bulb crops should be 3 or more years after ground keepers have been removed and no susceptible plant hosts such as bluebells, onions, scillas, snowdrops and strawberries should be grown during the interval. It may be considered worthwhile to sterilize small areas of ground with 'D-D mixture' at 300–400 lb/acre ($25–33\frac{1}{2}$ gal/acre) *or* metham-sodium (33%) 110 gal/acre.

DISEASES

****Fire** (*Sclerotinia polyblastis*) causes a rapid leaf decay during and following flowering, especially in wet weather, starting first on the flowers as small light brown spots, which appear first about picking time and soon destroy the flowers. The leaves then show elongated reddish-brown blotches and are killed very rapidly. Remove plants showing primary lesions and, when 1 in high, spray with thiram at $2\frac{1}{2}$–4 lb a.i. (3–5 lb 80% w.p.)/acre. Repeat at 10-day intervals but do not spray once the flower buds have emerged from the leaf sheaf. Diseased foliage should be raked up and burnt.

****Leaf scorch or tip burn** (*Stagonospora curtisii*). The fungus attacks the emerging leaves so that they have a scorched or burnt appearance at the tips. Spray the foliage soon after it emerges with zineb at 21 oz a.i. (2 lb 65% w.p.)/100 gal and repeat at 10–14 day intervals as long as necessary. Remove and burn diseased tips.

****Smoulder** (*Sclerotinia narcissicola*) causes a decay of stored bulbs but

in cold wet seasons it can rot foliage and flowers so that plants are malformed and affected tissues are covered with grey masses of spores. In
warmer drier conditions infection is usually restricted to one or two leaves,
and the plants grow away from the disease and flower normally. Spray as
soon as the disease is seen with zineb at 21 oz a.i. (2 lb 65% w.p.)/100 gal
and repeat at 10-day intervals.

*White mould (*Ramularia vallisumbrosae*) appears as dark-greenish
or yellowish-brown patches on the leaves, especially near the tips and these,
in damp weather, soon become covered with a whitish mass of fungus
spores. The foliage in these conditions is prematurely killed with a consequent check to bulb growth.

Spray with zineb at 21 oz a.i. (2 lb 65% w.p.)/100 gal and repeat at
monthly intervals. The first spray requires 120 gal/acre with foliage 3 in
high; the second spray 200 gal/acre and so on. Very susceptible cultivars
may need fortnightly spraying.

OAK

PESTS

**Caterpillars** of the **green oak tortrix** (*Tortrix viridana*), the winter
moth (*Operophtera brumata*) and many other species of moth feed on
leaves and may cause severe defoliation. Control measures are the same as
those given for tortrix caterpillars and winter moths in Chapter 9, p. 176.

DISEASES

**Powdery mildew** (*Microsphaera alphitoides*). Leaves and shoot tips
become covered with a floury-white coating and young shoots especially
can be severely injured and crippled in growth. Control is effected by
spraying either dinocap at 2 oz a.i. (8 oz 25% w.p.)/100 gal *or* colloidal
sulphur (40–50%) at 2 pints/100 gal *or* wettable sulphur $2\frac{1}{2}$–4 lb 75%
w.p./100 gal.

PAEONY

PESTS

**Leaf and bud eelworm** (*Aphelenchoides ritzemabosi*)—see Chapter 10,
p. 241.

DISEASES

**Wilt** (*Botrytis paeoniae*). The bases of the shoots develop a brown area
and soon wither. Leaves of other shoots can also develop brown, angular
patches usually at the tips and even the flower buds may turn brown and
die. Dust the crowns in early spring with a Bordeaux powder dust and
remove wilting stems by cutting them off (below ground level) for burning.

Good results have recently been obtained by spraying with dichlofluanid at 1 lb a.i. (2 lb 50% w.p.)/100 gal at 10–14 day intervals from emergence. Tree paeonies should be sprayed with captan at 1 lb a.i. (2 lb 50% w.p.)/100 gal *or* thiram at 25½ oz a.i. (2 lb 80% w.p.)/100 gal *or* zineb at 21 oz a.i. (2 lb 65% w.p.)/100 gal, commencing soon after foliage opens and repeating at 10–14-day intervals until flowering. Dichlofluanid could also be tried on this type of paeony.

## PANSY

### DISEASES

***Leaf spot** (*Ramularia* spp.). Small circular spots develop on the leaves and these are either greenish or yellowish, or whitish or straw-coloured with a brown margin. Spray with a copper-containing fungicide, e.g. Bordeaux mixture equivalent to 1–1½ lb metallic copper/100 gal *or* copper oxychloride at 6 oz metallic copper (12 oz 50% w.p.)/100 gal *or* maneb 2 lb a.i. (2½ lb 80% w.p.)/100 gal *or* zineb 21 oz a.i. (2 lb 65% w.p.)/100 gal.

***Stem rot** (*Mycothecium roridum*). This fungus is considered to be responsible at least in part for a brown rotting at the stem bases of violas and other plants. Apart from proper attention to soil drainage, etc., it is advised that 4% calomel dust be sprinkled in the planting holes.

## PEACH

See Prunus, p. 321.

## PELARGONIUM

### PESTS

***Common green capsid** (*Lygocoris pabulinus*). These active insects grow up to ¼ in long and are yellowish-green when young, becoming more green later. The injuries caused are similar to those of the tarnished plant bug but damage to flowers is less common. Punctured leaves of geranium pucker and tear as they grow, taking on a ragged lace-like appearance. Damage may occur from May onwards. To control this pest, keep the land clear of weeds and debris. Apply DDT 15 oz a.i. (3 pint 25% e.c.)/100 gal *or* diazinon at 3·2 oz a.i. (16 fl. oz 20% e.c.)/100 gal *or* gamma-BHC at 2 oz a.i. (½ pint 20%)/100 gal *or* nicotine at 8 fl. oz a.i./100 gal *or* TDE at ¾ lb a.i. (1½ lb 50% w.p.)/100 gal, giving 2–3 applications at 14-day intervals, starting as early as possible. Also spray the ground around the plants.

### DISEASES

****Rust** (*Puccinia pelargonii-zonalis*) appears on the undersurfaces of leaves as brown spore masses often arranged roughly in rings. Spray at fortnightly intervals with thiram at 25½ oz a.i. (2 lb 80% w.p.)/100 gal *or* zineb at 32 oz a.i. (3 lb 65% w.p.)/100 gal *or* preferably thiram and zineb alternately at the above rates.

## PETUNIA

DISEASES

**Foot rot** (*Phytophthora cryptogea*) causes a rot of the roots and stem bases. Seed dressing with captan 75% *or* thiram 50% at 2 oz to 28 lb seed has recently been advised. The soil can also be watered with captan at 2 oz a.i. (4 oz 50% w.p.)/10 gal *or* with Cheshunt compound at 1 oz metallic copper/2 gal *or* with zineb at 2 oz a.i. (3 oz 65% w.p.)/10 gal.

## PHLOX (Perennial)

PESTS

****Stem eelworm** (*Ditylenchus dipsaci*). The eelworms feed in stem and leaf tissues. On infested plants apical leaves often narrow and strap-like while lower leaves may be rolled, swollen and puffy or crinkled. Stems may be stunted, swollen or spindly, with a tendency to crack longitudinally. Infested stools may be given hot-water treatment (1 hour at 110°F). Clean stock may be raised from infested plants by propagating from root cuttings or seed. Eelworms on infested land must be starved out by denying them alternative host-plants, e.g. aubretia, gypsophila, helenium, annual phlox and certain weeds, and infested sites should be kept free of such host-plants for 2 or 3 years.

## PHLOX DRUMMONDII (Annual)

PESTS

**Aphids,** mainly peach-potato aphid (*Myzus persicae*)—control as for aphids on Dahlia (p. 305).

DISEASES

****Leaf spot** (*Septoria drummondii*). Brown spots with a pale centre develop on the leaves and eventually coalesce to give a scorched effect. Treat the seed with a captan 75% seed dressing at 2 oz/28 lb. Spray early with zineb at 21 oz a.i. (2 lb 65% w.p.)/100 gal.

**Powdery mildew** (*Sphaerotheca fuliginea*). A white, powdery coating on leaves, controlled by sprays of dinocap at 2 oz a.i. (8 oz 25% w.p.)/100 gal.

## PLANE

DISEASES

**Leaf scorch** (*Gloeosporium nervisequum*). Brown discoloration on the leaves roughly following the main veins causes the leaves and sometimes young shoots to wither. The fallen leaves should be collected and burnt in autumn. Spray with Bordeaux mixture equivalent to 1–1½ lb metallic copper/100 gal if possible in April and again before bud burst.


## POLYANTHUS

### PESTS

***Angleshades moth** (*Phlogophora meticulosa*)—see Dahlia (p. 305).

****Bryobia mite** (*Bryobia* spp.). These yellowish-green to reddish-brown mites suck the sap, causing a yellow mottling on the leaves which may spread until the whole leaf is yellowed. To control them, apply demeton-S-methyl 3·4 oz a.i. (6 fl. oz 58% e.c.)/100 gal *or* derris 0·65 oz a.i. (13 fl. oz 5% rotenone e.c.)/100 gal *or* dimethoate 3·6 oz (9 fl. oz 40% e.c.)/100 gal *or* malathion 18 oz a.i. (1½ pint 60% e.c.)/100 gal *or* tetradifon 2 oz a.i. (1¼ pint 8% conc.)/100 gal, repeating as necessary.

****Red spider mite** (*Tetranychus urticae*)—see entry under Dahlia (p. 305).

****Stem eelworm** (*Ditylenchus dipsaci*). Attacked flower stems become stunted, twisted and disfigured with pale or brown blisters and longitudinal furrows. The leaves become blotched with yellow and later turn brown and die, and the crown of the plant may also decay. No specific control measure is available but valuable stock may be worth treating with hot water as for Phlox (p. 319).


### DISEASES

**Leaf spots** (*Ramularia* spp. and *Phyllosticta* spp.). Various sized discoloured spots appear on the leaves making them unsightly and in some cases damaging or killing the leaves. These fungi are controlled by spraying with copper oxychloride at 6 oz metallic copper (12 oz 50% w.p.)/100 gal outdoors, but at 3 oz a.i. (6 oz)/100 gal under glass.


## POPLAR

### DISEASES

***Leaf spots** (*Marssonina* spp.). Small irregular blackish-brown spots occur on the leaves which fall prematurely. Control may be difficult but dead shoots should be cut out where possible, and smaller trees can be sprayed not less than three times in the spring and at least once in summer with Bordeaux mixture, equivalent to 1–1½ lb metallic copper/100 gal *or* copper oxychloride at 6 oz metallic copper (12 oz 50% w.p.)/100 gal.

***Yellow leafblister** (*Taphrina populina*). Affected leaves are distorted and bear large blisters which are bright yellow on the lower surface but remain green on the upper side. Where the disease is troublesome and it is feasible, spray in January or February with Bordeaux mixture equivalent to $1$–$1\frac{1}{2}$ lb metallic copper/100 gal *or* lime sulphur 3 gal/100 gal.


## PRIVET

PESTS

**Lilac leaf miner** (*Caloptilia syringella*)—see Lilac (p. 312).

**Privet thrips** (*Dendrothrips ornatus*). These tiny elongate insects are yellow when young but the adult has brown bars across the body. They feed on the sap of the foliage, causing silvery speckling and distortion. The affected foliage dies and falls during the winter. To control, spray with DDT at 15 oz a.i. (3 pints 25% e.c.)/100 gal *or* gamma-BHC 2 oz a.i. ($\frac{1}{4}$ pint 20% e.c.)/100 gal *or* malathion 18 oz a.i. ($1\frac{1}{2}$ pint 60% e.c.)/100 gal; repeat as necessary.


DISEASES

***Honey fungus** (*Armillaria mellea*) (A.L. 500). This fungus attacks the roots and kills the bushes especially in hedges. Spread is mainly by means of runner-like strands called 'rhizomorphs'. The fungus only shows above ground as toadstools growing at soil level on the bark of dead trees and bushes. The removal and burning of dead bushes with as much root as possible is recommended, together with soil disinfection using 2% formaldehyde (1 pint 38% formalin/49 pints water) to soak the freshly forked infected site.


## PRUNUS (Ornamental)

PESTS

The pests found on Cherry, Damson or Plum, see Chapter 9, p. 194, also attack ornamental Prunus.


DISEASES

**Bacterial canker** (*Pseudomonas mors-prunorum*). Lesions appear on leaves and branches in early spring with often some gumming on the branches. As these cankers develop, the shoots die back and some branches tend to be flattened. Leaves may show small brown spots. Spraying with Bordeaux mixture $1$–$1\frac{1}{2}$ lb/100 gal at end of August and end of September effects control.

***Peach leaf curl** (*Taphrina deformans*) (A.L. 81) causes the leaves to develop large blisters, at first red, which swell up and turn white often involving whole leaves and large numbers of them. To control, spray the bare branches *before* the flower buds burst in January or early February, again a fortnight later and once again just before leaf fall with Bordeaux mixture, equivalent to $1$–$1\frac{1}{2}$ lb metallic copper/100 gal *or* lime sulphur 3 gal/100 gal.

Y

**Silver leaf** (*Stereum purpureum*) (A.L. 246), so called from the silvering of foliage though the fungus is present inside branch or main stem. Apart from domestic plums it has been reported on ornamental species such as *Prunus spinosa, P. triloba, P. laurocerasus, P. avium.* Fungus fruit bodies appear only on dead branches, stems or roots. As it is a wound parasite, all wounds made in Prunus trees should be painted with a wound sealing preparation containing mercury or of a bituminous type. Infected dead trees should be grubbed before July 15 and are best kept dry until burnt.

## PYRACANTHA

PESTS

Many of the pests of apple, including green apple aphid (*Aphis pomi*), brown scale (*Parthenolecanium corni*), lackey moth (*Malacosoma neustria*) and vapourer moth (*Orgyia antiqua*) cause trouble and may be controlled by the methods given in Chapter 9, p. 160. *et seg*

DISEASES

**Scab** (*Fusicladium pyracanthae*) covers the berries with an olive-brown to black coating so spoiling their appearance. The leaves can also be attacked and defoliation caused. As copper fungicides are likely to cause damage, spray with captan 1 lb a.i. (2 lb 50% w.p.)/100 gal *or* lime sulphur (1 in 40) three times in March/April and twice (1 in 80) in June.

## PYRETHRUM

PESTS

**Leaf and bud eelworm** (*Aphelenchoides ritzemabosi*)—see Chrysanthemum, Chapter 10, p. 241.

DISEASES

**Grey mould** (*Botrytis cinerea*). Greyish furry mould attacking flower stalks, which can be controlled by spraying with captan at 1 lb a.i. (50% w.p.)/100 gal *or* thiram at $25\frac{1}{2}$ oz a.i. (2 lb 80% w.p.)/100 gal H.V.

## RHODODENDRON

PESTS

**Caterpillars,** e.g. March moth (*Alsophila aescularia*), winter moth (*Operophtera brumata*) fruit tree tortrix (*Archips podana*), etc., feed on young leaves and buds and can occasionally cause considerable damage. Control measures are the same as those given for caterpillars on Roses, p. 324.

***Rhododendron bug** (*Stephanitis rhododendri*) (A.L. 206). These shiny, yellowish to dark brown insects, up to ⅛ in long, feed in groups on the lower surface of leaves which show characteristic rusty or chocolate-brown spotting. The upper surface becomes mottled with yellow giving the shrub a sickly appearance. This damage is first seen in May or June and may be prevented by sprays of DDT at 8 oz a.i. (1 lb 50% w.p.)/100 gal *or* gamma-BHC at 3 oz a.i. (¾ pint 20% e.c.)/100 gal *or* malathion 18 oz a.i. (1½ pints 60% e.c.)/100 gal *or* nicotine 8 fl. oz 95–98%/100 gal *or* parathion 1·6 oz a.i. (8 fl. oz 20% e.c.)/100 gal. 2–3 applications at intervals of 3 weeks starting in mid-June or when infestations first seen should be made with the spray directed to the underside of the leaves.

*****Rhododendron leafhopper** (*Graphocephala coccinea*). This strikingly marked green and red leafhopper is one of the most important pests of rhododendron as it is mainly responsible for the infection of flower buds by bud blast disease (see below). Chemical control of the leafhopper reduces bud blast infections and may be achieved by spraying with DDT at 25 oz a.i. (5 pints 25% e.c.)/100 gal *or* malathion at 18 oz a.i. (1½ pints 60% e.c.)/100 gal. Sprays should be applied on two or three occasions during August and September, the object being to reduce the local leaf-hopper population and limit egg-laying as it is the wounds made in the bud scales by the female's ovipositor that provide the infection sites for the fungus.

***Rhododendron whitefly** (*Dialeurodes chittendeni*). The young stages appear as small yellowish-green 'scales', difficult to distinguish, on the underside of the leaves. The adults resemble miniature, white-winged moths and cluster on the foliage. Both stages suck the sap and may severely affect the vigour of the shrub. In some cases the foliage may become disfigured with yellow mottling. Honeydew is produced in copious quantities leading to further disfigurement by sooty moulds. Spray the underside of leaves with DDT at 15 oz a.i. (3 pints 25% e.c.)/100 gal *or* malathion 18 oz a.i. (1½ pints 60% e.c.)/100 gal, giving 2–3 applications at 14-day intervals.

***Weevils,** e.g. **clay-coloured weevil** (*Otiorhynchus singularis*) and **vine weevil** (*O. sulcatus*) (A.L. 57). These small wingless beetles are about ½ in long and black or brown. They feed on leaves, petioles and stems, usually at night, and hide in leaf litter, etc., during daylight. Spray foliage with DDT at 20 oz a.i. (4 pints 25% e.c.)/100 gal *or* apply DDT wettable powder (20 or 25%) as a dust to the soil under affected plants at about 1 oz/yd².


DISEASES

***Gall or False bloom** (*Exobasidium vaccinii*). The young leaves and sometimes the flower buds swell up into galls which soon turn white. Small-leaved rhododendrons are often severely attacked. The galls should be removed and burnt before they turn white; spray with zineb at 21 oz a.i. (2 lb 65% w.p.)/100 gal.

**Bud blast** (*Pycnostysanus azaleae*). The killed buds turn brown, black or silvery in spring and black bristle-like spore heads protrude from them. The fungus is a wound parasite and it is thought to enter wounds made by the Rhododendron Leafhopper (*Graphocephala coccinea*) as it deposits its eggs in the flower buds. Apart from picking off as many infected buds as possible, spray to control the hopper (see Insect Pests, p. 323).

ROSE

PESTS

***Aphids,** e.g. rose aphid (*Macrosiphum rosae*), feed on the foliage, shoots and buds, causing stunting and distortion. For their control, spray as necessary, with demeton-S-methyl 3·4 oz a.i. (6 fl. oz 58% e.c.)/100 gal *or* diazinon 3·2 oz a.i. (16 fl. oz 20% e.c.)/100 gal *or* dimethoate 4·8 oz a.i. (12 fl. oz 40% e.c.)/100 gal *or* formothion 7·5 oz a.i. (16 fl. oz 43% e.c.)/100 gal *or* malathion 18 oz a.i. (1½ pints 60% e.c.)/100 gal *or* nicotine 8 fl. oz 95–98%/100 gal *or* oxydemeton-methyl 3·4 oz a.i. (6 fl. oz 57% e.c.)/100 gal *or* parathion 1·2 oz a.i. (6 fl. oz 20% e.c.)/100 gal.

**Caterpillars,** e.g. tortrix moths (*Archips podana*, etc.), winter moths (*Operophtera brumata*, etc.), lackey moth (*Malacosoma neustria*), vapourer moth (*Orgyia antiqua*), etc. Caterpillars of very many different species feed on leaves, buds and blossoms. Many are fully exposed while others may gain protection by rolling or tying leaves together, by spinning silk coverings or by tunnelling into buds. Hand-picking can be effective on a small scale. On large-scale infestations spray with DDT at 15 oz a.i. (3 pints 25% e.c.)/100 gal or TDE at 12 oz a.i. (1½ lb 50% w.p.)/100 gal. Trichlorphon might also be effective.

**Red bud borer** (*Thomasiniana oculiperda*). These small reddish maggots feed in groups on the sap under new bud grafts and prevent the graft 'taking'. To prevent their attack, smear the graft with petroleum jelly after tying or apply DDT at 15 oz a.i. (3 pints 25% e.c.)/100 gal before or immediately after grafting.

**Red spider mite** (*Tetranychus urticae*)—see Chapter 10, page 264.

*Rose leafhopper** (*Typhlocyba rosae*). These pale yellow insects, up to ⅛ in long, feed on the underside of the foliage, producing a distinctive yellow spotting on the upper surface. The leaves are also disfigured by honeydew and sooty moulds. The adults are active, jumping insects. Damage is noticeable from May onwards and may be prevented by the methods given for rhododendron leafhopper (see p. 323).

**Rose thrips** (*Thrips fuscipennis*). These tiny, elongated, yellowish to dark-brown insects feed on the sap of the petals in opening buds causing them to be distorted and streaked with brown when the blooms open. They also feed on foliage which becomes disfigured with silvery mottling, and may be controlled by the methods cited for privet thrips (see p. 321).

**Sawfly larvae** are common rose pests, e.g. ***Leaf-rolling rose sawfly (*Blennocampa pusilla*), the pale green larvae of which shelter inside longitudinally rolled leaves, eating away the margin, and are found from June to August. Roses are disfigured and their vigour may be affected by severe attacks. The pest is difficult to control but applications of nicotine dust (3–4%) when the air temperature is 65°F or over, are partially effective against the larvae. Alternatively, to prevent egg-laying, DDT 15 oz a.i. (3 pint 25% e.c.)/100 gal may be applied at 14-day intervals from early May to mid June. The **rose slug sawfly (*Endelomyia aethiops*) feeds, as yellowish-green larvae, on one surface of the leaf, leaving the other intact. The injured area turns brown giving the foliage a mottled and scorched appearance; damage occurs mainly in June. Spray, as necessary, with DDT 15 oz a.i. (3 pints 25% e.c.)/100 gal *or* derris 0·65 oz a.i. (13 fl. oz 5% rotenone e.c.)/100 gal *or* gamma-BHC 2 oz a.i. ($\frac{1}{2}$ pint 20% e.c.)/100 gal *or* malathion 18 oz a.i. ($1\frac{1}{2}$ pints 60% e.c.)/100 gal *or* nicotine at 8 fl. oz 95–98%/100 gal (with wetter). Other sawfly pests include the antler sawflies (*Cladius pectinicornis* and *C. difformis*), the banded rose sawfly (*Allantus cinctus*) and the large rose sawfly (*Arge ochropus*). These yellowish-green or greyish-green larvae eat irregular holes out of the leaves and may cause severe defoliation. The large rose sawfly also lays its eggs in young shoots and the stalks of the blooms causing them to become blackened and twisted. They are controlled by the methods given above for rose slug sawfly.

Trichlophon at 12·8 oz a.i. (1 lb 80% s.p.)/100 gal has given good control of sawflies.

**Common green capsid** (*Lygocoris pabulinus*)—see Pelargonium, p. 318.

**Scale insects,** e.g. *brown scale (*Parthenolecanium corni*) and *mussel scale (*Lepidosaphes ulmi*), see Chapter 9, p. 173; *scurfy scale (*Aulacaspis rosae*) appears as small, dirty-white scales on stems causing disfigurement and poor growth. Their control is obtained by spraying 2–3 times, at 14-day intervals, with diazinon 6·4 oz a.i. (32 fl. oz 20% e.c.)/100 gal *or* malathion 30 oz a.i. ($2\frac{1}{2}$ pints 60% e.c.)/100 gal *or* parathion 1·2 oz a.i. (6 fl. oz 20% e.c.)/100 gal.


DISEASES

*****Black spot** (*Diplocarpon rosae*). Circular black spots on the leaves and in some species the fungus can be found in scabby places on the shoots. Start spraying immediately after pruning and repeat at fortnightly intervals throughout the season using captan at 1 lb a.i. (2 lb 50% w.p.)/100 gal *or* maneb 2 lb a.i. (2½ lb 80% w.p.)/100 gal. Promising results have recently been obtained with dichlofluanid at 1 lb a.i. (2 lb 50% w.p.)/100 gal used at 10–14 day intervals commencing when the leaves begin to unfold. Folpet is also effective but is currently available only in mixture with dinocap for combined black spot and mildew control.

**Powdery mildew** (*Sphaerotheca pannosa*). The fungus is the white powdery coating on shoots and young leaves which are crippled. Spray early with dinocap at 2 oz a.i. (8 oz 25% w.p.)/100 gal, though some petal-spotting may result on white-flowered cultivars. Repeat at 10–14 day intervals as necessary. Folpet when used for the control of black spot (see above) gives in addition some control of mildew.

**Rust** (*Phragmidium mucronatum*) causes orange coloured pustules on the leaves in summer and in autumn darker brown ones often on the same spots. Spraying with maneb 2 lb a.i. (2½ lb 80% w.p.)/100 gal *or* thiram at 3·2 lb a.i. (4 lb 80% w.p.)/100 gal *or* zineb at 21 oz a.i. (2 lb 65% w.p.)/100 gal.

SEEDLINGS (Various)

DISEASES

***Damping-off** (*Pythium* and *Phytophthora* spp.). These fungi cause a shrinking and rotting of stems at soil level and the plants soon collapse. The seed should be treated with captan 75% seed dressing at 2 oz/28 lb. Use enough just to give the seed a fine coating of the dressing. The soil can also be watered with captan at 2 oz a.i. (4 oz 50% w.p.)/10 gal *or* with zineb at 2 oz a.i. (3 oz 65%)/10 gal. Cheshunt compound at 1 oz metallic copper/2 gal is also recommended for this purpose.

SNOWDROP

PESTS

**Large narcissus fly** (*Merodon equestris*)—see Narcissus (p. 315).
**Stem eelworm** (*Ditylenchus dipsaci*)—see Narcissus (p. 316).
**Tulip bulb aphid** (*Dysaphis tulipae*)—see Tulip (p. 329).

DISEASES

**Grey mould** (*Botrytis galanthina*). The leaves and flower stalks are soon destroyed by a grey mould while the sclerotia formed on the bulbs contaminate the soil. Destroy infected clumps and disinfect the soil by mixing a quintozene dust (20%) in the soil at ratio of about 2 oz/yd². Spray the remaining healthy clumps with captan at 1 lb a.i. (2 lb 50% w.p.)/100 gal.

## SPRUCE

### PESTS

****Conifer spinning mite** (*Oligonychus ununguis*). These reddish-orange mites with black markings are found among fine webbing on young shoots together with round reddish-orange eggs. They suck the sap of the foliage causing it to assume an unhealthy greyish or yellowish-brown appearance. Damage is seen from May onwards, and may be prevented by spraying in late May or when damage is seen, with demeton-S-methyl 3·4 oz a.i. (6 fl. oz 58% e.c.)/100 gal *or* malathion 30 oz a.i. (2½ pints 60% e.c.)/100 gal *or* oxydemeton-methyl 3·4 oz a.i. (6 fl. oz 57% e.c.)/100 gal *or* parathion 1·2 oz a.i. (6 fl. oz 20% e.c.)/100 gal, repeating the application 2–3 weeks later. Other organophosphorus compounds used to control fruit tree red spider mite (see p. 170) may also be effective. If regular spraying is to be carried out, other types of chemicals should be alternated with the above to avoid a possible build-up of resistant strains of mite, e.g. dicofol 6·4 oz a.i. (32 fl. oz 20% e.c.)/100 gal *or* quinomethionate 2 oz a.i. (8 oz 25% w.p.)/100 gal *or* tetradifon 2 oz a.i. (1¼ pints 8% conc.)/100 gal. There are, however, no data on the efficiency of these materials.

*****Green spruce aphid** (*Elatobium abietinum*). This small red-eyed greenfly feeds on the undersides of the needles of spruce throughout the year except for a short period in summer. *Picea sitchensis* and *P. pungens* var. *glauca* are outstandingly susceptible but few spruce species are immune. Sucking of the sap causes yellow mottling and loss of all but the current year's foliage. Damage is associated with mild winters and normally occurs in early spring, and also from October onwards when autumn and winter are particularly mild. Control by spraying in March/April or when the aphid is first noticed, using malathion at 18 oz a.i. (1½ pints 60% e.c.)/100 gal.

****Spruce gall adelges** (*Adelges abietis* and *A. viridis*) are tiny brownish-black insects which form distinctive 'pine-apple' galls on the shoots, inside which they suck the sap. *A. abietis* also feeds on the bark and needles under patches of white 'wool' and causes yellowing of the needles. Both types of damage may seriously affect the vitality of the tree, and may be prevented by spraying with gamma-BHC 8 oz a.i. (2 pints 20% e.c.)/100 gal, with a wetter added, in early April.

### DISEASES

*****Damping-off of seedlings** (mainly *Pythium* spp.), causes damage in seedbeds and occurs in two phases of seedling development. In the pre-emergence phase the germinating seed is attacked below soil level, leaving bare patches in the nursery and, in the post-emergence phase, the seedling collapses due to attack at soil level, often whilst in the cotyledon stage. To control the disease a post-emergence drench should be applied immediately symptoms are observed using captan at 2 oz a.i. (4 oz 50% w.p.)/10 gal water/10 yd² of seedbed. If damping-off is a recurrent problem the nursery beds should be sterilized with formalin, 1 gal 38% formaldehyde/15 gal water/15 yd² of seedbed at least 3 weeks before sowing.

## STOCK

DISEASES

**Clubroot** (*Plasmodiophora brassicae*) (A.L. 276). Irregular swellings develop on the roots and the plants remain stunted. For long-term control, lime the soil using 1 to 2 ton hydrated lime/acre. For immediate control in affected beds, rake in 4% calomel dust at 1½ oz/yd² *or* dip the roots when planting in a 'sludge' made of 4% calomel dust (1 lb/⅓ pint water) or pure calomel (2 oz/1 pint water) and a little clay.

**Downy mildew** (*Peronospora parasitica*). The leaves develop yellow blotches which on the underside show a greyish furry coating and the plants may be crippled. It is important to keep plants grown for seed free from this disease by spraying with zineb at 28 oz a.i. (2½ lb 65% w.p.)/100 gal.

## SWEET PEA

PESTS

***Aphids,** e.g. peach-potato aphis (*Myzus persicae*)—control as for aphids on Dahlia, p. 305, but do not use malathion.

## SWEET WILLIAM

DISEASES

**Leaf spot** (*Didymellina dianthi*) appears as pale coloured circular spots on the leaves which can soon be destroyed. Do not grow plants too soft, and spray, at first sign of the disease, with Bordeaux mixture equivalent to 1–1½ lb metallic copper/100 gal *or* with copper oxychloride at 6 oz metallic copper (12 oz 50% w.p.)/100 gal and repeating as required.

*Rust (*Puccinia arenariae*). On the leaves brown pustules appear arranged roughly in rings. In autumn this disease may spread rapidly and cause great destruction. Do not encourage soft lush growth which is easily attacked and spray, at the first sign of the disease, with a copper-containing spray as given for Leaf spot above *or* with thiram at 3·2 lb a.i. (4 lb 80% w.p.)/100 gal *or* zineb at 28 oz a.i. (2½ lb 65% w.p.)/100 gal.

## SYCAMORE

See Acer, p. 298.

## SYRINGA

See Lilac, p. 312.

## THUJA

### PESTS

**Conifer spinning mite** (*Oligonychus ununguis*)—see Spruce, p. 327.

### DISEASES

***Needle blight** (*Didymascella* (*Keithia*) *thujina*) (F.C.L. 43). This disease can cause serious damage to *Thuja plicata* in the nursery, particularly in transplant lines, and occasionally attacks *T. occidentalis*. Individual needles turn brown and later small dark-brown spots about $\frac{1}{25}$ in across appear on the upper surface of the needle. In a heavy attack most of the needles may become infected on the lower shoots and considerable dieback occurs. (These symptoms should not be confused with winter bronzing, which is a natural occurrence during severe winters, when the whole plant turns reddish-brown recovering its natural colour in the following spring.) There is no fungicide at present marketed which will control this disease. Excellent control has been achieved using cyclo-heximide at 85 ppm/100 gal/acre, and this may become commercially available in the future.

## TULIP

### PESTS

*Bulb mite** (*Rhizoglyphus echinopus*)—see Narcissus, p. 314.
***Stem eelworm** (*Ditylenchus dipsaci*). Infested plants show fine cracks and pale or purplish streaks on the flower stem. In severe infestations the stem is distorted, split or swollen and the flowers may show pale or green streaking. On forced plants the leaves may also show pale streaks and tendency to split. Infested bulbs may have grey or brownish patches just above the basal plate. Thorough rogueing, strict attention to hygiene and care in cultural practices and in buying in new stocks will help to limit eelworm attacks. Hot water treatment of the bulbs has not proved satisfactory but, as an alternative, a cold dip in 0·23% thionazin (4 pints 46% e.c./100 gal) for 2 hours during July or August will give good control.
**Tulip bulb aphid** (*Dysaphis tulipae*). The yellowish or brownish aphids feed under the dry outer scales of stored bulbs and reduce their vitality. When the bulbs are planted the insects attack the young shoots, severely checking their growth and causing distortion. Infested bulbs should be immersed for 15 minutes in a solution of gamma-BHC 1 oz a.i. ($\frac{1}{4}$ pint 20% e.c.)/10 gal water to which a wetter has been added *or* fumigated with nicotine or gamma-BHC. On growing plants spray as for

aphids on Dahlia (see p. 305). For other aphids, e.g. peach-potato aphid (*Myzus persicae*) and potato aphid (*Macrosiphum euphorbiae*), the control is as for aphids on Dahlia (p. 305).

DISEASES

*****Fire disease** (*Botrytis tulipae*) (A.L. 536) is a most common and serious disease of tulips and is carried over by resting bodies (sclerotia) in the soil or on the bulbs. Young leaves are scorched brown at their tips and young shoots crippled. Older leaves show small spots and the flower petals are spoiled by similar spots. In some cases the shoot is injured at a leaf axil and the flower stem falls over. To control the disease, mix quintozene dust (20%) with the top soil at planting time at 2 oz/yd² (5½ cwt/acre) *or* the bulbs can be dusted with this powder at 10 lb for each ton of bulbs to reduce underground infection of the shoots. The growing plants should be sprayed in spring from the time they reach 1 in high with captan at 1 lb a.i. (2 lb 50% w.p.)/100 gal *or* maneb at 1·2–1·6 lb a.i. (1½–2 lb 80% w.p.)/acre L.V. *or* thiram at 2–3·2 lb a.i. (2½–4 lb 80% w.p.)/acre *or* zineb at 28 oz a.i. (2½ lb 65% w.p.)/100 gal.

****Grey bulb rot** (*Sclerotium tuliparum*). Unlike Fire (see above) which is serious on the foliage, this disease attacks the growing points and usual prevents the shoot getting far above ground. The nose portion is affected by a greyish rot and the bulb is soon destroyed. Resting bodies (sclerotia) are formed on the rotting portions and released into the soil where they are scattered by cultivation and will infect other bulbs the next season. The result is that bare patches are seen in the beds. Soil treatment is with quintozene dust (20%) at 2 oz/yd² or 5½ cwt/acre.

TURF

PESTS

****Chafer grubs** (*Phyllopertha horticola*, etc.) are whitish, fleshy grubs, with brown heads and curved bodies. *P. horticola* grows to about ½ in length but other species up to 1½ in long may also cause damage. They feed on the grass roots causing withered patches to appear. For control, use insecticides as for leatherjackets (below).

**Earthworms** (*Allolobophora nocturna* and *A. longa*). These species are responsible for depositing casts on the surface of turf, particularly during the spring and autumn, making it unsightly and causing unevenness. The turf should be dressed with chlordane (25% e.c.) 2 fl. oz/2 gal/12½ yd² *or* derris dust (0·5% rotenone) 2 oz/yd², watered well in, *or* lead arsenate (32%) 2 oz/yd² watered in *or* mowrah meal, 6–8 oz/yd² watered in under pressure *or* water with potassium permanganate solution, ½ oz/gal/yd². All except chlordane and lead arsenate bring the worms to the surface where they should be swept up. Of these materials only potassium permanganate should be used if there is risk of the contamination of ponds or water containing fish.

**Leatherjackets** (*Tipula paludosa, T. oleracea*, etc.). The greyish-brown or greyish-black grubs with tough wrinkled skin, feed on the roots of the grass causing it to wither. Damage is more severe if the autumn of the previous year has been wet. Water with DDT $7\frac{1}{2}$ oz a.i. ($1\frac{1}{2}$ pints 25% e.c.)/ 100 gal/200 yd² *or* apply a dressing of DDT $\frac{1}{2}$ oz (5%) dust/yd² *or* gamma-BHC (25% w.p.) $\frac{1}{2}$ lb/$\frac{1}{2}$ cwt bran/acre *or* lead arsenate $1\frac{1}{2}$–2 oz/yd². Treatment is most successful if given in mild, humid weather in autumn or spring.

DISEASES

**Dollar spot** (*Sclerotinia homoeocarpa*). Small brown patches, about 2 in across at first brown and then rather bleached, appear. The disease can be controlled by the calomel treatment recommended below for Snow mould.

***Fairy rings** (*Marasmius oreades* and other fungi). In summer the ring is composed of three zones, an outer and an inner in both of which the grass is green and vigorous while between there is a middle zone where the grasses are brown and killed by the fairy ring fungus. The fruit bodies of the fungi appear in summer and autumn especially in wet weather and towards the outer parts of the ring. This type of disease is often difficult to control because the fungus forms so dense a mat of mycelium in the soil that it is difficult to get a fluid to penetrate the affected area; the mycelium can be present to a depth of up to 18 in in some soils. It is usually necessary to strip the turf from the affected zones and this should be destroyed. The soil beneath should be forked or 'spiked' and then soaked with a fungicide; it has been suggested that corrosive sublimate (2 oz/22 gal) could be used but formalin (6 pints 38% formaldehyde plus $\frac{1}{2}$ pint non-ionic wetting agent/10 gal water/10 yd²) has also been recommended.

**Ophiobolus patch** (*Ophiobolus graminis*). This disease shows as depressed circular patches a few inches in diameter in which the grass has a light straw-coloured or bronzed appearance. Affected areas may increase in size and coalesce to form large irregularly-shaped patches. As the disease is usually restricted to the bents (*Agrostis* spp.), the centres of the patches become colonized by undesirable grasses and weeds. The disease may be confused with Snow mould (see below) but it is normally found only where the soil surface has a high pH and in wet conditions. Microscopic examination of the diseased grasses is required for confirmation of the disease as it cannot be controlled by the usual turf fungicides. Applications of phenylmercury acetate at 2 oz fungicide (containing 2·5% mercury) in 6 gal water/40 yd² at 10-day intervals in conjunction with sulphate of ammonia give some control. It has also been reported in America that chlordane applied periodically at 4 fl. oz (75% e.c.) in 10 gal water/1,000 ft$^2$ will control the disease.

****Red thread** (*Corticium fuciforme*). This fungus causes pinkish patches from 3–5 in in diameter on the turf and these are composed of dead grass with coral-red horn-like fungus growths showing amongst it. Control measures as for Snow mould below, are recommended.

***Seedling diseases** (*Pythium* spp., *Fusarium* spp. and *Helmintho-sporium* spp.). Various fungi can cause newly-sown seed to rot or prevent the emergence of seedlings and the same fungi can also cause a collapse of seedlings at ground level once they have emerged. To prevent trouble in newly-sown turf, care should be taken in the preparation of the seed bed. In addition the seed can be treated with captan 75% seed dressing at 1 oz/28 lb *or* thiram 50% seed dressing at 1½ oz/28 lb.

****Snow mould or Fusarium patch** (*Fusarium nivale*). Large brown patches covered with a whitish fungus mycelium appear on lawns especially after a snowfall. Water the affected patches with pure calomel at 3 oz in 30 gal/100 yd$^2$ of turf, repeating at monthly intervals. Addition of wetter is advisable. A mixture of 2 oz calomel and 1 oz corrosive sublimate/ 1,000 ft$^2$ may also be used. Experiments have recently shown that quinto-zene gives a good control of this disease used at a rate of 1–1·3 g a.i. (5–6·5 g 20% dust)/yd$^2$/month *or* 4 oz a.i. (6 oz 72% w.p.)/2 gal water/ 100 yd$^2$/month. It is really desirable to use chemical control as a preventive rather than a curative measure and the first application of fungicide should be given in autumn repeating at monthly intervals during conditions favouring fungal infection.

VIOLA

See Pansy, p. 318.

VIOLET

PESTS

****Leaf and bud eelworm** (*Aphelenchoides fragariae*) causes few definite symptoms apart from malformation of leaves and poor growth generally. Infested stock should be destroyed and the land allowed to remain fallow over the winter.

****Red spider mite** (*Tetranychus urticae*)—see Chapter 10, p. 231.

****Violet leaf midge** (*Dasyneura affinis*). The small, whitish or pale orange maggots feed on the leaves which are rolled inwards towards the mid-rib and usually greatly swollen. When badly infested the plant is dwarfed and flower production reduced or prevented altogether. The control of the midge is difficult but applications of DDT 15 oz. a.i. (3 pint 25% e.c.)/100 gal in early May, early July, mid-August and mid-October will reduce the numbers of egg-laying adults. On the continent the application of schradan to infested plants is reported to have controlled the larvae.

## WALLFLOWER

PESTS

**Cabbage root fly** (*Erioischia brassicae*). The white maggots eat away the lateral roots and tunnel in the main stem causing wilting and death of the plant. For control, see Chapter 11, p. 281.

**Caterpillars,** e.g. **angleshades moth** (*Phlogophora meticulosa*)—see under Dahlia (p. 305).

DISEASES

****Clubroot** (*Plasmodiophora brassicae*) (A.L. 276). The roots develop irregular swellings and the plants remain stunted. For long-range control, lime the soil using 1–2 ton hydrated lime per acre. For immediate control in affected beds, rake in 4% calomel dust at about $1\frac{1}{2}$ oz/yd$^2$ or dip the roots when planting in a 'sludge' made of 4% calomel dust ($1$ lb/$\frac{1}{3}$ pint water) or pure calomel ($2$ oz/$1$ pint water) and a little clay.

****Downy mildew** (*Peronospora parasitica*). Yellow blotches develop on the leaves and on the under surfaces a greyish furry coating can be seen. Affected plants may be crippled. Plants grown for seed should be sprayed with zineb at 28 oz a.i. ($2\frac{1}{2}$ lb 65% w.p.)/100 gal.

## WILLOW

PESTS

**Aphids,** e.g. willow stem aphid (*Tuberolachnus salignus*), large dark brown aphids feeding on shoots. Spring or summer infestations are controlled by demeton-S-methyl 3·4 oz a.i. (6 fl. oz 58% e.c.)/100 gal *or* diazinon 3·2 oz a.i. (16 fl. oz 20% e.c.)/100 gal *or* dimethoate 4·8 oz a.i. (12 fl. oz 40% e.c.)/100 gal *or* gamma-BHC 2 oz a.i. ($\frac{1}{2}$ pint 20% e.c.)/100 gal *or* malathion 18 oz a.i. ($1\frac{1}{2}$ pint 60% e.c.)/100 gal *or* oxydemeton-methyl 3·4 oz a.i. (6 fl. oz 58% e.c.)/100 gal *or* parathion 1·2 oz a.i. (6 fl. oz 20% e.c.)/100 gal.

****Willow ermine moth** (*Yponomeuta rorrella*). The small grey black-headed caterpillars of this moth feed in colonies on the foliage. The stems and leaves are enclosed in a web 'tent' which protects the caterpillars as they feed. Their control is as for small ermine moths on fruit (p. 173). For the control of other caterpillars, including lackey moth (*Malacosoma neustria*), tortrix moths (*Cacoecia* spp. and *Pandemis cerasana*), vapourer moth (*Orgyia antiqua*) and winter moths (*Operophtera brumata, Alsophila aescularia*, etc.)—see Chapter 9, p. 176.

****Willow scale** (*Chionaspis salicis*). The whitish scales encrust the stems and may be controlled by spraying in spring and summer with diazinon 6·4 oz a.i. (32 fl. oz 20% e.c.)/100 gal *or* malathion 30 oz a.i.

(2½ pint 60% e.c.)/100 gal *or* parathion 1·2 oz a.i. (6 fl. oz 20% e.c.)/ 100 gal.

### DISEASES

****Anthracnose** (*Marssonina salicicola*) appears as small brown spots on leaves and small blackish cankers on shoots of various *Salix* spp. The leaves fall prematurely and should be raked up and burnt. On large trees control is often not possible and feeding to encourage vigour and resistance may be tried. Where possible, cutting out badly affected shoots and spraying with Bordeaux mixture (1–1½ lb metallic copper)/100 gal or copper oxychloride at 6 oz metallic copper (12 oz 50% w.p.)/100 gal is advised, one application being given as the leaves unfold and at least one more in summer.

## YEW

### PESTS

****Yew scale** (*Parthenolecanium pomeranicum*). The oval, very convex, chestnut-brown scales on the stems, cover the foliage with honeydew on which grow sooty moulds. Severe infestations may cause bronzing of the foliage and defoliation. In January or February apply a tar oil wash at 5 gal 80% concentrate/100 gal. Spring and summer infestations are controlled as for scale on Willow (p. 333).

## YUCCA

### DISEASES

***Leaf spot** (*Coniothyrium concentricum*). The large brown well-defined spots with grey centres caused by this disease often show small dots in concentric rings on their surfaces. These appear on the leaves and spoil the appearance. Remove severely attacked leaves and spray with Bordeaux mixture equivalent to 1–1½ lb copper/100 gal *or* copper oxychloride at 6 oz metallic copper (12 oz 50% w.p.)/100 gal to protect young develop- ing leaves.

## ZINNIA

### DISEASES

*****Grey mould** (*Botrytis cinerea*). Even outdoors, zinnias are likely in wet seasons to suffer from rotting of the flowers and even collapse through stem rot caused by grey mould. To control the disease, remove all dead and dying leaves regularly and dust repeatedly with tecnazene dust (5%) at 1 lb for each 60 yd².

****Seedling blight** (*Alternaria zinniae*). This disease is seed-borne and causes reddish-brown spots with greyish centres on the leaves and dark brown canker-like areas on the stems. The seed should be treated with an organomercury *or*, preferably, a captan 75% seed dressing using 2 oz/28 lb. Even in the boxes the seedlings may need to be protected by spraying with Bordeaux mixture equivalent to $1-1\frac{1}{2}$ lb copper/100 gal *or* copper oxychloride at 6 oz metallic copper (12 oz 50% w.p.)/100 gal *or* zineb at 22 oz a.i. (2 lb 65% w.p.)/100 gal.


### ZONAL PELARGONIUM

See Pelargonium, p. 318.

# CHAPTER 13

# PESTS AND DISEASES OF FOREST CROPS

Commercial forest crops have, relatively speaking, a very low per acre value, and the forester tends therefore to rely on the cultural and ecological approach to his pest problems rather than the chemical one. Traditionally he has reserved insecticide and fungicide application as a means of saving his trees from death, on the basis that the cost of treatment could only be justified when his actual capital (or growing stock) is at risk. The same consideration does not apply, however, in forest nurseries where the per acre value can be very high indeed and losses correspondingly costly. There is, too, the special case of Norway spruce which is often planted out at high density for early thinning to supply the Christmas tree market and for this reason can economically carry comparatively high costs of treatment. Many forest trees are, of course, grown as ornamentals but, as such, will be judged by the yardstick of amenity rather than that of timber producer (see Chapter 12). Here, prescriptions for control have been strictly confined to those which apply to normal forest practice.

Routine chemical prophylactic treatments are seldom considered appropriate in forestry, the two outstanding exceptions to this rule being the measures taken against the fungus *Fomes annosus* and the weevil *Hylobius abietis*. The former organism can be regarded as the most important enemy of forest trees in Britain, not only on account of its ubiquity and for the continuing losses it causes, but for its increasing potential as a serious fungal pest in second and subsequent rotations. There are no known means of eradicating the disease once forest land has become infected and protective measures against invasion should be carried out after all thinning and felling operations.

It may be noted that the range of chemicals recommended against any one pest in this chapter may be rather more restricted than those given in others. This is often due to the choice having been limited on grounds of risks to workers; forests tend to be in isolated areas and distant from aid; they are also rough in terrain and provide conditions where men sweat and are inclined to strip off protective clothing. In general then, less toxic materials are chosen when possible.

It is not possible to give recommendations for rates of application, other than in nurseries, due to variations in age and therefore size of plants. The practice of spot-treating rather than applying to the whole area further confuses the issue, as does variation in planting density. A general injunction to 'visibly wet' foliage when using a manually-operated sprayer with fine nozzle is a useful guide with most materials, although definitely not a rule which should be applied with motorized knapsack mist-blowers. In this case rates can only be controlled by the speed of walking down the rows of trees, a single pass from one side being made in the case of small trees and a double pass from two sides in the case of larger ones.

## 13.1 FOREST NURSERIES
### PESTS

**Aphids.** A number of aphid species feed on forest nursery stock but only a few ever require measures to be taken against them.

   *__Beech woolly aphis__ (*Phyllaphis fagi*). A small yellow to green aphis which feeds on the underside of beech leaves in May to July, causing them to wilt, turn brown and fall. Control may be obtained by spraying, in spring with malathion 0·1% (1½ pints 60% e.c.)/100 gal/acre.
   *__Cherry blackfly__ (*Myzus cerasi*). See p. 194.
   *__Oak Phylloxera__ (*Phylloxera punctata*). An orange-yellow aphis which feeds on the underside of oak leaves from May to July, causing the leaves to wilt and turn a dirty mottled grey-brown. For control, spray in spring with malathion 0·1% (1½ pints 60% e.c.)/100 gal/acre.
   ***__Green spruce aphis__ (*Elatobium abietinum*). This small, red-eyed, greenfly feeds on the undersides of the needles of spruce throughout the year except for a short period in summer. *Picea sitchensis* and *P. pungens* var. *glauca* are outstandingly susceptible but few spruce species are immune. Sucking of the needle sap causes yellow mottling and loss of all but current year's foliage. Damage is associated with mild winters and normally takes place in early spring from March to June; damage may also occur from October onwards when autumn and winter are particularly mild. For control, spray when the aphis is noted, or, in March/April with malathion 0·1% (1½ pints 60% e.c.)/100 gal/acre.
   *__Pine root aphis__ (*Prociphilus* spp.). A small greyish-white aphis which covers itself and the root system of pine transplants with waxy wool. Severe infestation can cause yellowing, wilting and even death of individual trees. Infected plants are best burnt and the ground they occupied spot-treated with BHC 0·5% dust.

   *__Beech jassid__ (*Typhlocyba cruenta* var *douglasi*). A yellow aphis-like

z

insect which jumps like a flea on disturbance. It causes yellowing of the foliage between May and July and is particularly common where beech hedges are growing in the nursery area. For control, spray in spring with malathion 0·1% (1½ pints 60% e.c.)/100 gal/acre.

***Chafer.** Chafers are soil-inhabiting beetle larvae of the May or June bug (*Melolontha melolontha*), the garden chafer (*Phyllopertha horticola*) and of *Serica brunnea*. Damage takes place throughout the summer months. Attack first becomes obvious through the sudden wilting and browning of individual plants in the bed. When these are examined it will be found they can be removed from the ground with the slightest of effort, the whole root system having been severed about ½ cm or so below ground level. Damage is most common in heavier soils, particularly where there are surrounding hardwood hedges or woodlands; conversely it is rare in heathland nurseries.

No control is possible to the growing crop though ground known to be infested can be treated prior to use by working 0·5% gamma-BHC dust into the top 4 in of soil at the rate of 1–1½ cwt/acre.

****Conifer spinning mite** (*Oligonychus* (*Paratetranychus*) *ununguis*) is tiny, being barely visible to the naked eye in both egg and adult stages. The egg is orange-red and may be found, with the aid of a hand lens, on the wood of the past few years growth. Sitka and Norway spruce are particularly prone to attack, Scots pine less so. The young and adult stages are brown-coloured and are found on the needles and shoots amongst fine silk. Damage to transplants through their sap-sucking appears as a dirty yellowing, and continues throughout the period spring to autumn. In severe cases, treatment should be carried out when the overwintering eggs have hatched in late May, by spraying dicofol 0·1% (1½ pints 20% e.c.)/100 gal/acre.

****Cutworms.** Cutworms are caterpillars mainly of the Turnip moth (*Agrotis segetum*), the Heart and dart moth (*Agrotis exclamationis*) and the Yellow under-wing (*Noctua pronuba*) which inhabit the soil during day-light hours. Damage to seedlings takes place above ground at night, and begins in July. It reaches a peak in early autumn and ceases as temperatures fall in September or October. A short spell of intense feeding again occurs in February to May, depending on weather conditions, until the insects are fully fed and ready to pupate. The attack on conifer seedlings is quite characteristic; the caterpillars bite through the stem, usually just above soil level, and either leave the remains lying on the ground, or pull them down into the soil a short distance, leaving a portion still above ground. As soon as damage appears, spray the seedbeds with gamma-BHC at 0·1% (4 pints 20% e.c.)/100 gal/acre.

****Poplar leaf beetles.** *Phyllodecta vitellinae* and *P. vulgatissima* are shiny, metallic blue or green beetles some 5 mm long. Their larvae are dirty grey-brown in colour and are black dotted. They feed, as a group, skeletonizing the underside of the leaves of most poplar species. The adult of *Chrysomela populi* is about 12 mm long and is brick-red in colour, the larvae are cream with black dots and are similar to those of the Colorado beetle in general appearance. This species completely skeletonizes leaves, particularly of *Populus tremula*, and also eats holes out of them. All species can be found from May throughout the summer, and are particularly damaging in poplar stool beds. For control, apply DDT 0·1% (3 pints 25% e.c.)/100 gal/acre.

***Sawflies.** The main species are *Trichiocampus viminalis*, the Birch sawfly (*Croesus septentrionalis*), Large willow sawfly (*Nematus salicis*) and *Nematus melanaspis*. These caterpillar-like insects display a great variety of coloration at different stages of development and between species. Their feeding habits include skeletonization and biting of leaves. One or other may be found from May until October. For control, apply DDT 0·1% (3 pints 25% e.c.)/100 gal/acre.

*****Springtails.** The springtail, *Bourletiella hortensis*, is an occasional severe pest of conifer seedlings and particularly of *Pinus contorta*. The insect is no more than the size of a speck of dust but becomes quite noticeable when disturbed through its curious flea-like jumping habit. It is most active in the first few hours of dawn, between the end of April to mid-June. Damage is to the emerging seedling and appears as minute lesions on hypocotyl and cotyledons. The effect in the first year of growth is to produce a dumpy little seedling with a bush of distorted, fattened, needles. In the second year the damaged seedling grows on to produce a multileadered useless plant. For control apply malathion 0·1% (1½ pints 60% e.c.)/100 gal/acre of seedbed.

**Weevils.** A number of weevil species can be most damaging both to conifer and hardwood nursery stock. This is particularly so when nurseries are surrounded by woodland in which populations may first increase and later invade the nursery. Other ground vegetation, such as heather or grass may also provide the primary breeding ground.

****Barypithes araneiformis** and *B. pellucidus* are two very small weevils, about 3 mm long and dark brown in colour. The adult beetles feed at night between the end of April to mid-July. Death of emergent seedlings results from feeding upon, and severing of, the hypocotyl and from ring-barking of later stage seedlings. Control may be obtained by spraying 0·125% DDT (4 pints 25% e.c.)/100 gal/acre.

*****Clay-coloured weevil** (*Otiorhynchus singularis*) and *Strophosomus melanogrammus*, are two speckled brown weevils which attack a great variety of tree species, particularly when the nursery is surrounded by trees. The latter weevil is very common on larch, and Western hemlock; *Tsuga heterophylla* seems particularly susceptible to both. The adults are active from April until October, and it is this stage only which is damaging through feeding on the fine shoots and needles.

Spray when damage first appears, in April–May, with 0·125% DDT (4 pints 25% e.c.), *or* 0·025% of BHC (1 ping 20% e.c.)/100 gal/acre.

**Otiorhynchus ovatus.* This weevil is damaging both in the adult and larval stages. The adult is about $\frac{1}{2}$ cm long, with chestnut-brown body and darker head. It is active from April–May onwards and feeds at night on the finer shoots and needles of most conifers. The larva, which is indistinguishable from that of other weevils, feeds on finer roots throughout the summer months.

Control measures can be taken against both adult and larval stages. Against adults a spray of 0·125% DDT (4 pints 25% e.c.)/100 gal should be applied in April–May or when the insects first appear; against larvae, soil known to be infested should be treated before use by working in 0·5% gamma-BHC dust into the top 4 in of soil at the rate of 1–1$\frac{1}{2}$ cwt/acre.


# DISEASES

***Damping-off** (mainly *Pythium* spp.). Damping-off causes damage in seedbeds and occurs in two phases of seedling development. In the pre-emergent phase, the germinating seed is attacked below soil level, leaving bare patches in the nursery; in the post-emergent phase, the seedling collapses due to attack at soil level, often whilst in the cotyledon stage.

A post-emergent drench of captan at 2 oz a.i./10 gal water/10 yd$^2$ of seedbed should be applied immediately symptoms are observed. If damping-off is a recurrent problem the nursery beds should be sterilized with formalin (1 gal 38% formaldehyde/15 gal water/15 yd$^2$ of seedbed) at least 3 weeks before sowing.

***Grey mould** (*Botrytis cinerea*) (Forestry Commission Leaflet No. 50). A sparse web of grey mycelium appears over affected portions of plants often in damp weather or after they have been weakened or killed by frost, or other predisposing factors. *Sequoia* and *Cupressus* are particularly susceptible to Grey mould but many conifer species are attacked in seedbeds or, less frequently, in transplant lines.

Spray with Bordeaux mixture, equivalent to 2$\frac{1}{2}$ lb metallic copper/ 100 gal, immediately the disease is seen and repeat if necessary at intervals of 10–14 days. Thiram at 3 lb a.i. (4 lb 80% w.p.)/100 gal/acre, or captan at 1 lb a.i./100 gal/acre, can also be used.

**Leaf cast of larch** (*Meria laricis*) (Forestry Commission Leaflet No. 21). The first sign of this disease is a discoloration or browning of the middle or outer end of the needles in early May. It may be confused with frost damage which tends to kill whole needles at once, and of course takes place immediately after the frost. The needles attacked by the disease fall prematurely after becoming brown whilst those killed by frost remain attached longer. The disease is most serious in plants which have stood over either in seedbeds or transplant lines.

For control, spray with either wettable or colloidal sulphur, equivalent to $1\frac{1}{2}$ lb sulphur (3 lb sulphur w.p. or colloidal sulphur)/100 gal/acre, at flushing time and at 2–3 week intervals until the end of July.

****Needle cast of pine** (*Lophodermium pinastri*) causes the death of needles on pine seedlings and transplants particularly when a pine nursery is established close to a pine plantation. Pine should not be sown in nurseries close to mature pine plantations but, where this is unavoidable and infection appears, zineb or maneb should be applied at the rate of $2\frac{1}{2}$ lb a.i. (3 lb 80% w.p.)/100 gal/acre. Three annual sprays are recommended at monthly intervals at the end of July, August and September. Spraying should begin earlier if signs of infection are present.

*****Needle blight of Western red cedar** (*Didymascella* (*Keithia*) *thujina*) (Forestry Commission Leaflet No. 43). This disease can cause serious damage to Western red cedar in the nursery, particularly in transplant lines. Individual needles or fronds turn brown and, at a later stage, dark brown spots appear on the upper surface of the needle. In a heavy attack the majority of the needles may become infected on the lower fronds and considerable dieback occurs.

These symptoms should not be confused with winter bronzing, which is a natural occurrence during severe winters, when the whole plant turns reddish-brown, recovering its natural colour in the following spring.

There is no fungicide at present marketed which will control this disease. Excellent control has been achieved using cycloheximide at 85 ppm/ 100 gal/acre, and this may become commercially available in the future.

****Oak mildew** (*Microsphaera alphitoides*). During the summer the leaves and young shoots are covered with the powdery white spores of this fungus. This disease can be controlled in the nursery by the use of wettable or colloidal sulphur, equivalent to $1\frac{1}{2}$ lb sulphur (3 lb sulphur w.p. or colloidal sulphur)/100 gal/acre, as soon as it appears.

## 13.2 FOREST PLANTATIONS

### PESTS

#### YOUNGER PLANTATIONS

**Aphids.** Forest plantation trees are well endowed with aphid species, but there is seldom economic justification for their chemical control. The obvious exceptions to this rule are those species which cause degrade of Christmas trees (*Picea abies*).

***Green spruce aphis** (*Elatobium abietinum*) (see under Nursery above). Control of this aphis is seldom considered necessary on a forest scale since, though some loss in growth increment will be associated with defoliation, recovery is good. Such defoliation may not, however, be tolerable to the Christmas tree grower in which case the control measures given under Forest Nurseries above should be applied.

***Spruce shoot aphis** (*Cinara pinicola* syn. *Cinaropsis pilicornis*) is a grey or brown colonial aphis which infests the new shoots of spruce in June–August. Even large populations appear not to harm the trees directly but their copious honeydew production and the subsequent sooty moulds which grow upon this exudation can cause Christmas trees to be unsaleable at the end of the year. For control apply malathion 0·1% (1½ pints 60% e.c./100 gal).

********Adelges*. Spruce gall, or Woolly aphis (Forestry Commission Leaflet No. 7). These aphids have extremely complex life cycles with up to five forms often involving alternation between a spruce host and one other conifer which may be larch, Douglas fir, a pine or a Silver fir. Most stages secrete waxy wool and can cause yellowing and needle loss. To the forester only *Adelges nuesslini* (or one of its close allies now recognized as separate species) can cause plant death and is of sufficient economic importance to deserve control. The control method described, however, would be effective against most of the other species. This method is best carried out in spring, with a first application in mid-April, before the rapid increase of the summer generations have begun. Spray with gamma-BHC at 0·05% (2 pints 20% e.c./100 gal) *or* malathion 0·1% (1½ pints 60% e.c./100 gal) in two or more applications, at 2–3 week intervals, to obtain an effective control.

On the spruce host, those *Adelges* species which produce the characteristic spruce gall are of importance to the Christmas tree grower only. The control is as above but the timing of the first spray in mid-April is critical, in that the aphids must be killed before their feeding initiates gall formation.

*****Black pine beetles,** *Hylastes* spp. (Forestry Commission Leaflet No. 3). The most common species to be found on pine are *Hylastes ater* in the south and *Hylastes brunneus* in the north, and on spruce, *Hylastes cunicularius*. These are bark beetles which share the breeding sites of the Pine weevil (see below), and normally commence breeding on stumps some 6–8 months after felling. They are small, black, and less than 0·5 cm long. The damage caused by the adults of *Hylastes* spp. is to the root collar and main roots of young transplants which are ring-barked and killed. For control, see below under Pine weevil, p. 344.

***Conifer spinning mite** (*Oligonychus* (*Paratetranychus*) *ununguis*) (see under Forest Nurseries above). Infestation in the field is more common in the south than the north, and is associated most often with Norway spruce grown as Christmas trees in sites dry for this tree species. For control, see under Forest Nurseries above.

***Pine** or **Fox-coloured sawfly** (*Neodiprion sertifer*) (Forestry Commission Leaflet No. 35) is a grey-green colonial feeding larva often with diffuse darker strips running longitudinally, and about 2·5 cm long. This species is very common on pines in May and June, feeding on all needles except those of the current year. The species is under excellent natural control by a virus which will normally decimate a population some 2 or 3 years after the first attack. *Pinus contorta* is particularly susceptible and when the species is grown in remote areas, where natural virus infection may be delayed, some artificial control action may become necessary. The virus suspension is unfortunately not commercially available but excellent control may be obtained by DDT 0·1% (3 pints 25% e.c./100 gal) *or* gamma-BHC at 0·025% (1 pint 20% e.c./100 gal).

***Pine shoot moth** (*Rhyacionia buoliana*) (Forestry Commission Leaflet No. 40) is a small caterpillar without marked features to distinguish it from near relatives. Single insects mine into buds and new developing shoots. The shoots are not completely severed but flop over and grow on, forming a large kink in the main axis (traditionally likened to a German posthorn!). Normally in Britain the proportion of stems affected is small and the economic impact is seldom great enough to warrant control measures. For control use DDT 0·25% (8 pints 25% e.c./100 gal) applied during spring.

***Spruce bell moth** or **Streaked pine bell** (*Epinotia tedella*) is a small moth whose caterpillar is barely distinguishable from many other members of its family. During August to October the caterpillars mine the needles of spruce and spin them, together with excrement, on to the shoot. These areas of brown spun needles are later removed by wind and rain, leaving bald patches on the branches. Severe infestation can cause complete defoliation. Recovery from attack is good but the plant is unsaleable by the Christmas tree grower. For control use DDT 0·1% (3 pints 25% e.c./100 gal).

**Weevils.** This group of insects is responsible for large losses to new transplants, and prophylactic measures against their visitations, particularly in felled-over areas, is essential if adequate final stocking is to be assured. *****Clay-coloured weevil** (*Otiorhynchus singularis*) and *Strophosomus melanogrammus*. See under Forest Nurseries above for details and control. The chemical dipping treatments described below for control of Pine weevil will also give protection against these species.

***Leaf weevils** (*Phyllobius argentatus*, *P. pyri*, etc.) are some $\frac{1}{2}$ cm or less long, often bright green in colour or a dark speckled brown. Damage is caused by adult feeding on the leaves of a wide range of hardwoods and sometimes also conifers. They are most numerous in June and July. For control apply DDT 0·125% (4 pints 25% e.c./100 gal) *or* BHC 0·025% (1 pint 20% e.c./100 gal).

*****Pine weevil** (*Hylobius abietis*) (Forestry Commission Leaflet No. 1) is a large black weevil speckled with cream coloured scales, some 1·5 cm long. It breeds in conifer stumps, attacking fairly soon after felling, and takes, on average, some 15–18 months to develop in this country. It will also breed in fresh-felled lying logs under that part of the bark in contact with the ground. After emergence the young adults feed on green-barked material, such as that provided by the main stem of young transplants, before themselves seeking a breeding site. The transplants are ring-barked and killed, and the whole crop may easily be lost in this way if prophylactic measures are not taken.

Treatment against both Pine weevil and Black pine beetle may be carried out in one operation: (1) Prior to planting. By complete dipping in a stable water-based suspension of finely ground gamma-BHC at 1·6% a.i. (16 pints 80% conc./100 gal). Control of the Pine weevil alone may also be effected by dipping tops only in DDT 5% (1 part 25% w.p. to 4 parts water). (2) After planting. If attack develops after planting, spot treatment should be carried out with DDT 1% (20 pints 25% e.c./100 gal).

***Winter moths,** mainly *Operophtera brumata*, Northern winter moth (*Operophtera fagata*) and Mottled amber (*Erannis defoliaria*), are all looper catepillars and may often be found in joint attack. The first two are basically green larvae and both have longitudinal stripes, the last is variable in colour but is usually reddish brown above and yellowish below and on the sides. They feed early in the year, in the south from April on to June and in the north into July. These species are well known pests of hardwoods and orchard trees but can also be damaging to conifers, particularly the spruces. Conifers underplanted beneath scrub hardwood are often severely defoliated by caterpillars originating in the crowns of the overstorey. There have also been a number of cases in north Scotland in which the first two species have occurred in outbreak densities, feeding on Sitka spruce and upon the heather natural vegetation. For control, use DDT at 0·1% (3 pints 25% e.c./100 gal).

OLDER PLANTATIONS

****Bark beetles.** There are many species of bark beetle attacking dead, dying or unhealthy trees and felled timber, but few of them will damage live, established tree crops. There are two main British exceptions, the Pine shoot beetle, *Tomicus* (*Myelophilus*) *piniperda*, on pine (Forestry

Commission Leaflet No. 3), and the Larch shoot beetle, *Ips cembrae*, on larch. Both these small, brown and black, beetles, some 0·5 cm long, bore into the shoots of their respective tree hosts, pruning the crowns and causing multiple leaders and reduced growth. After feeding in this fashion they require thick-barked stem material for breeding. In the case of *T. piniperda* this may be logs felled within the previous 6 months or trees heavily defoliated, or weakened by fire, fungal attack, etc. Successful primary attack for breeding purposes is rare but can occur in conditions marginal for the tree species, such as sand dunes. *Ips cembrae*, so far confined to east Scotland, is potentially a rather more dangerous pest as it seems that a primary attack can take place on apparently healthy larch. Breeding commences in spring, often in March, reaches a peak in May–June and falls off in July. A second generation is rare in Britain.

The basic control of these species lies in management of fellings so that all suitable breeding material is either removed from the forest or has the bark removed within 6 weeks, during the period from April to August. This is not always possible when calamities, e.g. large-scale windblow, occur.

Chemical control is only economic if timber is first collected in sizeable stacks. It is best done either as a log protection treatment immediately before the attack can take place, or as early as possible after initiation of attack in order to reach the beetles before their progeny have eaten their way deep under the bark. Gamma-BHC at 0·5% (20 pints 20% e.c./ 100 gal) is made up in diesel oil and sprayed, preferably by mist-blower, to treat not only the log surfaces but obtain maximum penetration into the stack. The dosage in gallons is calculated on the superficial area of the stack (i.e. S.A. = Area in ft² of two sides, two ends and top). Dosage = (S.A. ÷ 70) gal.

****Douglas fir seed wasp** (*Megastigmus spermatrophu*) (Forestry Commission Leaflet No. 8). This small, yellow insect attacks the immature seed of Douglas fir and can cause serious losses in crops where seed collections are required to be made. The adult attacks the seed throughout the month of June and protective measures must be taken to cover this period. For this purpose, apply malathion 0·5% (7½ pints 60% e.c./100 gal) in three applications at intervals of 10 days.

*****Pine looper** or **Bordered white moth** (*Bupalus piniaria*) (Forestry Commission Leaflet No. 32) is a pine needle-feeding looper caterpillar green in colour with a number of white longitudinal stripes, and some 3 cm long. The moth is on the wing in June and July and the larvae may be found from mid-June until October or even later. The damage, at its worst, may be complete defoliation, and subsequent secondary attack by Bark beetles can cause the death of trees. Infestation typically occurs in regions of low rainfall (20–25 in), on sandy soils, and in large areas of pure

pine plantations. Control is only practicable by aerial application of
½–1 lb DDT/1 gal of water/acre, using an e.c. formulation.

FELLED PRODUCE

***Ambrosia beetle** or **Pinhole borer** (*Trypodendron lineatum*). This
small black beetle, about 3 mm long, bores deep into the wood of conifers,
particularly that of spruce and larch. Typically it is most harmful in the
wetter western parts of Britain. The adult attacks from the end of April
until mid-July. Timber felled during the period November to January of
the previous year is most susceptible, and that in the period February–
March rather less so. The bore tunnels become blackened with fungi and
the damage is essentially of a technical nature.

Chemical control is only economic if the timber is first collected into
sizeable stacks. It is carried out as a protective measure immediately prior
to the start of the attack period in mid-April. Dosage and technique of
spraying as for Bark beetles (see under Older Plantations above), but the
0·5% of gamma-BHC (20 pints 20% e.c./100 gal) is diluted with water.

*****Bark beetles.** The felled produce of most tree species may accommodate
one species of Bark beetle or another. The loss of bark resulting from
attack may lead to accelerated drying and shake or cracking of the timber.
For control see under Older Plantations, Bark beetles, above.

## DISEASES

*****Fomes** (*Fomes annosus*) (Forestry Commission Leaflet No. 5 and 13).
*Fomes* root and butt rot is the most serious disease of both young and
mature conifers in Britain. The fungus may kill pines on dry, alkaline
soils with pH 6 and above; pines on more acid sites are not affected. In
other species, young trees may also be killed, but more commonly the
fungus rots the tree roots (as in Douglas fir) or proceeds from the roots into
the stem, where it produces an extensive butt rot.

Fructifications of *Fomes* often appear at the base of trees killed by the
disease or on stumps of previously infected ones. The upper surface of the
growing fructification is reddish-brown with a white margin and the
under-surface is white and pierced by fine pores. Immature fructifications
are commonly found on infected stumps or roots; these are small white
pustules often bearing half-formed pore surfaces.

The fungus enters the crop by means of airborne spores from *Fomes*
fructifications that infect exposed woody tissues of conifers, particularly
cut stump surfaces immediately after thinning or clear-felling. The fungus
grows into the stump tissue and down into the root system, and if the roots

of healthy trees are in contact with such infected tissues, they in turn become infected.

Control in mature plantations is based on the protection of cut stump surfaces, thus preventing infection by airborne spores. Stump protectants should be applied immediately after all thinning and felling operations.

Sodium nitrite is the standard material used for stump protection, and it is used as a 10% solution with the addition of the marker dye Disulphine Blue at 1 gram/gal. The solution is liberally applied to the freshly cut stump surface with a brush immediately after felling. A dye is always included to assist in complete application and so that it is clear that the treatment has been done.

Local Forestry Commission offices will advise on the purchasing and application of this chemical.

Sodium nitrite is poisonous (see Chapter 2) and must therefore be handled, used and stored in accordance with Ministry of Agriculture recommendations. Where there is doubt about the use of sodium nitrite on grounds of its poisonous nature (e.g. where there is danger of contaminating water sources, or in areas where unattended domestic animals may graze) either disodium octaborate or urea may be used instead. Disodium octaborate should be used as a 10% solution with the marker dye Naphthalene Red JS added at the rate of 1 teaspoonful/gal. Urea is used as a 20% solution with the same marker dye at the same rate.

In pine plantations only, there is a fourth possible stump treatment. In this case, spore suspensions of a competing fungus, *Peniophora gigantea*, can be applied to the freshly cut stump surface. This method of biological control is very efficient in pine plantations but with no other species. Advice on the use of this method of stump protection should be sought from the Forestry Commission.

If stump protectants can be applied throughout the life of a crop, it can be kept substantially free from *Fomes* and danger to subsequent crops is then much reduced. This is particularly important as it is very difficult to eradicate *Fomes* once it becomes established within a crop.

If an infected crop is felled and replanted with conifers, rapid infection of the new plants may take place from the stumps of the previously infected crop.

No firm recommendations can be given for the control of *Fomes* in crops planted on infested sites. The mechanical removal of the stumps of the previous crop appears to be an effective control measure, but this is an expensive operation only justified where the disease is severe. The use of *Peniophora gigantea* for stump treatments in pine areas already infected by *Fomes* may help to reduce infection in the subsequent crop.

# PROPERTIES OF INSECTICIDAL AND FUNGICIDAL COMPOUNDS

| Common name | Chemical name | Structure<br>Me = $-CH_3$<br>Et = $-C_2H_5$ | Solubility in water | Volatility mm. Hg | Mammalian toxicity* Acute oral LD50 for rats (single dose) |
|---|---|---|---|---|---|
| **A. Naturally-occurring compounds** | | | | | |
| 1. Derris (rotenone) | | 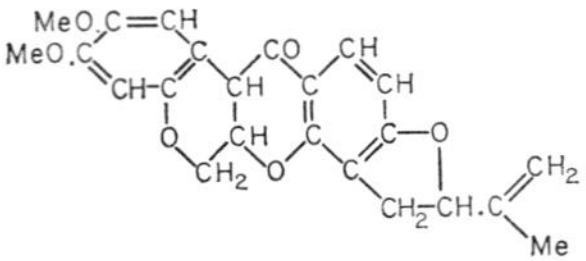 | 15 ppm at 100°C | Negligible | — |
| 2. Nicotine | *l*-3-(1-methyl-2-pyrrolidyl) pyridine | 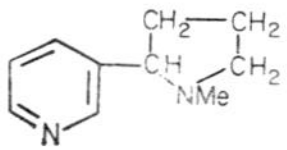 | Miscible below 60°C | 0.0425 at 25°C | 50–60 mg/kg |
| 3. Pyrethrum | mixed esters of pyrethrolone and cinerolone with chrysanthemic and pyrethric acids. | 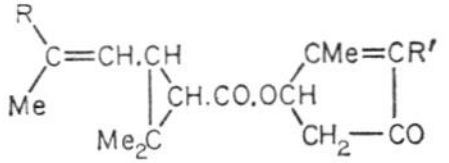<br>When R = —Me or —CO.OMe and R′ = $—CH_2.CH{=}CH.CH{=}CH_2$ or $—CH_2.CH{=}CH.Me$ | Negligible | Negligible | 800 mg/kg |
| **B. Organophosphorus compounds** | | | | | |
| 4. Azinphos-methyl | *S*-(3, 4-dihydro-4-oxo-benzo [*d*]-[1, 2, 3]-triazin-3-ylmethyl) dimethyl phosphorothiolothionate | 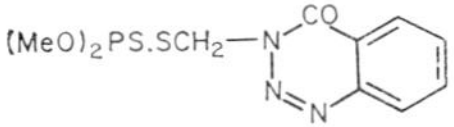 | 33 ppm | Below $3{\cdot}8 \times 10^{-4}$ at 20°C | 16·4 mg/kg, female rats |

| | Name | Chemical name | Formula | | | |
|---|---|---|---|---|---|---|
| 5. | Chlorfen-vinphos | 2-chloro-1-(2, 4-dichlorophenyl)-vinyl diethyl phosphate | $(EtO)_2PO.O-C=CHCl$ (structure) | 145 ppm at 23°C | $1·7 \times 10^{-7}$ at 23°C | 10–39 mg/kg |
| 6. | Demephion | mixture of demephion-O and demephion-S | | — | — | 138 mg/kg |
| | Demethion-O | dimethyl 2-(methylthio) ethyl phosphorothionate | $(MeO)_2PS.OCH_2CH_2SMe$ | 300 ppm | — | — |
| | Demethion-S | dimethyl $S$-[(2-methylthio)ethyl] phosphorothiolate | $(MeO)_2PO.SCH_2CH_2SMe$ | 3,000 ppm | — | — |
| 7. | Demeton-S-methyl | $S$-[2-(ethylthio)ethyl] dimethyl phosphorothiolate | $(MeO)_2PO.SCH_2CH_2SEt$ | 3,300 ppm | $3·6 \times 10^{-4}$ at 20°C | 40 mg/kg |
| 8. | Demeton-S-methyl sulphone | $S$-[2-(ethylsulphonyl) ethyl] dimethyl phosphorothiolate | $(MeO)_2PO.SCH_2CH_2SO_2Et$ | — | $0·5 \times 10^{-5}$ at 20°C | 37·5 mg/kg |
| 9. | Diazinon | diethyl (2-iso-propyl-6-methyl-4-pyrimidinyl) phosphorothionate | $(EtO)_2PS.O$ (structure with $C.CHMe_2$, $N$, $HC$, $C.Me$) | 40 ppm at 20°C | $1·4 \times 10^{-4}$ at 20°C | 100–150 mg/kg |
| 10. | Dichlorvos | 2, 2-dichlorovinyl di-methyl phosphate | $(MeO)_2PO.OCH = CCl_2$ | About 1% at 20°C | $1·2 \times 10^{-2}$ at 20°C | 80–107 mg/kg |
| 11. | Dimefox | $NNN'N'$-tetramethyl-phosphorodiamidic fluoride | $Me_2N$–$P(=O)(F)$–$Me_2N$ (structure) | Miscible | 0·36 at 25°C | 5 mg/kg |
| 12. | Dimethoate | dimethyl $S$-($N$-methylcarbamoylmethyl) phosphorothiolothionate | $(MeO)_2PS.SCH_2CONHMe$ | 2·5g/100 ml at room temperature | about $10^{-4}$ at 40°C | 245 mg/kg |

*As the actual hazard to people exposed to these pesticides depends on many factors such as manner of use, duration of exposure, safety precautions observed, nature of poisonous effect, ease of diagnosis of poisoning at an early reversible stage and availability of antidote, no simple assumptions should be made that the dangers are directly or solely related to the acute oral LD50 for experimental animals.

| Common name | Chemical name | Structure<br>Me = $-CH_3$<br>Et = $-C_2H_5$ | Solubility in water | Volatility mm. Hg | Mammalian toxicity*<br>Acute oral LD50 for rats (single dose) |
|---|---|---|---|---|---|
| 13. Disulfoton | diethyl S-2-(ethylthio)ethyl phosphorothiolothionate | $(EtO)_2PS.SCH_2CH_2SEt$ | 25 ppm at room temperature | $1.8 \times 10^{-4}$ at 20°C | 12.5 mg/kg |
| 14. Ethion | tetraethyl SS'-methylene bis(phosphorothiolothionate) | $(EtO)_2PS.SCH_2S.PS(OEt)_2$ | Negligible | $1.5 \times 10^{-6}$ at 25°C | 208 mg/kg (pure) 96 mg/kg (tech.) |
| 15. Ethoate-methyl | S-(N-ethyl-carbamoylmethyl) dimethyl phosphorothiolothionate | $(MeO)_2PS.SCH_2CONHEt$ | — | — | 125 mg/kg |
| 16. Fenchlorphos | dimethyl 2,4,5-trichlorophenyl phosphorothionate | $(MeO)_2PS.O$—(2,4,5-trichlorophenyl ring) | 40 ppm at 25°C | $8 \times 10^{-4}$ at 25°C | 1740 mg/kg |
| 17. Formothion | S-(N-formyl-N-methyl-carbamoylmethyl) dimethyl phosphorothiolothionate | $(MeO)_2PS.SCH_2.CO.N(Me_3)CHO$ | — | — | 330 mg/kg |
| 18. Malathion | S-[1,2-di (ethoxycarbonyl)ethyl] dimethyl phosphorothiolothionate | $(MeO)_2PS.SCH.COOEt$<br>$\mid$<br>$CH_2COOEt$ | 145 ppm at room temperature | $4 \times 10^{-5}$ at 30°C | 2,800 mg/kg |
| 19. Mecarbam | S-[N-ethoxycarbonyl-N-methylcarbamoylmethyl) diethyl phosphorothiolothionate | $(EtO)_2PS.SCH_2CON(Me)COOEt$ | Negligible | Negligible | 31–35 mg/kg |
| 20. Menazon | S-(4,6-diamino-1,3,5-triazin-2-ylmethyl) dimethyl phosphorothiolothionate | $(MeO)_2PS.SCH_2$—(4,6-diamino-1,3,5-triazin-2-yl ring with two $NH_2$ groups) | 250 ppm at 25°C | Negligible | 1950 mg/kg |

| | | | | | |
|---|---|---|---|---|---|
| 21. Mevinphos | 2-methoxycarbonyl-1-methylvinyl dimethyl phosphate | $(MeO)_2PO.OC(Me){=}CH.COOMe$ | Miscible | $2.9 \times 10^{-3}$ at 70°C | 6.8 mg/kg |
| 22. Morphothion | dimethyl $S$-(morpholino-carbonyl-methyl) phosphorothiolothionate | $(MeO)_2PS.SCH_2CO.N(CH_2CH_2)_2O$ | 0.5% | Negligible | 190 mg/kg |
| 23. Naled | 1,2-dibromo-2,2-dichloroethyl dimethyl phosphate | $(MeO)_2PO.OCHBr.CBrCl_2$ | Negligible | $2 \times 10^{-3}$ at 20°C | 430 mg/kg |
| 24. Oxydeme-ton-methyl | $S$-[2-(ethylsulphinyl) ethyl] dimethyl phos-phorothiolate | $(MeO)_2PO.SCH_2CH_2.SO.Et$ | Miscible | — | 65–75 mg/kg |
| 25. Parathion | diethyl 4-nitrophenyl phosphorothionate | $(EtO)_2PS.O{-}C_6H_4{-}NO_2$ | 24 ppm at 25°C | $3.78 \times 10^{-5}$ at 20°C | 6.4 mg/kg |
| 26. Phenkapton | $S$-(2,5-dichlorophenyl-thiomethyl) diethyl phosphorothiolothionate | $(EtO)_2PS.SCH_2S{-}C_6H_3Cl_2$ | Negligible | $4.1 \times 10^{-8}$ at 20°C | 182 mg/kg |
| 27. Phorate | diethyl $S$-(ethylthio-methyl) phosphoro-thiolothionate | $(EtO)_2PS.SCH_2.SEt$ | 50 ppm at 20°C | $8.4 \times 10^{-4}$ at 20°C | 3.7 mg/kg |
| 28. Phosalone | $S$-(6-chloro-2-oxobenzoxazolin-3-yl) methyl diethyl phosphoro-thiolothionate | $(EtO)_2PS.SCH_2{-}N{<}$ (6-chloro-2-oxobenzoxazolin ring) | Negligible | Negligible | 120 mg/kg |



*As the actual hazard to people exposed to these pesticides depends on many factors such as manner of use, duration of exposure, safety precautions observed, nature of poisonous effect, ease of diagnosis of poisoning at an early reversible stage and availability of antidote, no simple assumptions should be made that the dangers are directly or solely related to the acute oral LD50 for experimental animals.

| Common name | Chemical name | Structure<br>Me = $-CH_3$<br>Et = $C_2H_5$ | Solubility in water | Volatility mm. Hg | Mammalian toxicity*<br>Acute oral LD50 for rats (single dose) |
|---|---|---|---|---|---|
| 29. Phosphamidon | 2-chloro-2-diethyl-carbamoyl-1-methyl-vinyl dimethyl phosphate | $(MeO)_2PO.OC=\!\!\overset{\displaystyle Cl}{\underset{\displaystyle Me}{C}}\!\!.CO.N(Et)_2$ | Miscible | $2.5 \times 10^{-5}$ at 20°C | 28·3 mg/kg |
| 30. Schradan | bis-*NNN'N'*-tetra-methyl phosphoro-diamidic anhydride | $(Me_2N)_2PO.O.PO(NMe_2)_2$ | Miscible | $1 \times 10^{-3}$ at 25°C | 9·1 mg/kg |
| 31. Sulfotep | bis-*OO*-diethylphos-phorothionic anhydride | $(EtO)_2PS.O.PS(OEt)_2$ | 25 ppm at 20°C | $1.7 \times 10^{-4}$ at 20°C | *ca.* 5 mg/kg |
| 32. TEPP | Bisdiethylphosphoric anhydride | $(EtO)_2PO.O.PO(OEt)_2$ | Miscible | $1.55 \times 10^{-4}$ at 20°C | 1 mg/kg |
| 33. Thionazin | diethyl *O*-2-pyrazinyl phosphorothionate | (EtO)₂PS.O—[pyrazine ring] | 1140 ppm at 25°C | $3 \times 10^{-3}$ at 30°C | 12 mg/kg |
| 34. Trichlorphon | dimethyl 2,2,2-tri-chloro-1-hydroxyethyl phosphonate | $(MeO)_2PO_2.CH(OH)CCl_3$ | 15·4 g/160 ml at 25°C | $7.8 \times 10^{-6}$ at 20°C | 630 mg/kg |
| 35. Vamidothion | dimethyl *S*-[2-(1-methyl carbamoylethylthio) ethyl] phosphorothiolate | $(MeO)_2PO.SCH_2CH_2.SCHMe.CO.NHMe$ | Readily soluble | Negligible | 64–100 mg/kg |

## C. Chlorinated hydrocarbons

| | | Structure | Solubility in water | Vapour pressure (mm Hg) | Acute oral LD50* |
|---|---|---|---|---|---|
| 36. | Aldrin | 1,2,3,4,10,10-hexachloro-1,4,4a,5,8,8a-hexahydro-*exo*-1,4-*endo*-5,8-dimethano-naphthalene 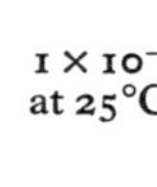 | 0.027 ppm at 25–29°C | $2.31 \times 10^{-5}$ at 20°C | 67 mg/kg |
| 37. | Carbon tetrachloride | — $CCl_4$ | 0.08% at 25°C | 114.5 at 25°C | 5,730–9,770 mg/kg |
| 38. | Chlordane | 1,2,4,5,6,7,10,10-octachloro-4,7,8,9-tetrahydro-4,7-methyleneindane 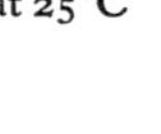 | Negligible | $1 \times 10^{-5}$ at 25°C | 457–590 mg/kg |
| 39. | DDT | 1,1,1-trichloro-2,2-di(4-chlorophenyl)ethane 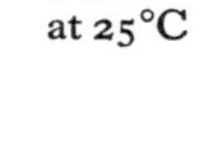 | Negligible | $1.9 \times 10^{-7}$ at 20°C | 113 mg/kg |
| 40. | Dieldrin | 1,2,3,4,10,10-hexachloro-6,7-epoxy-1,4,4a,5,6,7,8,8a-octahydro-*exo*-1,4-*endo*-5,8-dimethano-naphthalene  | 0.186 ppm at 25–29°C | $1.8 \times 10^{-7}$ at 25°C | 46 mg/kg |
| 41. | Endosulfan | 6,7,8,9,10,10-hexachloro-1,5,5a,6,9,9a-hexahydro-6,9-methano-2,4,3,-benzodioxa-thiepin-3-oxide 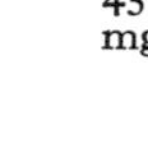 | Negligible | Negligible | 40–50 mg/kg |

*As the actual hazard to people exposed to these pesticides depends on many factors such as manner of use, duration of exposure, safety precautions observed, nature of poisonous effect, ease of diagnosis of poisoning at an early reversible stage and availability of antidote, no simple assumptions should be made that the dangers are directly or solely related to the acute oral LD50 for experimental animals.

| Common name | Chemical name | Structure<br>Me = −CH$_3$<br>Et = −C$_2$H$_5$ | Solubility in water | Volatility mm. Hg | Mammalian toxicity*<br>Acute oral LD50 for rats (single dose) |
|---|---|---|---|---|---|
| 42. Endrin | 1,2,3,4,10,10-hexa-chloro-6,7-epoxy-1,4, 4a,5,6,7,8a-octa-hydro-*exo*-1,4-*exo*-5,8-dimethanonaph-thalene | 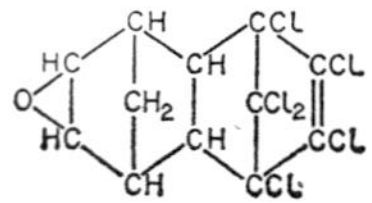 | Negligible | $2 \times 10^{-7}$ at 25°C | 17·5 mg/kg |
| 43. Ethylene dichloride | 1,2-dichloroethane | CH$_2$Cl . CH$_2$Cl | 0·43 g/100 ml at room temperature | 78 at 20°C | 670–890 mg/kg |
| 44. Gamma-BHC | $\gamma$-1,2,3,4,5,6-hexa-chlorocyclohexane | 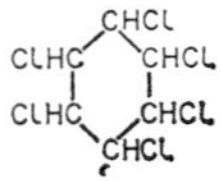 | 10 ppm at room temperature | $9·4 \times 10^{-6}$ at 20°C | 88 mg/kg |
| 45. Hexachloro-benzene | — | 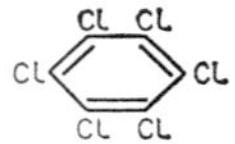 | Negligible | $1·089 \times 10^{-5}$ at 20°C | 10,000 mg/kg |
| 46. Paradichloro-benzene | *p*-dichlorobenzene | 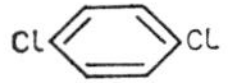 | 80 ppm at 25°C | 1·0 at 25°C | 500–5,000 mg/kg |
| 47. TDE | 1,1-dichloro-2,2,-di-(4-chlorophenyl)ethane | 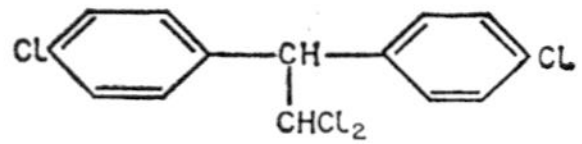 | Negligible | Negligible | 3,400 mg/kg |

## D. Bridged diphenyl acaricides

| No. | Common name | Chemical name | | | |
|---|---|---|---|---|---|
| 48. | Azobenzene | diphenyl diimide | Negligible | Appreciable | — |
| 49. | Chlorbenside | 4-chlorobenzyl 4-chlorophenyl sulphide | Negligible | $2 \cdot 59 \times 10^{-6}$ at 20°C | >2,000 mg/kg |
| 50. | Chlorfenson | 4-chlorophenyl 4-chlorobenzenesulphonate | Negligible | Negligible | *ca.* 2,000 mg/kg |
| 51. | Chlorobenzilate | ethyl 4,4′-dichlorobenzilate | Negligible | $2 \cdot 2 \times 10^{-6}$ at 20°C | 1,000–5,000 mg/kg |
| 52. | Dicofol | 2,2,2-trichloro-1,1-di-(4-chlorophenyl) ethanol | Negligible | Negligible | 809 mg/kg |
| 53. | Fenson | 4-chlorophenyl benzenesulphonate | Negligible | Negligible | 1,560–1,740 mg/kg |
| 54. | Tetradifon | 2,4,5,4′-tetrachlorodiphenyl sulphone | 0·02% at 50°C | $2 \cdot 4 \times 10^{-10}$ at 20°C | >5,000 |

*As the actual hazard to people exposed to these pesticides depends on many factors such as manner of use, duration of exposure, safety precautions observed, nature of poisonous effect, ease of diagnosis of poisoning at an early reversible stage and availability of antidote, no simple assumptions should be made that the dangers are directly or solely related to the acute oral $LD_{50}$ for experimental animals.

| Common name | Chemical name | Structure<br>Me = $-CH_3$<br>Et = $-C_2H_5$ | Solubility in water | Volatility mm. Hg | Mammalian toxicity*<br>Acute oral LD50 for rats (single dose) |
| --- | --- | --- | --- | --- | --- |
| 55. Tetrasul | 2,4,5,4′-tetrachloro-diphenyl sulphide | 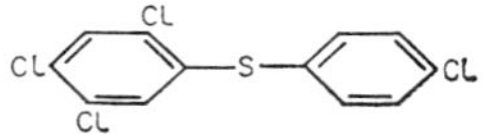 | Slightly soluble | $7.5 \times 10^{-7}$ at 20°C | 6,810 mg/kg |
| **E. Dithiocarbamates** | | | | | |
| 56 Dazomet | tetrahydro-3,5-di-methyl-2H-1,3,5-thia-diazine-2-thione | 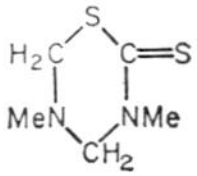 | 0.12% at 30°C | Volatile degradation products | 500 mg/kg |
| 57. Mancozeb | complex of zinc ion and maneb | — | Negligible | Negligible | 5,000 mg/kg |
| 58. Maneb | manganese ethylene-1,2-bisdithiocarbamate |  | Negligible | Negligible | 7,500 mg/kg |
| 59. Metham-sodium | sodium *N*-methyl-dithiocarbamate | Me.NH.CS.SNa | 72.2 g/100 ml at 20°C | Volatile degradation products | 820 mg/kg |
| 60. Metiram | complex of zineb and polyethylene thiuram disulphide | — | Negligible | Negligible | 10,000 mg/kg |
| 61. Nabam | Disodium ethylene-1,2-bisdithiocarbamate | $CH_2.NH.CS.SNa$<br>$\mid$<br>$CH_2.NH.CS.SNa$ | 20% at 20°C | Negligible | 395 mg/kg |

| No. | Name | | | | |
| --- | --- | --- | --- | --- | --- |
| 62. | Propineb | zinc propylene-bisdithiocarbamate | 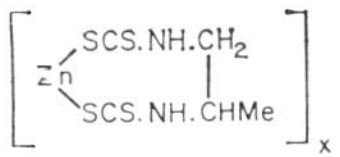 $\left[ Zn \left\langle \begin{array}{l} SCS.NH.CH_2 \\ SCS.NH.CHMe \end{array} \right. \right]_x$ | Negligible | Negligible | 8,500 mg/kg |
| 63. | Thiram | bis(dimethyl-dithiocarmoyl)disulphide | $Me_2N.CS.S$ — $Me_2N.CS.S$ | 30 ppm at room temperature | Negligible | Of low mammalian toxicity |
| 64. | Zineb | zinc ethylene 1,2-bis-dithiocarbamate |  $CH_2.NH.CS.S$ / $CH_2.NH.CS.S$ Zn | 10 ppm at 25°C | Negligible | More than 5,200 mg/kg |

**F. Dinitro compounds**

| 65. | Binapacryl | 2-(1-methyl-n-propyl)-4,6-dinitrophenyl 2-methylcrotonate | 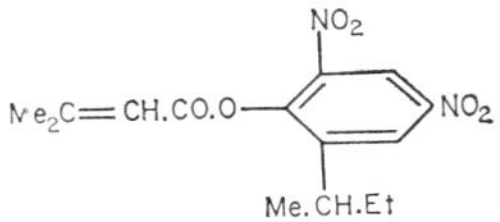 $Me_2C=CH.CO.O$—ring $NO_2$, $NO_2$, $Me.CH.Et$ | Negligible | $1 \times 10^{-4}$ at 60°C | 120–165 mg/kg |
| 66. | Dinobuton | isopropyl 2,4-dinitro-6-s-butylphenyl carbonate | 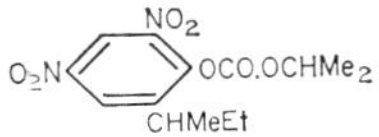 $O_2N$—ring $NO_2$, $OCO.OCHMe_2$, $CHMeEt$ | Negligible | Negligible | 140 mg/kg |
| 67. | Dinocap | 2-(1-methyl-n-heptyl)-4,6-dinitrophenyl crotonate (see p. 41) | 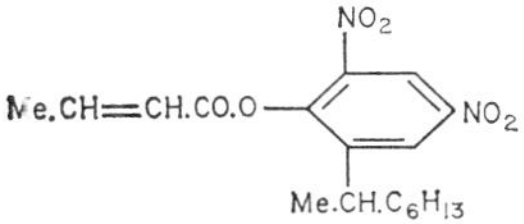 $Me.CH=CH.CO.O$—ring $NO_2$, $NO_2$, $Me.CH.C_6H_{13}$ | Negligible | Negligible | 980 mg/kg |



*As the actual hazard to people exposed to these pesticides depends on many factors such as manner of use, duration of exposure, safety precautions observed, nature of poisonous effect, ease of diagnosis of poisoning at an early reversible stage and availability of antidote, no simple assumptions should be made that the dangers are directly or solely related to the acute oral LD50 for experimental animals.

| Common name | Chemical name | Structure<br>Me = −CH₃<br>Et = −C₂H₅ | Solubility in water | Volatility mm. Hg | Mammalian toxicity* Acute oral LD50 for rats (single dose) |
|---|---|---|---|---|---|
| 68. DNOC | 2-methyl-4,6-dinitro-phenol | | 130 ppm at 15°C | $1.05 \times 10^{-4}$ at 25°C | 40 mg/kg |

**G. Chloronitrobenzenes**

| Common name | Chemical name | Structure | Solubility in water | Volatility mm. Hg | Mammalian toxicity |
|---|---|---|---|---|---|
| 69. Quintozene | pentachloronitro-benzene | | Negligible | $1.33 \times 10^{-4}$ at 25°C | > 12,000 mg/kg |
| 70. Tecnazene | 1,2,4,5-tetrachloro-3-nitrobenzene | | Negligible | — | Of low mammalian toxicity |

**H. Miscellaneous compounds**

| Common name | Chemical name | Structure | Solubility in water | Volatility mm. Hg | Mammalian toxicity |
|---|---|---|---|---|---|
| 71. Calomel | mercurous chloride | Hg₂Cl₂ | 2.0 ppm at 18°C | Negligible | 210 mg/kg |
| 72. Captan | N-(trichloromethylthio)-cyclohex-4-ene-1,2,-dicarboximide | | Negligible | Negligible | 9,000 mg/kg |

| 73. | Carbaryl | 1-naphthyl $N$-methyl-carbamate | OCO.NH.Me (naphthyl carbamate structure) | Less than 0·1% at 25°C | Less than 0·005 at 26°C | 850 mg/kg |
| 74. | Chloropicrin | trichloronitromethane | $CCl_3 . NO_2$ | 0·2% at 0°C | 23·8 at 25°C | Lethal at 0·8 mg/litre for 30 minutes |
| 75. | Copper sulphate | — | $CuSO_4 . 5H_2O$ | Soluble | Negligible | 300 mg/kg |
| 76. | Corrosive sublimate | mercuric chloride | $HgCl_2$ | 6·9 g/100 ml at 20°C | $1·4 \times 10^{-4}$ at 35°C | 1–5 mg/kg |
| 77. | Dichlofluanid | $N'$dichlorofluoro-methylthio-$NN$-dimethyl-$N'$-phenyl-sulphamide | $Me_2 NSO_2$—N—$SCCl_2F$ (phenyl sulphamide structure) | Negligible | $1 \times 10^{-6}$ at 20°C | 1,000 mg/kg |
| 78. | Dicloran | 2,6-dichloro-4-nitro-aniline | 2,6-dichloro-4-nitroaniline structure ($NH_2$, $Cl$, $Cl$, $NO_2$) | Negligible | $1·2 \times 10^{-6}$ at 20°C | 1,500–4,000 mg/kg |
| 79. | Dimethirimol | 5-butyl-2-dimethylamino-4-hydroxy-6-methylpyrimidine | pyrimidine structure ($(CH_2)_3Me$, $Me$, $OH$, $NMe_2$) | 0·2g/100 ml at 25°C | $1·1 \times 10^{-5}$ at 30°C | 4,000mg/kg |
| 80. | Disodium octaborate | — | $Na_2B_8O_{13} . 4H_2O$ | — | — | 5,330 mg/kg (guinea pigs) |

*As the actual hazard to people exposed to these pesticides depends on many factors such as manner of use, duration of exposure, safety precautions observed, nature of poisonous effect, ease of diagnosis of poisoning at an early reversible stage and availability of antidote, no simple assumptions should be made that the dangers are directly or solely related to the acute oral $LD_{50}$ for experimental animals.

| Common name | Chemical name | Structure<br>Me = $-CH_3$<br>Et = $-C_2H_5$ | Solubility in water | Volatility mm. Hg | Mammalian toxicity*<br>Acute oral LD50 for rats (single dose) |
|---|---|---|---|---|---|
| 81. Dithianon | 2,3-dicyano-1,4-dithia-anthraquinone | | Negligible | Negligible | 1,015 mg/kg |
| 82. Dodine | dodecylguanidine acetate | $C_{12}H_{25}NH . CNH_2 . CH_3COOH$ (=NH) | Soluble in hot water | Negligible | 1,000–2,000 mg/kg |
| 83. Fentin acetate | triphenyltin acetate | [C6H5]3—SnOCOMe | 20 ppm at 20°C | $1.33 \times 10^{-6}$ at 30°C | 125 mg/kg |
| 84. Fentin hydroxide | triphenyltin hydroxide | [C6H5]3—SnOH | Negligible | — | 108 mg/kg |
| 85. Folpet | N-(trichloromethyl-thio)phthalimide | | Negligible | Negligible | >10,000 mg/kg |
| 86. Lead arsenate | diplumbic hydrogen arsenate | $PbHAsO_4$ | Negligible | Negligible | 10–50 mg/kg fatal |
| 87. Lime sulphur | calcium polysulphides | $CaS . Sx$ | Miscible | — | No evidence of toxicity |

| | | | | | |
|---|---|---|---|---|---|
| 88. Methiocarb | 4-methylthio-3,5-xylyl $N$-methylcarbamate | 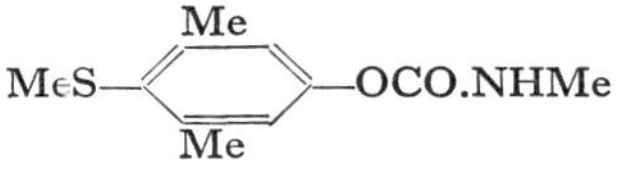 $\text{MeS}$—⟨3,5-Me$_2$C$_6$H$_2$⟩—$\text{OCO.NHMe}$ | Negligible | Negligible | 100 mg/kg |
| 89. Methyl bromide | — | MeBr | 1·34% at 25°C | Gaseous | Respirators must be worn at conc. over 17 ppm |
| 90. Paris green | copper aceto-arsenite | $(\text{MeCOO})_2\text{Cu} . 3\text{Cu}(\text{AsO}_3)_2$ | Negligible | Negligible | 22 mg/kg |
| 91. Phenyl-mercury acetate | — | 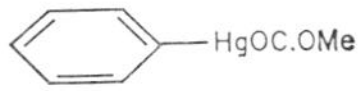 Ph—HgOC.OMe | 0·437 g/100 ml at room temperature | $9 \times 10^{-6}$ at 35°C | ca. 60 mg/kg |
| 92. Quinome-thionate | 6-methyl-2-oxo-1,3-dithiolo[4,5-b]-quinoxaline | 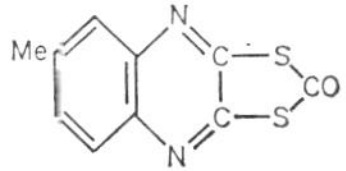 6-methyl-2-oxo-1,3-dithiolo[4,5-b]quinoxaline | Negligible | $2 \times 10^{-7}$ at 20°C | 2,500–3,000 mg/kg |
| 93. Salicyl-anilide | — | 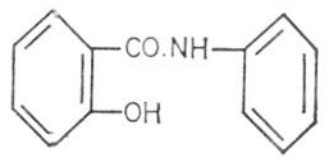 ⟨C$_6$H$_4$(OH)⟩—CO.NH—Ph | 55 ppm at 25°C | Negligible | — |
| 94. Sodium nitrite | — | NaNO$_2$ | Readily soluble | — | 90 mg/kg (cattle) |
| 95. Urea | — | $\text{CO(NH}_2)_2$ | Readily soluble | — | — |

*As the actual hazard to people exposed to these pesticides depends on many factors such as manner of use, duration of exposure, safety precautions observed, nature of poisonous effect, ease of diagnosis of poisoning at an early reversible stage and availability of antidote, no simple assumptions should be made that the dangers are directly or solely related to the acute oral LD50 for experimental animals.

361

# INDEX OF SCIENTIFIC NAMES

(other than host plants)

*Abraxas grossulariata* 207
*Acarus siro* 110
*Aceria essigi* 213
*Acherontia atropos* 114
*Acleris comariana* 218
*Aclypea opaca* 125
*Acyrthosiphon pisum* 142, 144, 293
*Adelges abietis* 327
*Adelges nuesslini* 342
*Adelges viridis* 327
*Adoxophyes orana* 174
*Agriolimax reticulatus* 99, 115, 131, 140, 143
*Agriotes obscurus* 270
*Agriotes* spp. 100, 131, 276
*Agrobacterium* spp 16
*Agrostis exclamationis* 338
*Agrostis segetum* 127, 273, 338
*Agrostis* spp 215, 219, 298
*Allantus cinctus* 325
*Allolobophora longa* 330
*Allolobophora nocturna* 330
*Alsophila aescularia* 173, 176, 304, 312, 322, 333
*Alternaria brassicae* 156, 157
*Alternaria brassicicola* 156, 286
*Alternaria dauci* 288
*Alternaria dianthi* 238
*Alternaria godetiae* 308
*Alternaria* spp 101, 105
*Alternaria tenuis* 253, 313
*Alternaria zinniae* 334
*Amaurosoma* spp 142
*Amelia paleana* 142
*Ametastegia glabrata* 170
*Amphimallon solstitialis* 113, 127
*Amphorophora rubi* 210
*Anisandrus dispar* 166
*Anothrips orchidaceous* 262

*Anthonomus pomorum* 162, 187
*Anthonomus rubi* 216
*Anuraphis farfarae* 186
*Apamea sordens* 140
*Aphelenchoides fragariae* 215, 332
*Aphelenchoides ritzemabosi* 206, 215, 241, 302, 317, 322
*Aphelenchoides* spp 259
*Aphelinus mali* 6
*Aphis craccivora* 142
*Aphis fabae* 126, 130, 134, 142, 144, 277, 278, 279, 295, 305, 306
*Aphis gossypii* 232, 234, 245
*Aphis grossulariae* 202
*Aphis idaei* 211
*Aphis nasturtii* 111
*Aphis pomi* 160, 322
*Aphis sambuci* 301
*Aphis schneideri* 202
*Aphrodes* spp 216
*Apion* spp 142
*Appelia schwartzi* 192
*Aptinothrips rufus* 140
*Aptinothrips stylifer* 140
*Archips podana* 174, 190, 322, 324
*Arge ochropus* 325
*Argyresthia conjugella* 164
*Argyresthia curvella* 195
*Arion ater* 115
*Arion hortensis* 115
*Armadillidium* spp 248
*Armillaria mellea* 321
*Ascochyta brassicae* 156, 157
*Ascochyta clematidina* 304
*Ascochyta fabae* 156
*Ascochyta imperfecta* 153, 154
*Ascochyta orobi* 154
*Ascochyta trifolii* 153
*Aspidiotus hederae* 263

*Atomaria linearis* 130
*Aulacaspis rosae* 265, 325
*Aulacorthum circumflexum* 235, 240, 245, 250, 251, 261, 263, 298, 313
*Aulacorthum solani* 111, 234, 235, 236, 245, 252, 256, 267

*Balaninus nucum* 201
*Barypithes araneiformis* 339
*Barypithes pellucidus* 339
*Blaniulus guttulatus* 115, 130
*Blennocampa pusilla* 325
*Botrytis anthophila* 153
*Botrytis cinerea* 153, 154, 156, 157, 220, 233, 235, 243, 245, 251, 252, 253, 254, 257, 263, 265, 266, 270, 291, 299, 303, 306, 308, 322, 334, 340
*Botrytis elliptica* 313
*Botrytis fabae* 154, 156
*Botrytis galanthina* 326
*Botrytis gladiolorum* 308
*Botrytis paeoniae* 317
*Botrytis tulipae* 330
*Bourletiella hortensis* 116, 339
*Brachycaudus helichrysi,* 197, 240, 301, 302, 305
*Brachycaudus persicaecola* 192
*Brachydesmus superus* 130
*Bremia lactucae* 256, 290
*Bremiola onobrychidis* 144
*Brevicoryne brassicae* 145
*Bryobia cristata* 187
*Bryobia ribis* 206
*Bryobia rubrioculus* 166, 187
*Bryobia* spp 320
*Bupalus piniaria* 345
*Byturus tomentosus* 211

*Cacoecimorpha pronubana* 237
*Cacoecia* spp 333
*Caenorhinus aequatus* 164
*Caenorhinus germanicus* 213, 217
*Caliroa cerasi* 189, 196
*Calocoris fulvomaculatus* 224
*Calocoris norvegicus* 113, 126, 241
*Caloptilia syringella* 312, 321
*Campylomyza ormerodi* 143
*Carulaspis* spp 311

*Caulomyia radifica* 141
*Cavariella aegopodii* 287
*Cecidophyopsis ribis* 202
*Cephalosporium lamellicola* 260
*Cerapteryx graminis* 139
*Cerataphis lataniae* 261
*Cercospora beticola* 132
*Cercospora zebrina* 153
*Cercospora zonata* 156
*Cercosporella brassicae* 157
*Cercosporella herpotrichoides* 102, 105, 108
*Cetema* spp 140
*Cetonia aurata* 113
*Ceutorhynchus assimilis* 146
*Ceutorhynchus pleurostigma* 147, 284
*Ceutorhynchus quadridens* 147, 283
*Chaetocnema concinna* 129
*Chaetosiphon fragaefolii* 214
*Chaetomium* spp 261
*Chionaspis salicis* 312, 333
*Chlorops pumilionis* 104
*Cholus cattleyae* 262
*Chrysomela populi* 339
*Cinara pinicola* 342
*Cinaropsis pilicornis* 342
*Cladius difformis* 325
*Cladius pectinicornis* 325
*Cladosporium cucumerinum* 249
*Cladosporium fulvum* 271
*Cladosporium herbarum* 101
*Claviceps purpurea* 102, 105, 108, 109, 150, 151
*Coccus hesperidum* 263
*Collembola* spp 248, 259
*Colletotrichum atramentarium* 272
*Colletotrichum lagenarium* 249
*Colletotrichum trifolii* 153, 154
*Coniothyrium concentricum* 334
*Coniothyrium hellebori* 309
*Contarinia dactylidis* 141
*Contarinia festucae* 141
*Contarinia humuli* 144, 224
*Contarinia lolii* 142
*Contarinia loti* 142
*Contarinia medicaginis* 143
*Contarinia merceri* 141
*Contarinia nasturtii* 148
*Contarinia onobrychidis* 143
*Contarinia pisi* 294
*Contarinia poae* 141
*Contarinia pyrivora* 189
*Corticium fuciforme* 332

Corticium solani 103, 106, 109, 153, 156, 157, 235, 256, 270
    (see also *Rhizoctonia solani*)
*Corynebacterium betae* 134, 280
*Corynebacterium rathayi* 150, 152
*Corynebacterium* spp 16
*Croesus septentrionalis* 339
*Cronartium ribicola* 209
*Cryptococcus fagi* 300
*Cryptomyzus galeopsidis* 202
*Cryptomyzus ribis* 202
*Cumminsiella mirabilissima* 313
*Cylindrocarpon radicicola* 251
*Cylindroiulus londinensis* 115
*Cymadothea trifolii* 153, 154
*Cystopus candidus* 156, 157

*Dactylium dendroides* 260
*Dasyneura affinis* 332
*Dasyneura alopecuri* 141
*Dasyneura brassicae* 148
*Dasueura dactylidis* 141
*Dasyneura festucae* 141
*Dasyneura gentneri* 143
*Dasyneura ignorata* 143
*Dasyneura leguminicola* 143
*Dasyneura mali* 164
*Dasyneura pyri* 189
*Dasyneura tetensi* 203
*Dasyneura trifolii* 143
*Dasyneura viciae* 143, 144
*Delia antiqua* 292
*Delia platura* 251, 278
*Dendrothrips ornatus* 312, 321
*Dialeurodes chittendeni* 323
*Dicliomyces microsporus* 261
*Dichomeris marginella* 312
*Dicyphus errans* 113
*Didymascella thujina* 329, 341
*Didymella lycopersici* 272, 296
*Didymellina dianthi* 301, 328
*Didymellina macrospora* 311
*Diplocarpon rosae* 265, 325
*Ditylenchus destructor* 114
*Ditylenchus dipsaci* 107, 114, 128, 138, 143, 215, 310, 316, 319, 320, 326, 329
*Ditylenchus* spp 259
*Drechslera avenae* 109
*Drechslera dictyoides* 149, 150, 151
*Drechslera graminea* 106
*Drechslera siccans* 149, 150, 151

*Dysaphis devecta* 160
*Dysaphis mali* 160
*Dysaphis pyri* 186
*Dysaphis tulipae* 304, 307, 311, 312, 326, 329

*Elatobium abietinum* 327, 337, 342
*Elsinoe veneta* 213, 214
*Empoasca decipiens* 115
*Empoasca* spp 143
*Enarmonia formosana* 167, 195
*Encarsia formosa* 6
*Endelomyia aethiops* 325
*Entomosporium maculatum* 192
*Entyloma calendulae* 306
*Epichloe typhina* 150, 151
*Epinotia tedella* 343
*Epipolaeus caliginosus* 224
*Erannis defoliaria* 173, 176, 344
*Erioischia brassicae* 148, 281, 333
*Erioischia floralis* 149
*Eriophyes pyri* 188
*Eriosoma lanigerum* 177, 304
*Erwinia amylovora* 179, 190
*Erwinia carotovora* 124, 249, 311
*Erwinia chrysanthemi* 238
*Erwinia* spp 16
*Erysiphe graminis* 103, 106, 109, 149, 158
*Erysiphe polygoni* 153, 154, 156, 157, 267, 304, 306
*Erysiphe* spp 133
*Erythroneura alneti* 172
*Eumerus strigatus* 315
*Eumerus tuberculatus* 315
*Eupteryx aurata* 115
*Eurytoma orchidearium* 262
*Euscellis* spp 216
*Euxoa nigricans* 127, 275
*Exobasidium vaccinii* 300, 323

*Fabraea maculata* 192
*Feronia* spp 217
*Fomes annosus* 336, 346
*Forficula auricularia* 170, 241, 302, 303, 305
*Fusarium lycopersici* 273
*Fusarium moniliforme* 251
*Fusarium nivale* 332
*Fusarium oxysporum* 238, 251, 272

*Fusarium roseum* 239
*Fusarium* spp 101, 105, 108, 124, 152, 233, 250, 332
*Fusicladium pyracanthae* 322

*Geoktapia pyraria* 186
*Geomyza* spp. 140
*Gibberella zeae* 103, 106, 109
*Gloeosporium album* 179
*Gloeosporium concentricum* 156, 157
*Gloeosporium fructigenum* 179
*Gloeosporium nervisequum* 319
*Gloeosporium perennans* 179
*Gloeosporium* spp 201
*Gloeotinia temulenta* 150, 152
*Glomerella cincta* 262
*Glyphipteryx cramerella* 141
*Graphocephala coccinea* 323, 324

*Harpalus rufipes* 217
*Hedya pruniana* 200
*Helicobasidium purpureum* 134, 153, 157
*Heliothrips haemorrhoidalis* 232, 233, 234, 250, 262
*Helminthosporium* spp 332
*Helophorus nubilus* 100
*Hemitarsonemus latus* 234, 253, 276
*Hepialus humuli* 116
*Hepialus lupulina* 116
*Hepialus* spp 148, 218, 276
*Hercinothrips femoralis* 232, 234, 262
*Heterodera avenae* 107
*Heterodera rostochiensis* 114, 268
*Heterodera schachtii* 127
*Heterosporium echinulatum* 301
*Heterosporium phlei* 150, 151
*Hoplocampa brevis* 189
*Hoplocampa flava* 199
*Hoplocampa testudinea* 164
*Hyalopteroides humilis* 141
*Hyalopterus amygdali* 192
*Hyalopterus pruni* 197
*Hydraecia micacea* 131, 225
*Hylastes ater* 342
*Hylastes brunneus* 342
*Hylastes curricularius* 342
*Hylobius abietus* 336, 344
*Hypera nigrirostris* 142

*Hypera variabilis* 142
*Hyperomyzus lactucae* 202
*Hypomyces rosellus* 260

*Ips cembrae* 345
*Itersonilia perplexans* 243, 303, 304

*Jaapiella medicaginis* 143

*Kabatiella caulivora* 153, 154
*Kakothrips robustus* 194
*Keithia thujina* 329, 341

*Lacanobia oleracea* 269
*Lampronia rubiella* 213
*Lasius flavus* 111
*Lasius niger* 111
*Laspeyresia funebrana* 199
*Laspeyresia nigricana* 294
*Laspeyresia pomonella* 167, 188
*Lema melanopa* 139
*Lepidosaphes ulmi* 325
*Leptohylemyia coarctata* 100
*Leptosphaeria avenaria* 108
*Leptosphaeria nodorum* 102
*Lilioceris lilii* 313
*Limothrips cerealium* 140
*Liriomyza bryoniae* 267
*Longidorus* spp 128
*Lophodermium pinastri* 341
*Loxostege sticticalis* 148
*Lygocoris pabulinus* 113, 169, 188, 192, 198, 205, 211, 215, 241, 305, 307, 310, 318, 325
*Lygus rugulipennis* 113, 126, 241, 302, 305

*Macrosiphoniella sanborni* 240, 302
*Macrosiphum avenae* 98
*Macrosiphum euphorbiae* 111, 232, 251, 299, 301, 307, 311, 330
*Macrosiphum fragariae* 211
*Macrosiphum rosae* 264, 324
*Malacosoma neustria* 309, 312, 322, 324, 333
*Mamestra brassicae* 148, 281, 307
*Marasmius oreades* 331
*Marssonina daphnes* 306

*Marssonina panattoniana* 291
*Marssonina salicicola* 334
*Marssonina* spp 320
*Mastigosporium cylindricum* 150, 151
*Mastigosporium rubricosum* 150, 151
*Mayetiola dactylidis* 141
*Mayetiola joannisi* 141
*Mayetiola poae* 141
*Mayetiola schoberi* 141
*Megaselia bovista* 258
*Megaselia halterata* 258
*Megaselia nigra* 258
*Megastigmus spermatrophus* 345
*Megoura viciae* 142, 144
*Meligethes* spp 145
*Meloidogyne hapla* 128, 246
*Meloidogyne incognita acrita* 246
*Meloidogyne naasi* 128
*Meloidogyne* spp 269
*Melolontha melolontha* 113, 127, 167, 338
*Meria laricis* 312, 340
*Merodon equestris* 315, 326
*Meromyza saltatrix* 140
*Mesapamea secalis* 139
*Mesographe forficalis* 148
*Metopolophium dirhodum* 98
*Metopolophium festucae* 98
*Microsphaera alphitoides* 317, 341
*Microsphaera berberidis* 313
*Microsphaera grossulariae* 209
*Microsphaera polonica* 255, 311
*Milax budapestensis* 115
*Milax gagates* 115
*Mycelia sterilia* 272
*Mycogone perniciosa* 260
*Mycosphaerella brassicicola* 156
*Mycosphaerella carinthiaca* 153
*Mycosphaerella ligulicola* 244
*Mycothecium roridum* 318
*Myelophilus piniperda* 345
*Mytilococcus ulmi* 173, 188, 193
*Myzus ascalonicus* 214, 307
*Myzus cerasi* 194, 337
*Myzus persicae* 11, 111, 130, 134, 145, 192, 232, 234, 235, 236, 240, 245, 251, 252, 256, 261, 263, 266, 298, 299, 301, 302, 306, 307, 319, 328, 330

*Nasonovia ribisnigri* 202, 256
*Nectria galligena* 177, 190

*Nematus melanaspis* 339
*Nematus olfaciens* 204
*Nematus salicis* 339
*Neodipron sertifer* 343
*Nephrotoma maculata* 115
*Nephrotoma* spp 98
*Noctua pronuba* 338
*Notocella uddmanniana* 211

*Oidium begoniae* 235, 300
*Oidium chrysanthemi* 243, 303
*Oidium euonymi-japonicae* 307
*Oidium* spp 238, 304
*Oligonychus ununguis* 311, 327, 329, 338, 342
*Oniscus* spp 248
*Onychiurus* spp 116
*Oospora pustulans* 124
*Operophtera brumata* 176, 304, 312, 317, 322, 324, 333, 344
*Operophtera fagata* 344
*Ophiobolus graminis* 103, 106, 109, 149, 150, 152, 331
*Opomyza florum* 140
*Opomyza germinationis* 140
*Orgyia antiqua* 309, 322, 324, 333
*Oryzaephilus surinamensis* 110
*Oscinella frit* 107, 138, 140
*Otiorhynchus claripes* 200, 213, 219
*Otiorhynchus ovatus* 340
*Otiorhynchus rugosostriatus* 219
*Otiorhynchus singularis* 167, 188, 205, 211, 221, 303, 323, 339, 343
*Otiorhynchus sulcatus* 219, 250, 253, 264, 323
*Oxidus gracilis* 246

*Pammene rhediella* 174
*Pandemis cerasana* 333
*Panonychus ulmi* 170, 188, 193, 196, 198
*Papulaspora byssina* 261
*Parallelodiplosis cattleya* 262
*Parthenolecanium corni* 193, 204, 253, 304, 322, 325
*Parthenolecanium pomeranicum* 334
*Pealius azaleae* 299
*Pectobacterium carotovorum* 156, 157, 232, 263
*Pediculopsis gramineum* 140

*Pegomya betae*   129, 279, 295
*Pemphigus bursarius*   290
*Penthaleus major*   141
*Peronospora antirrhini*   299
*Peronospora arborescens*   314
*Peronospora destructor*   292
*Peronospora effusa*   295
*Peronospora farinosa*   133
*Peronospora ficariae*   298
*Peronospora parasitica*   156, 157, 286, 328, 333
*Peronospora sparsa*   265
*Peronospora trifoliorum*   153
*Peronospora viciae*   154, 156
*Pestalotia guepini*   300
*Phaedon cochleariae*   146
*Phialophora cinerescens*   239
*Philopedon plagiatum*   131
*Philophylla heraclei*   289
*Phlogophora meticulosa*   111, 239, 298, 305, 307, 320, 333
*Phoma exigua*   124
*Phoma lingam*   156, 157, 158, 285
*Phoma spp*   244
*Phomopsis sclerotioides*   249
*Phorodon humuli*   197, 222
*Phragmidium mucronatum*   326
*Phyllachora graminis*   150, 151
*Phyllachora sylvatica*   150, 151
*Phyllaphis fagi*   300, 337
*Phyllobius abietis*   336
*Phyllobius argentatus*   173, 174, 343
*Phyllobius oblongus*   166, 173
*Phyllobius pyri*   139, 169, 173, 343
*Phyllobius spp*   173, 188, 196
*Phyllodecta vitellinae*   338
*Phyllodecta vulgatissima*   338
*Phyllopertha horticola*   113, 167, 330, 338
*Phyllosticta antirhini*   299
*Phyllosticta spp*   320
*Phyllotreta spp*   146, 284, 298
*Phylloxera punctata*   337
*Phytomyza atricornis*   241, 245, 302
*Phytomyza ilicis*   309
*Phytomyza rufipes*   148
*Phytophthora cactorum*   178
*Phytophthora cryptogea*   157, 270, 302, 319
*Phytophthora infestans*   117, 295
*Phytophthora megasperma*   156, 157
*Phytophthora parasitica*   270
*Phytophthora richardiae*   232

*Phytophthora spp*   228, 276, 326
*Phytophthora syringae*   178
*Pieris brassicae*   148
*Pieris napi*   148
*Pieris rapae*   148
*Pieris spp*   283
*Planococcus citri*   234, 253, 266
*Planococcus spp*   254, 255, 262
*Plasmodiophora brassicae*   156, 157, 285, 328, 333
*Plasmopara viticola*   221
*Pleospora betae*   279
*Pleospora bjoerlingii*   132
*Pleospora herbarum*   153, 154, 156
*Plesiocoris rugicollis*   163
*Plutella xylostella*   148, 283
*Podosphaera leucotricha*   179
*Podosphaera oxyacanthae*   309
*Polydesmus angustus*   115
*Porcellio spp*   248
*Pratylenchus spp*   315
*Prociphilus spp*   337
*Procus strigilis*   140
*Prodenia litura*   240
*Pseudobalsamia microspora*   261
*Pseudococcus adonidum*   234, 263
*Pseudococcus citri*   263
*Pseudomonas coronafaciens*   108, 150
*Pseudomonas delphinii*   306
*Pseudomonas marginata*   308
*Pseudomonas mors-prunorum*   196, 321
*Pseudomonas spp*   16
*Pseudomonas syringae*   153, 312
*Pseudomonas tolaasi*   259
*Pseudoperonospora humuli*   225
*Pseudopeziza medicaginis*   153
*Pseudopeziza ribis*   209
*Pseudopeziza trifolii*   153
*Psila nigricornis*   242
*Psila rosae*   286, 288
*Psylla mali*   165
*Psylla simulans*   189
*Psylliodes affinis*   115
*Psylliodes attenuata*   224
*Psylliodes chrysocephala*   146
*Pteronidea leucotrocha*   207
*Pteronidea ribesii*   207
*Pterstichus spp*   317
*Puccinia arenariae*   328
*Puccinia chrysanthemi*   244, 303
*Puccinia coronata*   150, 151
*Puccinia cyani*   304
*Puccinia dispersa*   150

*Puccinia glumarum* 150
*Puccinia graminis* 150, 151, 313
*Puccinia hordei* 105
*Puccinia malvacearum* 310
*Puccinia menthae* 291
*Puccinia pelargonii-zonalis* 253, 318
*Puccinia pringsheimiana* 209
*Puccinia striiformis* 104, 106
*Pycnostysanus azaleae* 324
*Pyrenophora avenae* 109
*Pyrenophora dicyoides* 149
*Pyrenophora graminea* 106
*Pyrenophora siccans* 149
*Pyrenophora teres* 106
*Pythium* spp 149, 152, 156, 157, 228, 233, 235, 253, 263, 276, 326, 327, 332, 340

*Quadraspidiotus ostreaeformis* 173, 188, 193

*Ramularia beticola* 133
*Ramularia onobrychidis* 154
*Ramularia* spp 318, 320
*Ramularia vallisumbrosae* 317
*Rhadinoceraea micans* 311
*Rhizoctonia solani* 103, 106, 109, 124, 228, 285
    (see also *Corticium solani*)
*Rhizoglyphus echinopus* 314, 329
*Rhopalosiphonus latysiphon* 111
*Rhopalosiphonus staphyleae* 134
*Rhopalosiphum insertum* 139, 160
*Rhopalosiphum maidis* 138
*Rhopalosiphum padi* 98, 105
*Rhopobota naevana* 310
*Rhyacionia buoliana* 343
*Rhynchites coeruleus* 166, 187
*Rhyncosporium orthosporum* 150, 151
*Rhyncosporium secalis* 106, 150, 151
*Rhytisma acerinum* 198

*Scaptomyza apicalis* 148
*Sciara* spp 246
*Schizoneura ulmi* 202
*Sclerotinia fructigena* 200
*Sclerotinia gladioli* 308
*Sclerotinia homoeocarpa* 331
*Sclerotinia laxa* 177, 200
*Sclerotinia narcissicola* 316

*Sclerotinia polyblastis* 316
*Sclerotinia sclerotiorum* 156, 157
*Sclerotinia* spp 253
*Sclerotinia trifoliorum* 153, 154, 156
*Sclerotium cepivorum* 293
*Sclerotium tuliparum* 310, 330
*Scolytus mali* 166
*Scolytus rugulosus* 166
*Scutigerella immaculata* 116, 242, 248, 269
*Scythropia crataegella* 304, 309
*Selenophoma donacis* 106, 150, 151
*Septogloeum oxysporum* 150, 151
*Septoria apii* 289
*Septoria avenaria* 108
*Septoria chrysanthemella* 242
*Septoria drummondii* 319
*Septoria gladioli* 308
*Septoria nodorum* 102
*Septoria orobina* 154
*Septoria passerinii* 106
*Septoria* spp 302
*Septoria tritici* 102
*Serica brunnea* 113, 338
*Siteroptes graminum* 140
*Sitobion avenae* 98
*Sitodiplosis cambriensis* 141
*Sitodiplosis dactylidis* 141
*Sitona lineatus* 11
*Sitona* spp 143, 145, 295
*Sitophilus granarius* 110
*Sitophilus* spp 110
*Sphaerotheca fuliginea* 245, 249, 314, 319
*Sphaerotheca macularis* 220, 226
*Sphaerotheca mors-uvae* 208, 210
*Sphaerotheca pannosa* 194, 266, 326
*Sphaerotheca saliginea* 249
*Sphaerulina trifolii* 153
*Spilonota ocellana* 175
*Spodoptora littoralis* 240
*Spongospora sunterranea* 296
*Sporendonema purpurascens* 260
*Stagonospora curtisii* 255, 316
*Stagonospora meliloti* 153
*Stemphylium radicinum* 288
*Steneotarsonemus laticeps* 254, 314
*Steneotarsonemus pallidus* 217, 234, 250, 253, 266
*Steneotarsonemus spirifex* 140
*Stenodiplosis geniculati* 141
*Stephanitis rhododendri* 323
*Stereum purpureum* 322

BB

*Streptomyces scabies*   157
*Strophosomus melanogrammus*   339, 343
*Synanthedon tipuliformis*   206

*Taeniothrips inconsequens*   190, 199
*Taeniothrips simplex*   307
*Taphrina deformans*   194
*Taphrina populina*   320, 321
*Tarsonemus confusus*   266
*Tetranychus cinnabarinus*   233, 234, 237
*Tetranychus urticae*   193, 207, 213, 216,
    224, 232, 233, 234, 237, 242, 247,
    251, 252, 254, 255, 262, 263, 264, 267,
    268, 305, 310, 320, 324, 332
*Thielaviopsis basicola*   153, 154, 156,
    235, 251, 253, 263
*Thomasiniana oculiperda*   173, 324
*Thomasiniana theobaldi*   212
*Thrips angusticeps*   131, 142, 248
*Thrips fuscipennis*   250, 324
*Thrips tabaci*   232, 233, 234, 237, 248,
    250, 269
*Tilletia caries*   102
*Tipula oleracea*   331
*Tipula paludosa*   129, 331
*Tipula spp*   98, 115, 143, 148, 275
*Tomicus piniperda*   345
*Tortrix viridana*   317
*Trialeurodes vaporariorum*   233, 235,
    242, 245, 248, 252, 253, 269
*Trichiocampus viminalis*   339
*Trichodorus spp*   128
*Trichoniscus spp*   248
*Trypodendron lineatum*   346
*Tuberolachnus saligenus*   333
*Typhlocyba cruenta*   337
*Typhlocyba froggatti*   172
*Typhlocyba jucunda*   115
*Typhlocyba rosae*   172, 324
*Tyrophagus dimidiatus*   234, 246

*Uncinula necator*   221, 254
*Urocystis cepulae*   292
*Urocystis occulta*   149

*Uromyces betae*   133
*Uromyces caryophyllinus*   238
*Uromyces dactylidis* 150, 151
*Uromyces dianthi*   238, 301
*Uromyces fabae*   154, 156
*Uromyces flectens*   153
*Uromyces onobrychidis*   154
*Uromyces striatus*   153
*Uromyces trifolii*   153
*Urophlyctis alfalfae*   153, 155
*Ustilago avenae*   109
*Ustilago hordei*   105, 108
*Ustilago maydis*   149
*Ustilago nuda*   103, 106

*Venturia inaequalis*   180, 314
*Venturia pirina*   191
*Verticillium albo-atrum*   153, 155, 244,
    250, 273
*Verticillium cinerescens*   239
*Verticillium dahliae*   153, 154, 273
*Verticillium malthousei*   260
*Verticillium psalliotae*   260
*Vespula germanica*   176
*Vespula spp*   190, 193, 196, 202
*Vespula vulgaris*   176

*Xanthomonas campestris*   156, 157, 158
*Xanthomonas pelargonii*   253
*Xiphinema diversicaudatum*   216, 220
*Xyleborus saxeseni*   166

*Yponomeuta cagnagella*   307, 309
*Yponomeuta malinella*   174
*Yponomeuta padella*   174
*Yponomeuta rorrella*   333
*Yponomeuta spp*   173

*Zygina pallidifrons*   235, 252, 264, 268
*Zygophiala jamaicensis*   238

# GENERAL INDEX

Acaricide 23
Acer, diseases of 298
Adelges, spruce gall 327, 342
Aerosol 27, 231
   application of 77
African violet, diseases of 266
    pests of 266
Agricultural Chemicals Approval
    Scheme 12, 93
Agriculture (Poisonous Substances)
    Act 86
Agriculture (Poisonous Substances)
    Regulations 86
Airblast sprayers 64
Aircraft spraying 64
Aldrin 29, 353
Alternaria blight, carnation 238
Amarylis, diseases of 255
    pests of 254
Ambrosia beetles 346
American blight 177
   gooseberry mildew 208
Amiton 87
Anemone, diseases of 298
    pests of 298
Angleshades moth 111, 239, 298, 305,
    307, 320, 333
Anionic surfactants 26
Anthracene oils 58
Anthracnose, cucumber 249
   lucerne 154
   orchid 262
   willow 334
Antirrhinum, diseases of 299
    pests of 299
Antler moth 139
   sawfly 325
Ants 111

Aphicide 23
Aphids, anemome 298
   antirrhinum 299
   apple 160
   apricot 192
   arum 232
   beans, glasshouse 234
   begonia 232
   blackberry 210
   calceolaria 235
   carnation 236, 301
   cherry 194
   china aster 301
   chrysanthemum 240, 303
   cineraria 245
   cucumber 245
   currant 201
   cyclamen 250
   dahlia 305
   damson 197
   field beans 144
   freesia 257
   gooseberry 201
   grasses 139
   legumes 142
   lily 312
   loganberry 210
   melon 245
   nectarine 192
   orchids 261
   peach 192
   pear 186
   plum 197
   potato 111
   primula 263
   raspberry 210
   rose 264
   strawberry 214

Aphids (*cont.*)
  sweet pea   266
  wheat   98
Apple, American blight   177
  aphids   160
  bitter rot   179
  blossom weevil   162
    wilt   177
  canker   177
  capsid   163
  collar rot   178
  diseases of   177
  fire blight   179
  fruit miner   164
    rhynchites   164
    tree red spider mite   170
  Gloeosporium rot   179
  leaf midge   164
  leafhoppers   172
  pests of   160
  powdery mildew   179
  replant problem   180
  sawfly   164
  scab   180
  sucker   165
  twig cutter   166
  woolly aphid   177
Apple and pear bryobia   166
Apple-grass aphid   139, 160
Approved products   93
Apricot, diseases of   194
  pests of   192
Arabis mosaic   219
Arsenic tolerance   85
Arum lily, diseases of   232
    pests of   232
Ascomycetes   15
Asparagus fern, diseases of   233
    pests of   232
Aster, China, pests of   301
Atomizing solution   27
Atomization   27, 231
Azalea, diseases of   233, 300
  false bloom   300
  gall   300
  grey mould   233
  lichen on   300
  pests of   233, 299
  red spider mite   233
  thrips   233
  whitefly   299
Azinphos-ethyl   87, 91
Azinphos-methyl   30, 87, 91, 348

Azobenzene   31, 355

Bacteria   15
Bacterial blotch, mushroom   259
  canker, prunus   321
  pit, mushroom   259
  soft rot, orchid   263
Banded rose sawfly   325
Bark beetles   166, 344, 346
Barley, diseases of   101
  pests of   104
  yellow dwarf virus   101, 105, 108
Basal stem rot, asparagus fern   233
    cucumber   249
Basic copper carbonate   121
    chloride   34, 121
Basidiomycetes   15
Beaumont period   119
Bean aphid   305, 306
    (see also Black bean aphid)
  seed fly   251, 278
Beans, glasshouse, pests of   234
    (see also Broad beans)
Beech aphid   300
  jassid   337
  pests of   300
  scale   300
  woolly aphid   337
Bees, spray hazards   4, 29, 85
Beet, Black leg   279
  carrion beetle   125
  cyst eelworm   127
  diseases of   279
  docking disorder   128
  eelworm   127
  Eelworm Orders   7, 127
  pests of   125
  'sickness'   127
  silvering   280
  webworm   148
  yellows virus   134
Begonia, diseases of   235, 300
  pests of   234
  powdery mildew   300
  spray damage   234
Benzene hexachloride   31
BHC   31
Big bud mite   202
Binapacryl   31, 357
Biological control   6
Birch sawfly   339
Bird-cherry aphid   98
Bitter rot, apple   179

Black bean aphid   126, 142, 144, 277, 278, 279, 295
Blackberry aphid   211
  diseases of 213
  pests of   210
  blotch, delphinium   306
    tares   154
    vetch   154
Blackcurrant gall midge   202
  leaf midge   203
    spot   209
  rust   209
  sawfly   204
Black fly   126
  leg, beet   279
    potato   124
    sugar beet   132
  millepede   115
  mildew, rose   265
  mould   101
  peach aphid   192
  pine beetles   342
  point   101
  root rot   249
  rot, carrot   288
    orchid   263
  scurf   124
  slug   115
  spot   265
  stem   154
Blight, lilac   312
  potato   117
  tomato   295
Blind seed disease, ryegrass   150, 152
Blossom beetles   145
  wilt, apple   177
    plum   200
Blotch, begonia   235
  chrysanthemum   242
Bluestone   32
Bordeaux mixture   32, 121
Bordered white moth   345
Botrytis rot, see Grey mould
Bramble shoot moth   211
Brassica crops, diseases of   285
    pests of   280
  pod midge   148
Bridged diphenyl acaricides   31, 355
Broad bean, pests of   277
Brown chafer   113
  foot rot, barley   105
      wheat   101
  leaf weevil   173

Brown (cont.)
  plaster mould   261
  rot, plum   200
  rust   105, 150
  scale   193, 200, 204, 253, 325
Bryobia mite   166, 187, 320
Bubbles, mushroom   260
Buck-eye rot   270
Buckthorn-potato aphid   111
Bud blast, rhododendron   324
Bud moth, apple   175
Bulb and potato aphid   111
  mite   314, 329
  scale mite   254, 314
Bunt   102
Burgundy mixture   32, 121

Cabbage aphid   280
  leaf miner   148
  moth   148, 281, 307
  root fly   148, 281, 333
  stem flea beetle   146
    weevil   283
  thrips   142
  white butterflies   283
Calceolaria, diseases of   236
  pests of   235
Calla lily, diseases of   232
    pests of   235
    root rot   232
Calomel   32, 358
Camellia leaf spot   300
Canker, apple   177
  brassica crops   285
  pear   191
Capsid, china aster   302
  dahlia   305
  potato   113
  sugar beet   126
  (see also Common green capsid)
Captan   32, 358
Carbaryl   33, 359
Carbon tetrachloride   33, 353
Carnation, diseases of   238, 301
  leaf spot   238
  mildew   238
  pests of   236, 301
  red spider mite   237
  ring spot   301
  rust   238, 301
  slow wilt   238

Carnation (*cont.*)
  stem rot   239
  thrips   237
  tortrix   237
  Verticillium wilt   238
Carrot, diseases of   288
  fly, carrot   286
    celery   288
  pests of   286
    -willow aphid   287
Cationic surfactants   26
Cattleya fly   262
  midge   262
  weevil   262
Cecid midges   257
Celery, diseases of   289
  fly   289
  pests of   288
Centrifugal pumps   67
Cercospora leaf spot   132
Cereal leaf aphid   138
    beetle   139, 153
  cyst eelworm   107
Cereals, diseases of   97, 149
  pests of   97, 138
Chafer beetle, apple   167
    crucifers   146
    grasses   139
  grubs, beet   127
    forest nurseries   338
    potato   113
    strawberry   215
    turf   330
Cherry bacterial canker   196
  bark tortrix   195
  black fly   194
  diseases of   196
  fruit moth   195
  pests of   194
  replant problem   197
Cheshunt compound   33
Chewing insects   3
China aster, diseases of   302
    foot rot   302
    pests of   301
Choke   151
Chlorbenside   33, 355
Chlordane   34, 353
Chlorfenson   34, 355
Chlorfenvinphos   34, 87, 91, 349
Chlorinated hydrocarbons   8, 353
Chlorobenzilate   34, 355
Chloronitrobenzenes   358

Chloropicrin   34, 229, 359
Chocolate spot, field beans   156
Chrysanthemum aphid   240, 302
  blotch   242, 302
  capsid   241
  diseases of   242
  eelworm   241
  grey mould   243
  leaf miner   241
  mildew   243
  pests of   239, 302
  petal blight   243
  Phoma root rot   244
  ray blight   244
  rust   244
  stool miner   242
  Verticillium wilt   244
Clarkia grey mould   303
Clay-coloured weevil   205, 323, 339, 343
Cineraria, diseases of   245
  pests of   245
Clematis, pests of   303
  powdery mildew   303
  wilt   303
Clover, diseases of   153
  flower midge   143
  leaf midge   143
    weevil   142
  phyllody   155
  rot   154
  seed midge   143
    weevils   142
  spot   154
Clubroot   157, 285, 328, 333
Cobnut, diseases of   201
  pests of   201
Cobweb, mushroom   260
Cockchafers   113
Cocksfoot aphid   141
  diseases of   150
  gall midges   141
  moth   141
  mottled virus   153
  pests of   141
  streak   153
  yellow slime disease   152
Code of conduct   29, 95
Codling moth, apple   167
    pear   188
Collar rot, apple   178
Colletotrichum root rot   270
Colloidal formulations   35

Colorado beetle  6
Common green capsid  113, 169, 188,
    198, 205, 305, 318
  leaf weevil  139, 169
  rustic moth  139
Compatability  26
Concentrate spraying  27
Cone nozzles  67
Conifer spinning mite  327, 338, 342
Contact poison  24
Conventional spraying  26
Copper acetoarsenite  361
  dusts  34, 121
  fungicides  34, 121
  oxychloride  34, 121
  sulphate  32, 359
Core rot, arum  232
    gladiolus  308
Corn flower, diseases of  304
Corrosive sublimate  32, 35, 359
Cotoneaster, pests of  304
Cottonseed oil-Bordeaux  196
Cover  63
Covered smut, barley  105
    oats  108
Cress germination test  48, 229
Cresols  35
Cresylic acid  35
Crocus, pests of  304
Crook root, watercress  296
Crop rotation  7, 97
Crown rust  150
  wart, lucerne  155
Cucumber anthracnose  249
  aphid  245
  basal stem rot  249
  black root rot  249
  diseases of  249
  'French fly'  246
  fungus gnat  246
  gumnosis  249
  pests of  245
  powdery mildew  249
  root-knot eelworm  246
  root rot  250
  Verticillium wilt  250
Cultural control  7
Currant aphid  202
  clearwing moth  206
  -sowthistle aphid  202
  root aphid  202
  diseases of  208
  pests of  201

Cuprous oxide  35, 121
Cutworms, anemone  298
  beet  127
  chrysanthemum  241
  forest nurseries  338
  potato  114
  strawberry  215
  vegetables  275
Cyclamen aphid  250
  diseases of  251
  grey mould  251
  mite  266
  pests of  250
  root rots  251
  tarsonemid mite  250
  vine weevil  250
Cyclodiene insecticides  29


Daffodil, see Narcissus
Dahlia, diseases of  306
  grey mould  306
  pests of  305
  red spider mite  305
  smut  306
Damping-off, brassica  285
  field beans  156
  forest nurseries  340
  grasses  152
  lettuce  256
  maize  149
  seedlings  276, 326
  spruce  327
  tomato  270
  vegetables  276
Damson, diseases of  200
  pests of  197
Damson-hop aphid  197, 222
Daphne, diseases of  306
  leaf spot  306
  pests of  306
Dark leaf spot  108
Dazomet  35, 229, 230, 356
D-D mixture  35, 229
DDT  36, 353
Deaths head hawk moth  114
Decontamination of sprayers  74
Delphinium black blotch  306
  diseases of  306
  powdery mildew  306
Demethion  37, 349
Demeton  87

Demeton-methyl   37, 87, 91
Demeton-S-methyl   37, 87, 91, 349
Demeton-S-methyl sulphone   37, 349
Demeton-S-methyl sulphoxide   51
Derris   8, 37, 348
Destructive Insects and Pests Acts   6
Diamond-back moth   148, 283
Diaphragm pumps   66
Diazinon   38, 349
Dichlofluanid   38, 359
Dichloropropane   35
Dichloropropene   35
Dichlorvos   38, 87, 92, 349
Dicloran   39, 359
Dicofol   39, 355
Didymella stem rot   296
Dieldrin   39, 353
Dimefox   40, 87, 89, 349
Dimethirimol   41, 359
Dimethoate   40, 349
Dinitro-o-cresol   42
Dinobuton   41, 357
Dinocap   41, 357
Dinoseb   87, 89
Diplumbic hydrogen arsenate   46
Disodium octaborate   359
Disulfoton   41, 87, 89, 350
Dithianon   42, 360
Dithiocarbamates   356
DNC   42
DNOC   42, 87, 89, 358
Dock sawfly   170
Docking disorder   128
Dodine   42, 360
Dollar spot   331
Dose   63
Douglas fir seed wasp   345
Downy mildew, anemone   298
    antirrhinum   299
    beet   132
    brassica   286
    hop   225
    lettuce   256
    meconopsis   314
    onion   292
    rose   265
    spinach   295
    stock   328
    sugar beet   132
    swede   157
    wallflower   333
Droplet spectrum   62
Dry bubble, mushroom   260

Dry rot, gladiolus   308
    potato   124
Dusts   27, 78
    application of   78
Dusting machines   78

Ear blight, barley   105
    oats   108
    wheat   101
Earthworms   330
Earwig, apple   170
    chrysanthemum   241, 302
    potato   114
Eelworm, leaf and bud, blackcurrant   206
    chrysanthemum   241, 302
    paeony   317
    pyrethrum   322
    strawberry   215
        (see also Nematode)
Elder aphid   301
EMP   132
Emulsifiable concentrate   25
Emulsions   25
Endosulfan   43, 87, 89, 353
Endothal   87
Endrin   43, 87, 89, 354
Eradicant fungicides   25
Ergot   109, 150, 151
Ethion   44, 87, 92, 350
Ethoate-methyl   44, 350
Ethylene dichloride   44, 354
Ethylene thiuram monosulphide   49
Ethylmercuric phosphate   132
Euonymus, pests of   306
    powdery mildew   307
European gooseberry mildew   209
Eyespot, oats   108
    wheat   102

Factory Acts 1937–39   50
Fairy rings, turf   331
False bloom, azalea   300
    rhododendron   323
    truffle   261
Fan nozzle   67
    spray   69
Fenchlorphos   44, 350
Fenson   44, 355
Fentin acetate   44, 87, 92
    hydroxide   44, 87, 92

Fescue aphid 98
Fescues, diseases of 150
  pests of 141
Field beans, diseases of 156
  crops, spraying of 75
Filbert, diseases of 201
  pests of 201
Fire, narcissus 316
  tulip 330
Fire blight, apple 179
    pear 190
Flat millepede 115, 130
Flea beetles 146, 284
Fluoroacetamide 87
Fodder beat, disease of 132
    pests of 125
Folpet 45, 360
Fomes 346
Food and Drugs Act 1955 46, 85
Foot-rot, china aster 302
  petunia 319
  ryegrass 149
  tomato 270
Forest nurseries, diseases of 340
    pests of 337
Formaldehyde 45, 230
Formalin 45, 230
Formothion 45, 350
Formulation 25
Forsythia, pests of 307
Fox-coloured sawfly 343
Freesia, diseases of 251
  pests of 251
French beans, pests of 278
'French-fly' 246
Frit fly, grasses 140
    maize 140
    oats 107
Fruit spot, quince 192
  tree red spider mite, apple 170
        cherry 196
        damson 198
        peach 193
        pear 188
        plum 198
    tortrix 174
Fruitlet mining tortrix 174
Fuchsia, pests of 252, 307
Fumigants 23
Fungi, classification of 15
Fungi Imperfecti 15
Fungus gnat 246

Fusarium patch 332
  wilt, carnation 238
  yellows 251

Gall, azalea 300
  midges 141, 142, 143
  rhododendron 323
gamma-BHC 45, 354
Gangrene, potato 124
Garden chafer beetle 113
  dart moth 127, 275
  pebble moth 148
  slug 115
  springtails 116
  swift moth 116
Gear pumps 66
Geranium, diseases of 253
  pests of 252
Ghost swift moth 116
Gill mildew, mushroom 260
Gladiolus, core rot 308
  diseases of 308
  dry rot 308
  hard rot 308
  pests of 307
  scab 308
  thrips 307
Glasshouse crops, spraying of 77
  hygiene 231
  millepede 246
  potato aphid 111, 267
  symphilid 116
Gloeosporium bud rot 201
  rot, apple 179
Godetia, diseases of 308
  grey mould 308
  stem rot 308
Gooseberry American mildew 208
  aphid 202
  cluster cup rust 209
  diseases of 208
  European mildew 209
  pests of 201
  red spider mite 206
  sawfly 207
Gout fly 104
Grain aphid 98
  mite 110
  weevils 110
Granules 27, 78

Grape vine, brown scale   253
   diseases of   221
   downy mildew   221
   grey mould   221
   mealy bug   254
   pests of   253
   powdery mildew   221
Grass aphid   98, 139
  crops, pests of   138
  thrips   140
Grass and cereal flies   140
     mites   140
Grasses, diseases of   149
  pests of   138
Greasy-blotch, carnation   238
Green apple aphid   160
  leafhoppers   115
  oak tortrix   317
  petal virus   216
  spruce aphid   327, 337
Green-veined white butterfly   148
Grey bulb rot, tulip   330
  field slug   115
  leaf, barley   105
   oats   108
   wheat   102
  mould, anemone   299
   antirrhinum   299
   azalea   233
   begonia   235
   chrysanthemum   243, 303
   cineraria   245
   clarkia   303
   cyclamen   251
   dahlia   306
   forest nurseries   340
   freesia   252
   geranium   253
   godetia   308
   grape   254
   hydrangea   255
   lettuce   257, 291
   pelargonium   253
   poinsettia   263
   pyrethrum   322
   rose   265
   saintpaulia   266
   snowdrop   326
   strawberry   220
   tomato   271
   zinnia   334
  sterile fungus   272
Ground crop sprayers   63

G.S.F.   272
Gumnosis   249

Halo blight   108
  spot   106
Hard rot, gladiolus   308
Haulm destruction   123
Hawthorn, pests of   309
  powdery mildew   309
  webber   309
Hazelnut, diseases of   201
  pests of   201
Heart and dart moth   338
Heat treatment   29, 228
Hellebore leaf spot   309
HETP   87
Hexachlorobenzene   46, 354
Hexachlorocyclohexane   31
High volume spraying   26, 62
Hippeastrum, diseases of   255
  pests of   254
Holly leaf miner   309
  leaf-tying moth   310
  pests of   309
Hollyhock, diseases of   311
  pests of   310
Honey fungus   321
Hop capsid   224
  -damson aphid   222
  diseases of   225
  downy mildew   225
  flea beetle   224
  mould   226
  pests of   221
  powdery mildew   226
  root weevil   224
  strig midge   224
Horticultural crops, spraying of   77
Hyacinth, diseases of   310
  pests of   310
Hydrangea, diseases of   255
  grey mould   255
  pests of   255
  powdery mildew   255
  spray damage   255

Insecticide resistance   11
Insecticides, classification of   23
Iris leaf spot   311
  pests of   311
  rhizome rot   311
  sawfly   311

Juniper, pests of 311
  scale 311
  webber 311

Kale, diseases of 156
Keeled slugs 115
Knapsack sprayers 65

Lackey moth 312, 324
Larch leaf cast 312, 340
  shoot beetle 345
Large narcissus fly 315
  rose sawfly 325
  white butterfly 148
  willow sawfly 339
Lead arsenate 46, 360
  tolerance 85
Leaf blight, carrot 288
   lily 312
   quince 192
  blotch, barley 106
   grasses 151
   ryegrass 149
  cast, larch 312, 340
  curling plum aphid 197, 305
  fleck 151
  miner, chrysanthemum 241
   holly 309
   lilac 312
   tomato 268
  mould 271
  roll, potato 124
  rolling rose sawfly 325
  spot, antirrhinum 299
   barley 106
   blackcurrant 209
   brassica 286
   camelia 238, 300
   carnation 238
   celery 289
   chrysanthemum 302
   daphne 306
   gernaium 253
   hellebore 309
   iris 311
   lobelia 313
   oats 109
   pansy 318
   pelargonium 253
   phlox 319
   polyanthus 320

Leaf spot (*cont.*)
   poplar 320
   sweet william 328
   wheat 102
   yucca 334
  scorch, amaryllis 255
   hippeastrum 255
   narcissus 316
   plane 319
  stripe 106
  -tying moth, holly 310
  weevil, apple 173
   cherry 196
   forest trees 343
   pear 188
Leafhoppers, apple 172
  calceolaria 235
  legumes 143
  potato 115
  primula 264
  rhododendron 323
  rose 324
  strawberry 216
  tomato 268
Leatherjackets, barley 105
  crucifers 148
  grasses 140
  legumes 143
  mangolds 128
  potato 115
  sugar beet 128
  turf 331
  vegetable crops 275
  wheat 98
Legumes, diseases of 153
Lettuce aphid 202, 256, 289
  damping-off 256
  diseases of 256, 290
  downy mildew 256, 290
  grey mould 291
  pests of 256, 289
  ring spot 291
  root aphid 290
Lilac blight 312
  leaf miner 312
  pests of 312
Lily beetle 313
  leaf blight 313
Lime sulphur 46, 360
Lindane 45
Linnets, as pests 217
Liver of sulphur 46
Lobelia, diseases of 313

Loganberry cane spot 213
  diseases of 213
  pests of 210
Lonchocarpus 8, 38
Loose smut, barley 106
  oats 109
  wheat 103
Low volume spraying 62, 159
Lucerne, diseases of 153
  flower midge 143
  leaf midge 143

Magpie moth 207
Mahonia, diseases of 313
Malathion 47, 350
Malus, pests of 313
  scab 314
Mangold clamp aphid 134
  diseases of 132
  fly 129, 279
  pests of 125
Mancozeb 47, 356
Maneb 47, 356
Manual knapsack sprayers 65
Maple 314
Marbled minor moth 140
March moth 176, 322
Marsh crane fly 129
Mazidox 87
Meadow foxtail, pests of 141
  grasses, pests of 141
Mealy bug, amaryllis 255
  begonia 234
  grape vine 254
  hippeastrum 255
  orchids 262
  cabbage aphid 145
  peach aphid 192
  plum aphid 197
Mecarbam 47, 87, 92, 350
Meconopsis downy mildew 314
Medinoterb 87
Melon aphid 245
  diseases of 249
  pests of 245
Menazon 47, 350
Mercuric chloride 32, 35, 359
Mercurous chloride 32, 360
Metaldehyde 48
Metham-sodium 48, 230, 231, 356

Methiocarb 48, 361
Methoxyethylmercuric silicate 50
Methyl bromide 48, 230, 361
  cellulose 293
  isocyanate 35, 48, 230
Methylmercuric dicyandiamide 50
Metiram 49, 356
Mevinphos 49, 87, 89, 351
Michaelmas daisy, mildew 314
Mildew, barley 106
  brassica 157
  carnation 238
  chrysanthemum 243
  grasses 150
  legumes 154
  oats 109
  ryegrass 149
  strawberry 220
  tares 154
  tomato 271
  wheat 103
Millepedes, cucumber 246
  melon 246
  potato 115
  sugar beet 130
  vegetables 276
Mills table 181
Mineral oils 52
Mint rust 291
Mirids 126
Miscible oil 25
Morphothion 49, 351
Motorized knapsack sprayer 65
Mottled arum aphid 298
  umber moth 176, 312
Mushroom, bacterial blotch 259
  pit 259
  bubbles 260
  cobweb 260
  diseases of 259
  dry bubble 260
  flies 257
  gill mildew 260
  houses, disinfection of 261
  pests of 257
  red geotrichum 260
  truffle 261
Mussel scale, apple 173
  peach 193
  pear 188
  plum 200
  rose 325
Mustard beetle 146

Nabam  49, 356
Naled  50, 351
Narcissus bulb mite  314
  bulb scale mite  314
  diseases of  316
  fire  316
  leaf scorch  316
  pests of  314
  root lesion eelworm  315
  smoulder  316
  tip burn  316
  white mould  317
Nectarine, diseases of  194
  pests of  192
Needle blight, thuja  329
    western red cedar  341
  cast, pine  341
Nematode, strawberry  259
      (see also Eelworm)
Net blotch, barley  106
    ryegrass  149
Nicotine  50, 87, 92, 348
Nonionic surfactants  26
Northern root-knot eelworm  128
Notification of Pesticides Scheme  53
Nozzle arrangement  75
Nozzles, spray  67
Nut weevil  201


Oak mildew  341
  pests of  317
  phylloxera  337
  powdery mildew  317
Oat spiral mite  140
Oats, cereal cyst eelworm  107
  frit fly  107
  stem eelworm  107
  tulip root  107
Olive green mould  261
Onion, diseases of  292
  downy mildew  292
  fly  292
  pests of  292
  smut  292
  white rot  293
Ophiobolus patch  331
Orchids, anthracnose  262
  aphids  261
  bacterial soft rot  263
  black rot  263
  diseases of  262
  pests of  261

Organomercury compounds  87, 192
Organophosphorus compounds  9, 348
Oxydemeton-methyl  51, 87, 92, 351
Oxytetracycline  57
Oxythioquinox  55
Ovicide  24
Oyster-shell scale, apple  173
  peach  193
  pear  188
  plum  200


Paeony, leaf and bud eelworm  317
  wilt  317
Paint spray gun  77
Pansy leaf spot  318
  stem rot  318
Paradichlorobenzene  51, 354
Paris green  52, 361
PCNB  55
Pea aphid  142, 144, 293
  midge  294
  moth  293
  pests of  293
Pea and bean thrips  294
      weevils  143, 144, 295
Peach aphid  192
  diseases of  194
  leaf curl  194, 321
  pests of  192
  powdery mildew  194
  scale  193, 200, 204
Peach-potato aphid  111, 130, 145, 192
Pear aphids  186
  bryobia mite  187
  canker  190
  diseases of  190
  fire blight  190
  leaf blister mite  188
    midge  189
  midge  189
  pests of  186
  sawfly  189
  scab  191
  slug sawfly  189
  sucker  189
  thrips  190
Pear-bedstraw aphid  186
Pear-coltsfoot aphid  186
Pelargonium, diseases of  253, 318
  pests of  252, 318
Pentachlorophenol  56
Permanent currant aphid  202

Pesticide resistance  11
Pesticides Safety Precautions Scheme
    83
Petal blight, chrysanthemum  243, 303
    corn flower  304
Petroleum oils  52
Petunia foot rot  319
Pharmacy and Poisons Act, 1933  35,
    93
Phenkapton  87, 92, 351
Phenylmercuric acetate  53
    chloride  53
    dimethyldithiocarbamate  53
    nitrate  53
    salicylanilide  53
    salicylate  53
    urea  53
Phlox, leaf spot  319
    pests of  319
    powdery mildew  319
Phoma root-rot, chrysanthemum  244
Phorate  54, 89, 351
Phorid flies  258
Phosalone  54, 351
Phosphamidon  54, 87, 92, 352
Phycomycetes  15
Phyllody  155
Phytotoxicity  13
Pine looper  345
    needle cast  341
    root aphid  337
    sawfly  343
    shoot beetle  345
        moth  343
    weevil  344
Pinhole borer  346
Plane leaf scorch  319
Plantation crops, spraying of  76
Plum blossom wilt  200
    diseases of  200
    fruit moth  199
    pests of  197
    red-legged weevil  200
    sawfly  199
    scale mite  200
    tortrix  200
Plunger pumps  66
Poinsettia, diseases of  263
    pests of  263
Pollen beetles  145
Pollinating insects  4
Polyanthus leaf spots  320
    pests of  320

Poplar beetles  338
    leaf spots  320
    yellow leaf blister  320
Positive displacement pumps  67
Potassium arsenite  87
Potato aphid  111
    blight  117
    capsid  126
    cyst eelworm  114
    diseases of  117
    flea beetles  115
    gangrene  124
    leafhoppers  115
    pests of  111
    root eelworm  268
    tuber eelworm  114
    virus diseases  124
Powdery mildew, apple  179
        beet  133
        begonia  235, 300
        blackcurrant  220
        chrysanthemum  243, 303
        cineraria  245
        clematis  304
        corn flower  304
        cucumber  249
        delphinium  306
        euonymus  307
        grape  221, 254
        hawthorn  309
        hop  226
        hydrangea  255, 311
        mahonia  313
        melon  249
        michaelmas daisy  314
        oak  317
        peach  194
        phlox  319
        rose  266, 326
        sugar beet  133
        sweet pea  267
            (see also Mildew)
Predatory insects  5
Primula, pests of  263
Privet, honey fungus  321
    pests of  321
    thrips  321
Propineb  55, 357
Protective clothing  88, 89
    fungicide  24
    insecticide  24
Prunus, diseases of  321
    pests of  321

Pumps, spray   66
Pygmy mangold beetle   130
Pyracantha, pests of   322
  scale   322
Pyrethrum   8, 55, 348
  grey mould   322
  pests of   322

Quince, diseases of   192
Quinomethionate   55, 361
Quintozene   55, 358

Ramularia leaf spot   133
Rape, diseases of   157
Raspberry aphid   211
  beetle   211
  cane midge   212
    spot   214
  diseases of   213
  moth   213
  pests of   210
Ray blight, chrysanthemum   244
'Redberry'   213
Red beet, pests of   279
  bud borer   173, 324
  clover gall gnat   143
    scorch   154
  geotrichum   260
  leaf aphid   160
  leg   291
    legged earth mite   141
    weevil   200, 219
  spider mite, amaryllis   255
    arum   232
    asparagus fern   233
    azalea   233
    beans, glasshouse   234
    carnation   237
    chrysanthemum   242
    cucumber   247
    currant   207
    dahlia   305
    gooseberry   207
    grape   254
    hippeastrum   255
    hop   224
    hydrangea   255
    melon   247
    orchids   262
    peach   193
    poinsettia   263

Red spider mite (*cont.*)
  primula   264
  rose   264
  strawberry   216
  sweet pea   267
  tomato   268
  thread, turf   332
Replant problem, apple   180
  cherry   197
Resin sticker   293
Rhizome rot, iris   311
Rhododendron bud blast   324
  bug   323
  diseases of   323
  leafhopper   323
  pests of   323
  white fly   323
Ring spot, carnation   301
  lettuce   291
Root-knot eelworm, cucumber   246
  tomato   269
  lesion eelworms   315
  rot, arum   232
    asparagus fern   233
    cucumber   250
    cyclamen   251
    geranium   253
    pelargonium   253
    tomato   272
Rose aphids   264, 324
  black spot   265, 325
  diseases of   265
  downy mildew   265
  grey mould   265
  leafhopper   324
  pests of   264, 326
  powdery mildew   266, 326
  rust   326
  scale   265
  slug sawfly   325
  thrips   324
Rose-grain aphid   98
Rosy apple aphid   160
  leaf-curling aphid   160
  rustic moth   131, 225
Rotary nozzles   68
Rotenoids   37
Rotenone   37, 348
Rubus aphid   210
  stunt virus   212
Runner bean, pests of   278
Rust, carnation   238, 301
  chrysanthemum   244, 302

Rust (*cont.*)
  cornflower  304
  geranium  253
  mahonia  313
  pelargonium  253, 318
  rose  326
  sugar beet  133
  sweet william  328
Rustic shoulder knot moth  140
Rusts, grasses  150
Rye, diseases of  109, 149
  pests of  109
Ryegrass, diseases of  149
  pests of  142
  mosaic virus  153


Safety precautions  28, 83
Sainfoin, diseases of  154
  flower midge  143
  leaf midge  143
Saintpaulia, diseases of  266
  pests of  266
Salicylanilide  55, 361
Sand weevil  131
Sawflies  339
Saw-toothed grain beetle  110
Scab, apple  180
  barley  106
  gladiolus  308
  malus  314
  oats  109
  pear  191
  pyracantha  322
  wheat  103
Scale insects, pear  193, 262
    poinsettia  263
Schradan  56, 87, 90, 352
Sciarid flies  258
Sclerotinia rot, field beans  156
Scurfy scale  325
Seed disinfectant  25
  dressing  25, 97
  weevil  146
Seedborne vegetable diseases  277
Seedling blight, zinnia  334
  damping-off  326
  diseases, turf  332
Severe mosaic, potato  124
Shallot aphid  214, 307, 312
Sharp eyespot  103
Shot-hole borers  166

Silver leaf  322
Silver-green leaf weevil  173, 174
Silvering  280
Skin spot, potato  124
Slow wilt, carnation  238
Slugs, legumes  143
  mushroom 259
  potato  115
  sugar beet  131
  tomato  269
  wheat  99
Small ermine moth  173
  narcissus flies  315
  white butterfly  148
Smokes  27, 231
Smoulder, narcissus  316
Smut, dahlia  306
  maize  149
  onion  292
  wheat  102
Snowdrop, grey mould  326
  pests of  326
Snow mould  332
Sodium arsenite  87
  nitrite  56, 347
  pentachlorophenate  56
Soil insecticides  12
  sterilants  228
  treatment  228
Soluble concentrate  28
Spinach, diseases of  295
  pests of  295
Spindle ermine moth  307
Spotted millepede  115, 130
Spraing disease  114
Spray drift  70
  pattern  67
  pumps  66
Sprayers, maintenance of  69
  types of  63
Spreaders  26
Springtails, cucumber  248
  forest nurseries  339
  lettuce  256
  melon  248
  potato  116
Spruce aphid  327
  bell moth  343
  damping-off  327
  gall adelges  327, 342
  pests of  327
  shoot aphid  342

Stem eelworm, daffodil 316
   hyacinth 310
   legumes 143
   narcissus 316
   oats 107
   phlox 319
   polyanthus 320
   potato 114
   strawberry 215
   sugar beet 128
   tulip 329
Stem rot, antirrhinum 299
   carnation 239
   field beans 156
   geranium 253
   godetia 308
   lobelia 313
   pansy 318
   pelargonium 253
   tomato 272
Stem weevil 147
Stinking smut 102
Stock, diseases of 328
  downy mildew 328
Stock emulsion 26
Stomach poison 24
Stool miner, chrysanthemum 242
Stored creal pests 110
Strawberry aphid 214
  arabis mosaic 219
  blossom weevil 216
  bud eelworm 215
  cutworms 215
  diseases of 219
  grey mould 220
  leaf eelworm 215
  leafhoppers 216
  mildew 220
  mite 217
  pests of 214
  rhyncites 213, 217
  root weevil 219
  seed beetle 217
  stem eelworm 215
  tortrix moth 218
Streaked pine bell 343
Streptomycin 56
Stripe smut 149
Sucking insects 4
Sugar beet, diseases of 132
  downy mildew 133
  mild yellowing virus 133
  pests of 125

Sugar beet (*cont.*)
  powdery mildew 133
  virus yellows 133
  yellows 133
Sulfotep 57, 87, 90, 352
Sulphur 57
Sulphur-shy varieties 57, 184, 203, 208
Summer chafer 113
  fruit tortrix 174
  petroleum oils 52
Surfactants 25
Swede midge 148
Sweet pea, diseases of 267
  pests of 266, 328
  spray damage 266
  william, diseases of 328
  leaf spot 328
  rust 328
Swift moths, crucifers 148
  strawberry 218
  vegetable crops 276
Swirl nozzle 67
  plate 67
Sycamore tar spot 298
Symphilids, cucumber 248
  potato 116
  tomato 269
Syringa blight 312
Systemic pesticides 9, 24

Take-all, grasses 152
  maize 149
  oats 109
  wheat 103
Tanks, sprayer 65
Tar distillate washes 58
  oils 58
  spot 298
Tares, diseases of 154
Tarnished plant bug, dahlia 305
  potato 113
Tarsonemid mite, cyclamen 250
  saintpaulia 266
TCNB 59
TDE 58, 354
Tecnazene 59, 358
TEPP 59, 87, 90, 352
Tetradifon 59, 355
Tetraethyl pyrophosphate 59
Tetrasul 50, 356
Thionazin 60, 87, 90, 352
Thiram 60, 357

Thrips, arum  232
  asparagus fern  233
  azalea  233
  begonia  234
  carnation  237
  cucumber  248
  cyclamen  250
  orchids  262
  privet  321
  rose  324
  tomato  269
Thuja needle blight  329
  pests of  329
Timothy, diseases of  150
  fly  142
  pests of  142
  stem rust  150
  tortrix moth  142
Tip burn, narcissus  316
Tobacco rattle virus  128
Tolerance, arsenic  85
  lead  85
Tomato blight  295
  blackring virus  127
  brown root rot  270
  buckeye rot  270
  Colletotrichum root rot  270
  corky root  270
  damping-off  270
  diseases of  270, 295
  foot rot  270
  leaf miner  267
  leaf mould  271
  leafhopper  268
  moth  269
  pests of  267
  root rots  272
  stem rot  272
Tortrix moths, apple  174
  rhododendron  322
  rose  324
  willow  333
Tree sprayers  64
Trefoil, diseases of  153
  flower midge  144
Trichlorphon  60, 352
Triphenyltin acetate  360
  hydroxide  360
Truffle  261
Tulip bulb aphid  312, 329
  diseases of  330
  fire  330
  grey bulb rot  330

Tulip (cont.)
  pests of  329
  root  107
  stem eelworm  329
Turf, diseases of  331
  dollar spot  331
  Fusarium patch  332
  Ophiobolus patch  331
  pests of  330
  red thread  332
  snow mould  332
Turnip, diseases of  157
  gall weevil  147, 284
  moth  127, 275, 338
  root fly  149
Twig canker, cobnut  201

Vamidothion  61, 87, 92, 352
Vane pumps  67
Vapourer moth  324
Verticillium wilt, carnation  239
  chrysanthemum  244
  cucumber  250
  lucerne  155
  tomato  273
Vetch aphid  142, 144
  diseases of  154
  leaf midge  144
Vine weevil, cyclamen  250
  primula  264
  rhododendron  323
  strawberry  219
Viola stem rot  318
Violet leaf midge  332
  pests of  332
  root rot  134
Virus diseases  4, 153
  yellows  134

Wallflower club root  233
  downy mildew  333
  pests of  333
Wasps  176
Watercress, crook root  296
Western red cedar, needle cast  341
Wettable powders  25
Wetters  26
Wetting agents  26
Wheat black mould  101
  point  101
  brown foot rot  101

Wheat (*cont.*)
  bulb fly 100
  bunt 102
  diseases of 101
  ear blight 101
  eyespot 102
  glume blotch 102
  grey leaf 102
  leaf spot 102
  loose smut 103
  mildew 103
  pests of 98
  scab 103
  sharp eyespot 103
  shoot beetle 100
  stinking smut 102
  take-all 103
  yellow rust 104
White mould, mushroom 260
    narcissus 317
  rot, onion 293
Whitefly, asparagus 233
  begonia 233
  calceolaria 235
  chrysanthemum 242
  cucumber 248
  fuchsia 252
  rhododendron 323
  tomato 269
Willow anthracnose 334
  ermine moth 333
  pests of 333
  scale 333
  stem aphid 333
Wilt, clematis 304
  paeony 317
  tomato 273
(see also Fusarium wilt and Verticillium
        wilt)
Wingless weevils 219

Winter moths, apple 176
    apricot 194
    cherry 196
    cobnut 201
    currant 208
    forest plantations 344
    nectarine 194
    peach 194
    plum 200
    rhododendron 322
    rose 324
Winter washes 52
Wirestem 285
Wireworms, hops 225
  potato 116
  strawberry 219
  sugar beet 131
  tomato 270
  wheat 100
Woodlice, cucumber 248
  mushroom 259
  orchids 262
  tomato 270
Woolly aphid, apple 177
    spruce 342

Yellow leaf blister 320
  rust, barley 106
    cocksfoot 150
    wheat 104
  slime disease, cocksfoot 152
  underwing 338
Yew scale 334
Yucca leaf spot 334

Zinc-PETD 49
  sulphate 49
Zineb 49, 61, 337
Zinnia grey mould 334
  seedling blight 334

In small gardens, in orchards and on thousands of acres of farmland, products from Baywood are at work, providing a brighter and more bountiful harvest every year.

Backed by the world-wide research of Bayer Germany, Baywood continue to provide the most effective means of pest and disease control.

# ®Metasystox 55

The best and most widely used systemic insecticide for the control of aphids on all crops.

# ®Disyston new formula

Granular systemic insecticide for prolonged protection against aphids on a wide range of field crops.

# ®Gusathion MS

The most effective protection yet for apples. Simultaneously controls, pre-blossom, aphids, tortrix and winter moth caterpillars, apple sucker and apple blossom weevil.

# ®Elvaron

A revolution in the control of botrytis on strawberries. Also controls botrytis on many other soft fruit crops, and downy mildew on cauliflower seedlings.

Full details on these and the many other insecticides, fungicides and herbicides from Baywood may be obtained on request from:

**BAYWOOD**  Baywood Chemicals Limited,
Eastern Way, Bury St. Edmunds, Suffolk.
® Registered Trade Marks of BAYER GERMANY.

# A unique service

Boots are amongst the leading commercial research organisations in Great Britain. Backed by many years' experience in both crop and animal research they have produced such outstanding compounds as CMPP and benazolin (weedkillers) dicloran (fungicide) picadex (anthelmintic) and butacarb (insecticide), and have pioneered the use of compounds such as di-allate, tri-allate, dazomet and binapacryl in the U.K market.

All these products are thoroughly tested under laboratory, field trial and commercial farming conditions—the trials being controlled centrally from Boots research stations at Lenton and Thurgarton near Nottingham.

Apart from their extensive research facilities, Boots have recently set up a new Subsidiary Company—Boots Farm Sales Ltd.—through which their own products and those of other leading manufacturers are sold *direct* to the farmer. Boots Farm Sales Limited now has over 200 trained and experienced sales staff operating from more than 100 distributive centres throughout the country. As new research developments take place, involving both new and established lines, so this information is fed systematically to the B.F.S. fieldsmen who are in direct contact with the farmer. With this and Boots long established reputation for quality and service, farmers and growers recognise that when a product is promoted by Boots then it has the backing of their wide experience.

Overseas, Boots agricultural compounds are available either through their Subsidiary Companies or through their specially appointed agents or licensees.

## Boots Farm Sales Limited
CROP AND ANIMAL PRODUCTS, NOTTINGHAM

CC*

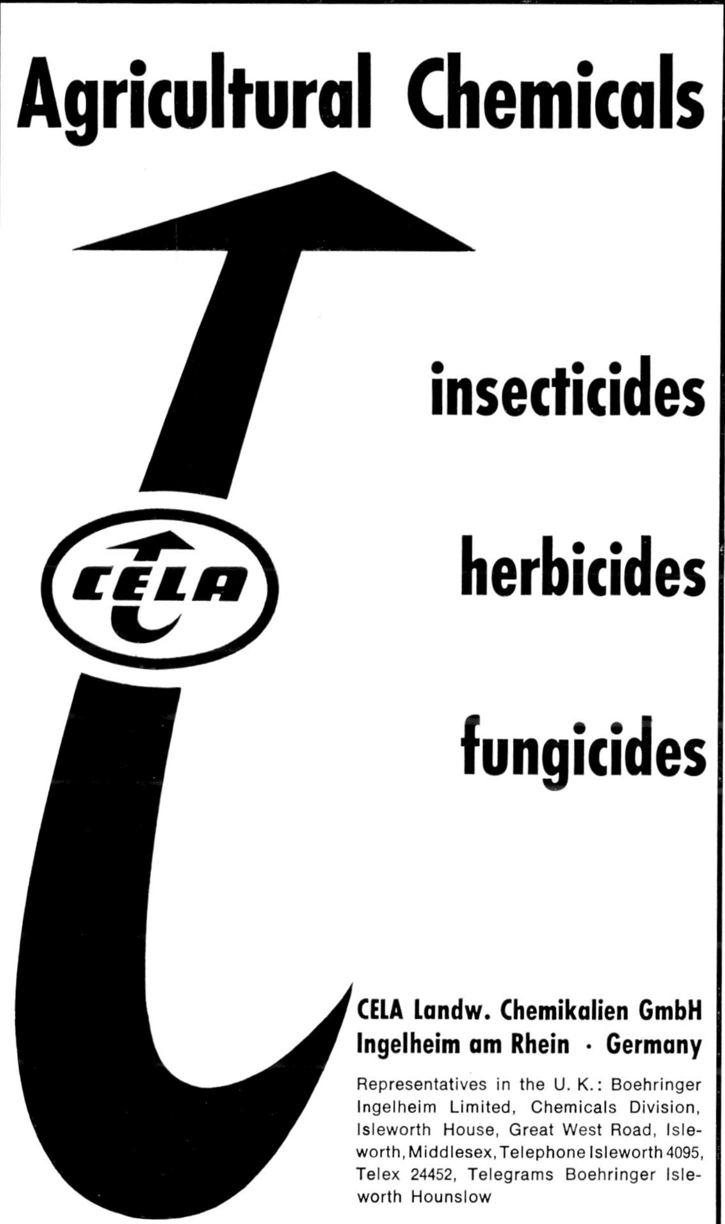
Agricultural Chemicals
CELA
insecticides
herbicides
fungicides
CELA Landw. Chemikalien GmbH
Ingelheim am Rhein · Germany
Representatives in the U. K.: Boehringer
Ingelheim Limited, Chemicals Division,
Isleworth House, Great West Road, Isle-
worth, Middlesex, Telephone Isleworth 4095,
Telex 24452, Telegrams Boehringer Isle-
worth Hounslow

# A Fragment

*T*HE beans no more by aphids shall be plagued,
   Nor lodging wreak its havoc in the wheat.
And scab, which often did the fruit degrade,
No more its costly visits will repeat.

And all the winged messengers of destruction,
The moulds and all the other troublous pests,
Shall meet their match with modern crop protection
And pass away forgotten, like the rest.

For them no more shall precious food play host,
Nor busy farmer lose his well earned sleep,
Nor housewives have to bear a higher cost
Now that more of what we sow we reap.

All that effort, all the research we freely gave
To find the dreadful secrets by Nature hid,
Now sends the pests to a sterile grave —
*A fruitful victory by courtesy of Cyanamid.*

  *    *    *    *    *    *    *

With apologies to Thomas Gray

We've kept
our ears to
the ground
to some effect
Study, research; learning,
developing; testing and proving.
Du Pont research teams have
produced over 100 top-flight
agricultural chemicals — all
developed and proven by the
test of years of successful use
by agriculture and industry
throughout the world.
Look for the Du Pont oval on
the weed and pest control
products you can rely on.
DU PONT

# Every year, Fisons Cambridge Division research synthesises over 1,000 new chemicals. One of them could make the world a little less hungry tomorrow.

# Which?

It's by no means certain that even one of these new chemicals will have a future. But if one does, then the real work begins.

It's one thing to find a new chemical, quite another to prove its potential— safe for farm use, safe to farming people, safe to animals and wildlife, safe to the consumers.

Fisons Cambridge Division take seemingly incredible pains to make sure it's all these things and more. It takes an endless and painstaking attention to detail and quality to prove fourteen new products in five years. This is Fisons Cambridge Division's record. It's a good one. You can see the results in healthy crops all around you.

Some problems are still unsolved-of these we say have hope. Tomorrow's answer may be just days away.

Fisons Cambridge Division

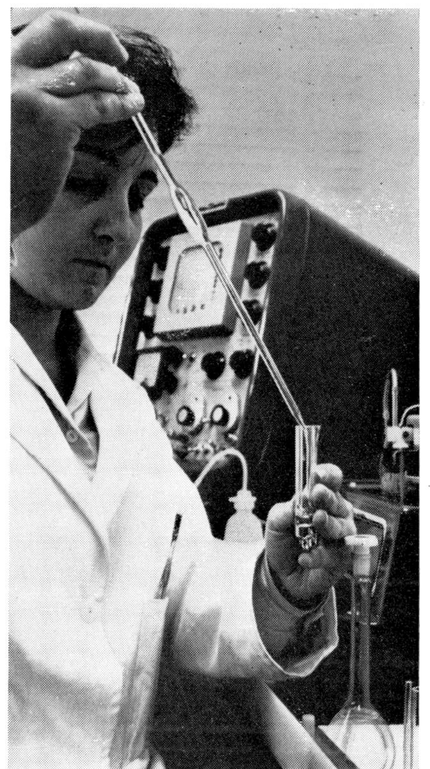 

# Research in the lab plus research in the field

## is still discovering new methods
## of pest and disease control

# Plant Protection



Administrative Headquarters: Fernhurst, Haslemere, Surrey, England

THIS
IS A FARM
WHERE
PROBLEMS GROW

# Crop Protection Products »Merck« for the control of pests, diseases and weeds

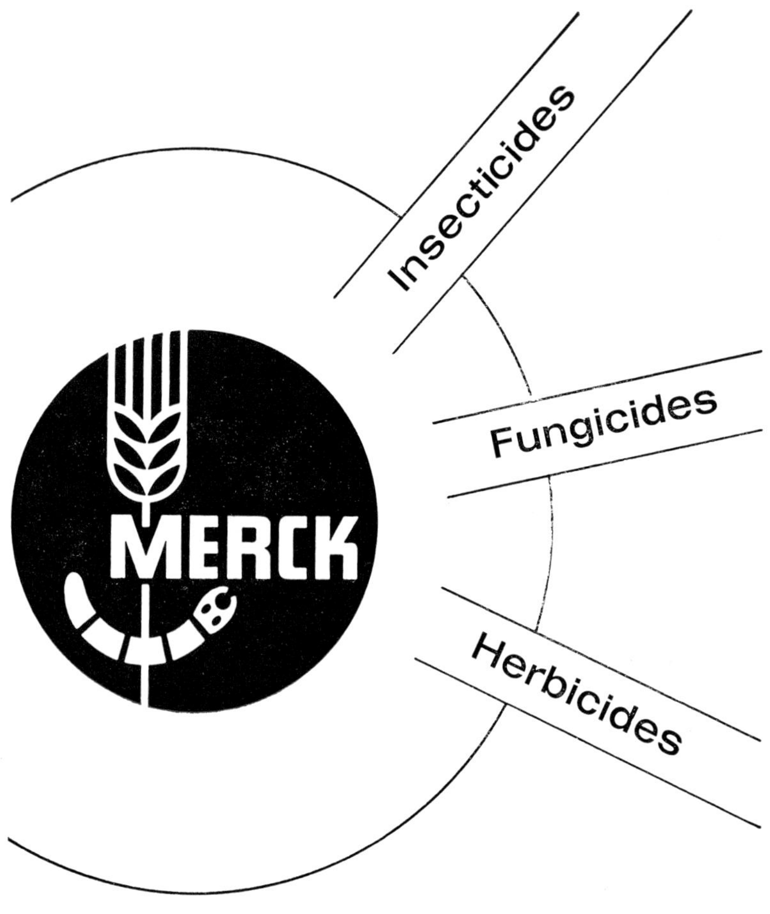

## Gone to ruin, every one

Thanks to the Murphy range of Insecticides and Fungicides.   Specialists in agricultural and horticultural pest and disease control, Murphy have developed products to combat most insects and fungi.   When you use Murphy you know you are getting the very best in protection.  Contact your local Murphy representative, or write to us for advice on your crop protection problem.

# Murphy

**The Murphy Chemical Co. Ltd., Wheathampstead, St. Albans, Herts. Phone: Wheathampstead 2001.**

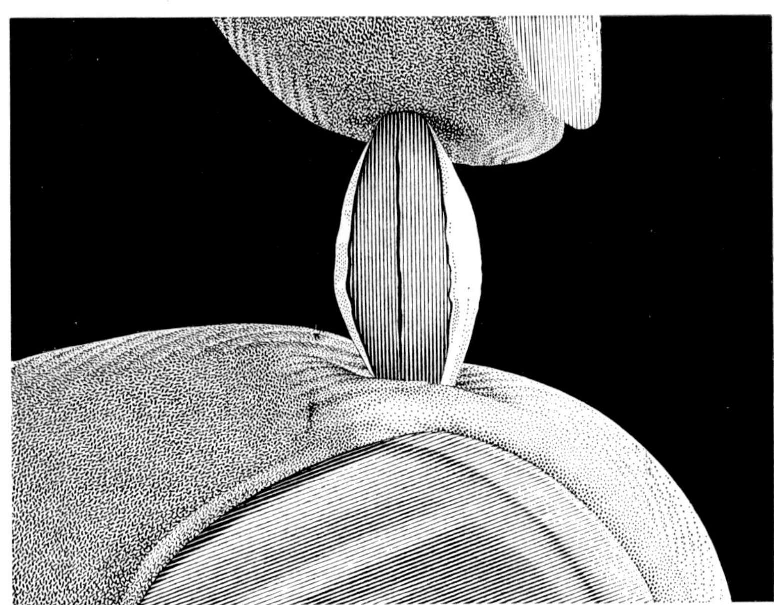

# Shellstar liquid seed dressings stay put–for good

With Shellstar Liquid Seed Dressings there's no dust. No dressing lost through handling or storage. No delays: the seed flows easily through the drill. And no mistakes: the treated seed is immediately recognisable by its red colouring, so there's no risk of it being fed to stock accidentally.

Give your cereals a great start in life. Insist on seed treated with:

**PANOGEN** – for the control of most seed-borne diseases in winter and spring cereals.

**BIRLANE** – applied with Panogen – for the most effective control of wheat bulb fly.

**KOTOL** – applied with Panogen – for the control of wireworm in winter and spring cereals.